APPLIED PYROLYSIS HANDBOOK

APPLIED PYROLYSIS HANDBOOK

edited by

THOMAS P. WAMPLER
CDS Analytical Inc.
Oxford, Pennsylvania

Marcel Dekker, Inc. New York • Basel • Hong Kong

Library of Congress Cataloging-in-Publication Data

Applied pyrolysis handbook / edited by Thomas P. Wampler.
p. cm.
Includes bibliographical references and index.
ISBN 0-8247-9446-X (acid-free)
1. Pyrolysis. I. Wampler, Thomas P.
TP156.P9A67 1995
543'.086—dc20 94-43929
CIP

The publisher offers discounts on this book when ordered in bulk quantities. For more information, write to Special Sales/Professional Marketing at the address below.

This book is printed on acid-free paper.

Marcel Dekker, Inc.
270 Madison Avenue, New York, New York 10016

Current printing (last digit):
10 9 8 7 6 5 4 3 2 1

PRINTED IN THE UNITED STATES OF AMERICA

Preface

Analytical pyrolysis is the study of molecules by applying enough thermal energy to cause bond cleavage, and then analyzing the resulting fragments by gas chromatography, mass spectrometry, or infrared spectroscopy. Pyrolysis has been employed for the analysis of organic molecules for most of this century. It was initially connected with investigations of vapor-phase hydrocarbons, and later became a routine technique for analyzing fuel sources and natural and synthetic polymers. Current applications include analysis of trace evidence samples in forensic laboratories, evaluation of new composite formulations, authentication and conservation of artworks, identification of microorganisms, and the study of complex biological and ecological systems.

This book is intended to be a practical guide to the application of pyrolysis techniques to various samples and sample types. To that end, general and theoretical considerations, including instrumentation and degradation mechanisms, have been consolidated in the first two chapters. The balance of the book describes the use of pyrolysis as a tool in specific fields. Synthetic polymers, forensic materials, and other samples with a long history of analysis by pyrolysis are covered here. In addition, we have been pleased to see some new areas of study, such as the analysis of surfactants, antiquities, and environmental materials, and these topics are presented as well.

The chapters examine the scope of work based on pyrolysis in particular fields of analysis and give specific examples of methods currently being used for the examination of representative samples. This book is intended to serve as a starting point for analysts who are adding pyrolysis to their array of analytical techniques, by providing concrete examples and suggesting additional reading.

I would like to thank all of the authors for their contributions. Each is actively involved in scientific pursuits, and the time that they have taken away from their busy schedules to contribute to this project was valuable and greatly appreciated.

Thomas P. Wampler

Contents

Contributors

NORBERT S. BAER Conservation Center, New York University, New York, New York

JOHN M. CHALLINOR Forensic Science Laboratory, Chemistry Centre (WA), East Perth, Western Australia

RANDOLPH C. GALIPO Department of Chemistry and Biochemistry, The University of South Carolina, Columbia, South Carolina

JOHN MADDOCK Horizon Instruments Ltd., Heathfield, East Sussex, England

STEPHEN L. MORGAN Department of Chemistry and Biochemistry, The University of South Carolina, Columbia, South Carolina

T. O. MUNSON Analytical and Testing Services Division, ECKENFELDER INC., Nashville, Tennessee

HAJIME OHTANI Department of Applied Chemistry, School of Engineering, Nagoya University, Nagoya, Japan

T. W. OTTLEY Horizon Instruments Ltd., Heathfield, East Sussex, England

ALEXANDER M. SHEDRINSKY Department of Chemistry, Long Island University, Brooklyn, New York, and Conservation Center, New York University, New York, New York

SHIN TSUGE Department of Applied Chemistry, School of Engineering, Nagoya University, Nagoya, Japan

THOMAS P. WAMPLER Research and Development, CDS Analytical Inc., Oxford, Pennsylvania

JOHN W. WASHALL CDS Analytical Inc., Oxford, Pennsylvania

BRUCE E. WATT Department of Chemistry and Biochemistry, The University of South Carolina, Columbia, South Carolina

1

Analytical Pyrolysis: An Overview

THOMAS P. WAMPLER
CDS Analytical Inc., Oxford, Pennsylvania

I. INTRODUCTION

Pyrolysis, simply put, is the breaking apart of chemical bonds by the use of thermal energy only. Analytical pyrolysis is the technique of studying molecules either by observing their behavior during pyrolysis or by studying the resulting molecular fragments. The analysis of these processes and fragments tells us much about the nature and identity of the original larger molecule. The production of a variety of smaller molecules from some larger original molecule has fostered the use of pyrolysis as a sample preparation technique, extending the applicability of instrumentation designed for the analysis of gaseous species to solids, and especially polymeric materials. As a result, gas chromatography, mass spectrometry, and Fourier-transform infrared (FTIR) spectroscopy may be used routinely for the analysis of samples such as synthetic polymers, biopolymers, composites, and complex industrial materials.

The fragmentation which occurs during pyrolysis is analogous to the processes that occur during the production of a mass spectrum. Energy is put into the system, and as a result, the molecule breaks apart into stable fragments. If the energy parameters (temperature, heating rate, and time) are controlled in a reproducible way, the fragmentation is characteristic of the original molecule, based on the relative strengths of the bonds between its atoms. The same distribution of smaller molecules will be produced each time an identical sample is heated in the same manner, and the resulting fragments carry with them much information concerning the arrangement of the original macromolecule.

The application of pyrolysis techniques to the study of complex molecular systems covers a wide and diversified field. Several books have been published that present theoretical as well as practical aspects of the field, including a good introductory text by Irwin [1] and a compilation of gas chromatographic applications by Liebman and Levy [2]. A recent bibliography [3] lists approximately 500 recent papers in areas as diverse as food and environmental and geochemical analysis, while the application to microorganisms has been examined by Morgan et al. [4]. This chapter will include only a few representative examples of the kinds of applications being pursued, with references for further reading. Specific areas of analysis are detailed in subsequent chapters.

II. DEGRADATION MECHANISMS

The degradation of a molecule which occurs during pyrolysis is caused by the dissociation of a chemical bond and the production of free radicals. The general processes employed to explain the behavior of these molecules are based on free radical degradation mechanisms. The way in which a molecule fragments during pyrolysis, and the identity of the fragments produced, depend on the types of chemical bonds involved and on the stability of the resulting smaller molecules. If the subject molecule is based on a carbon chain backbone, such as that found in many synthetic polymers, it may be expected that the chain will break apart in a fairly random fashion to produce smaller molecules chemically similar to the parent molecule. Some of the larger fragments produced will preserve intact structural information snipped out of the polymer chain, and the kinds and relative abundances of these specific smaller molecules give direct evidence of macromolecular structure. The traditional degradation mechanisms generally applied to explain the pyrolytic behavior of macromolecules will now be reviewed, followed by some general comments on degradation via free radicals.

A. Random Scission

Breaking apart a long-chain molecule such as the carbon backbone of a synthetic polymer into a distribution of smaller molecules is referred to as random scission. If all of the C—C bonds are of about the same strength, there is no reason for one to break more than another, and consequently the polymer fragments to produce a wide array of smaller molecules. The polyolefins are good examples of materials that behave in this manner. When polyethylene (shown as structure I with hydrogen atoms left off for simplicity) is heated sufficiently to cause pyrolysis, it breaks apart into hydrocarbons, which may contain any number of carbons, including methane, ethane, propane, etc.

I —C—C—C—C—C—C—

II —C—C— C·   ·C—C—C—

Chain scission produces hydrocarbons with terminal free radicals (structure II), which may be stabilized in several ways. If the free radical abstracts a hydrogen atom from a neighboring molecule, it becomes a saturated end, and creates another free radical in the neighboring molecule (structure III), which may stabilize in a number of ways. The most likely of these is beta scission, which accounts for most of the polymer backbone degradation by producing an unsaturated end and a new terminal free radical.

III $-C-C-\dot{C}-C-C-C-$

Beta scission $\downarrow$

IV $-C-C-C = CH_2 + \cdot C-C-$

This process continues, producing hydrocarbon molecules which are saturated and have one terminal double bond or a double bond at each end. When analyzed by gas chromatography, the resulting pyrolysate looks like the bottom chromatogram in Figure 1. Each triplet of peaks represents the diene, alkene, and alkane containing a specific number of carbons and eluting in that order. The next set of three peaks contain one more carbon, etc. It is typical to see all chain lengths from methane to compounds containing 35–40 carbons, limited only by the upper temperature of the gas chromatography (GC) column.

When polypropylene is pyrolyzed, it behaves in much the same manner, producing a series of hydrocarbons that have methyl branches indicative of the structure of the original polymer. The center pyrogram in Figure 1 shows poly(propylene), revealing again a recurring pattern of peaks, with each group now containing three more carbons than the preceding group. Likewise, when a polymer made from a four-carbon monomer such as isobutylene is pyrolyzed, it produces yet another pattern of peaks, with oligomers differing by four carbons, as seen in the top pyrogram in Figure 1. The relationships of specific compounds produced in the pyrolyzate to the original polymer structure have been extensively studied by Tsuge et al. [5], for example, in the case of poly(propylenes). The effects of temperature and heating rate have also been studied [6].

B. Side Group Scission

When polyvinyl chloride is pyrolyzed, no such oligomeric pattern occurs. Instead of undergoing random scission to produce chlorinated hydrocarbons, PVC produces aromatics, especially

---

Figure 1 Polyolefin pyrograms (750°C for 10 s). Top: polyisobutylene, center: polypropylene, bottom: polyethylene.

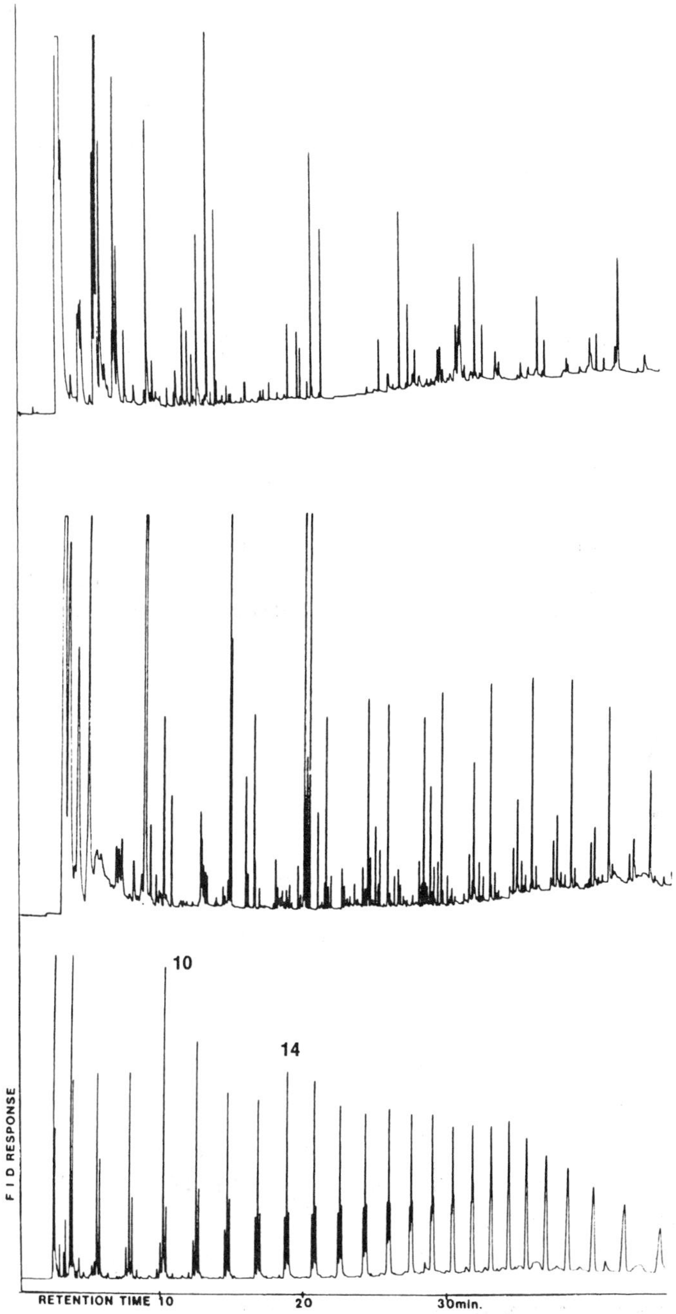

5

benzene, toluene, and naphthalene, as shown in Figure 2. This is the result of a two-step degradation mechanism which begins with the elimination of HCl from the polymer chain (structure V), leaving the polyunsaturated backbone shown as structure VI.

V

VI

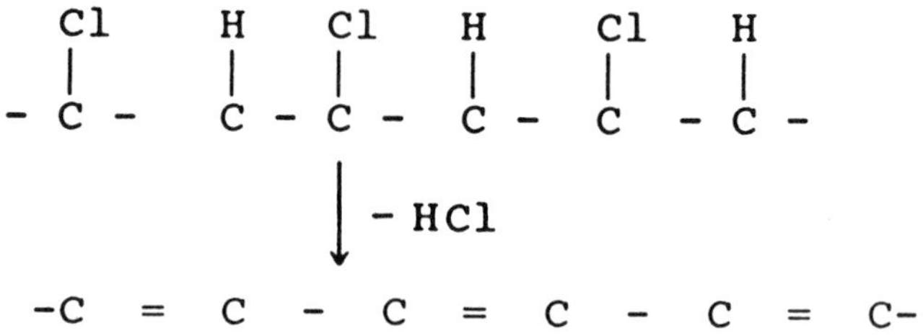

Upon further heating, this unsaturated backbone produces the characteristic aromatics seen in the pyrogram. This mechanism has been well characterized and the occurrence of chlorinated

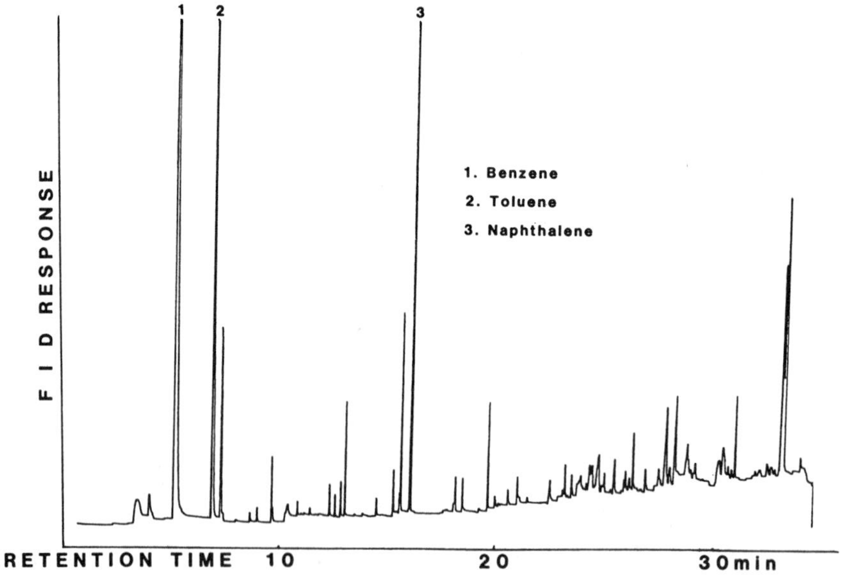

FIGURE 2 Pyrogram of poly(vinyl chloride) (600°C for 10 s).

aromatics is used as an indication of polymer defect structures, as in the work of Lattimer and Kroenke [7].

C. Monomer Reversion

A third pyrolysis behavior is evidenced by polymers such as poly(methyl methacrylate). Because of the structure of methacrylate polymers (structure VII), the favored degradation is essentially a reversion to the monomer.

VII

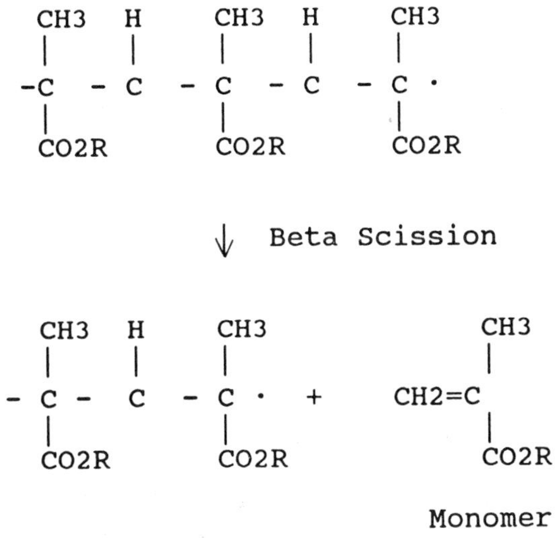

Monomer production is for the most part unaffected by the R group, so that poly(methyl methacrylate) will revert to methyl methacrylate, and poly(ethyl methacrylate) will produce ethyl methacrylate, etc. This proceeds in copolymers as well, with the production of both monomers in roughly the original polymerization ratio. Figure 3 shows a pyrogram of poly(ethyl methacrylate), with the ethyl methacrylate monomer peak by far the predominant product. A pyrogram of a copolymer of two or more methacrylate monomers would contain a peak for each of the monomers in the polymer.

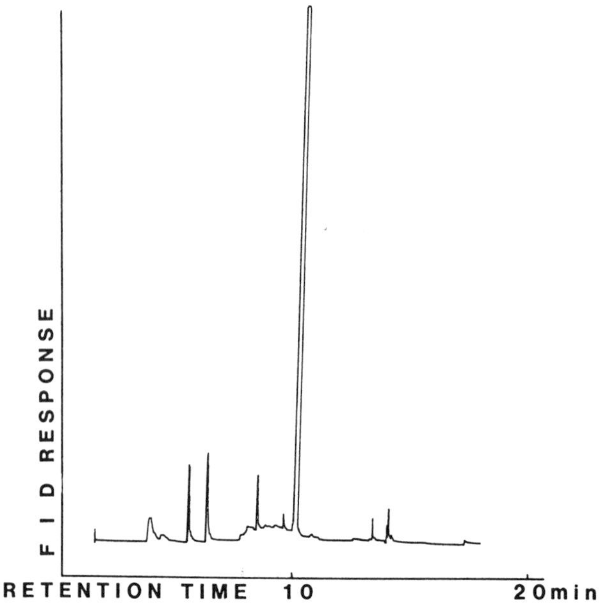

FIGURE 3 Pyrogram of poly(ethyl methacrylate) showing large monomer peak (600°C for 10 s).

D. Relative Bond Strengths

The question of which degradation mechanism a particular polymer will be subjected to—random scission, side group scission, or monomer reversion, or a combination of these—is simplified by considering the nature of thermal degradation as a free radical process. All of the degradation products shown, as well as minor constituents, and deviations from the simplified rules are consistent with the following general statements:

Pyrolysis degradation mechanisms are free radical processes and are initiated by breaking the weakest bonds first.

The composition of the pyrolyzate will be based on the stability of the free radicals involved and on the stabilities of the product molecules.

Free radical stability follows the usual order of 3° > 2° > 1° > CH_3, and intramolecular rearrangements which produce more stable free radicals play an important role, particularly the shift of a hydrogen atom.

A quick review of the previous degradation examples will help show how each of the above categories is in reality just one aspect of the general rule of free radical processes.

1. Polyolefins

Poly(ethylene) and the other polyolefins contain only C—C bonds and C—H bonds. Since an average C—C bond is about 83 kcal/mole and a C—H bond 94 kcal/mole, the initiation step involves breaking the backbone of the molecule, with subsequent stabilization of the free radical. The original free radicals formed are terminal or primary. Hydrogen abstraction from a neighboring molecule creates a C—H bond (stable product) and a new, secondary free radical, which may then undergo beta scission to form an unsaturated end. In addition, transfer of a hydrogen atom from the carbon five removed from the free radical (via a six-membered ring) transforms a primary free radical to a secondary, increasingthe free radical stability.

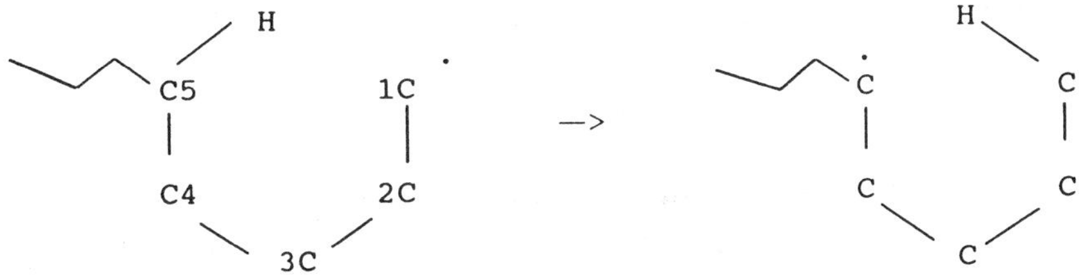

This new secondary free radical will either undergo beta scission or another 1–5 H shift. This stabilization by 1–5 hydrogen shifting explains the increased abundance of products in the pyrolyzate of poly(ethylene) containing 6, 10, and 14 carbons (see again Fig. 1 in which 10 and 14 carbon species are marked).

2. Vinyl Polymers

Poly(vinyl chloride) contains C—C, C—H, and C—Cl bonds, with the C—Cl bonds weakest at about 73 kcal/mole. Consequently, the first step is the loss of Cl·, which subsequently combines with hydrogen to form HCl, leaving the unsaturated polymer backbone, with formation of the very stable aromatic products upon further heating.

3. Methacrylates

Carbons in the chain of a methacrylate polymer are bonded to the CO_2R side group, the CH_3 side group, to hydrogens, and to other chain carbons (structure VII), with C—C bonds being weaker than C—H bonds. Of the C—C bonds, the ones making up the chain are the weakest and produce the most stable free radicals, since breaking C—CH_3 produces CH_3·, the least stable free radical. Consequently, beta scission with an unzipping back to the monomer represents the most stable product formed by the most stable free radical.

III. EXAMPLES AND APPLICATIONS

Analytical pyrolysis is frequently considered to be a technique mainly applied to the analysis of polymers, which may at first seem fairly limited. However, when one considers that proteins, polysaccharides, plastics, adhesives, paints, etc. are included in this general category of "polymers," the list of applications becomes much longer. Natural and synthetic polymers, in the forms of textile fibers, wood products, foods, leather, paints, varnishes, plastic bottles and bags, and paper and cardboard make up the bulk of what we come into contact with every day. In fact, it is difficult to sit in a room and touch something—paint, paneling, carpet, clothing, countertop, telephone, upholstery, books—that is not made of some sort of polymer. Consequently, the study of materials using pyrolysis has become a very broad field, including such diverse topics as soil nutrients, plastic recycling, criminal evidence, bacteria and fungi, fuel sources, oil paintings, and computer circuit boards. The examples in this chapter will review in only a very general way some of the applications of analytical pyrolysis. Subsequent chapters treat some of these areas in greater depth.

A. Forensic Materials and Paints

The application of pyrolysis techniques to the study of forensic samples has a long and well-documented history, including a review of pyrolysis-mass spectrometry as a forensic tool in 1977 by Saferstein and Manura [8], and a general review by Wheals [9]. A wide variety of sample types has been investigated, including chewing gum, rubber and plastic parts from automobiles, drugs, and bloodstains.

Perhaps the best known forensic application of pyrolysis is the analysis of paint flakes from automobiles, a standard practice in many laboratories backed by substantial libraries of pyrograms and sample materials. Munson et al. [10] describe their work using pyrolysis-capillary GC-MS for the analysis of paint samples, and Fukuda [11] has published results on nearly 80 paints used in the Japanese automobile industry. The same techniques may be applied to paint samples recovered from nonautomotive sources, including house paints and tool and machine coatings, as well as varnishes from furniture and musical instruments.

Most automotive finishes are applied in layers, which may be removed selectively and analyzed individually, or pyrolyzed intact. An advantage provided by pyrolysis is that the inorganic pigment material is left behind and only the organic pyrolyzate transferred to the analytical instrument. Because of the great variety of polymeric materials used as paints and coatings, including acrylics, urethanes, styrenes, epoxies, etc., the resulting pyrograms may be quite complex. It is not always necessary to identify all of the constituents involved, however, to make comparisons among related paints. Figure 4 shows a comparison of three paint samples of similar monomer composition. Although many of the peaks are very similar for all three formulations, the inversion on the relative peak height in the second and third largest peaks makes it relatively straightforward to see that paints A and B are the same formulation and paint C is not a match.

Frequently samples like paint flakes present a problem to the analytical lab because they are small, nonvolatile and opaque

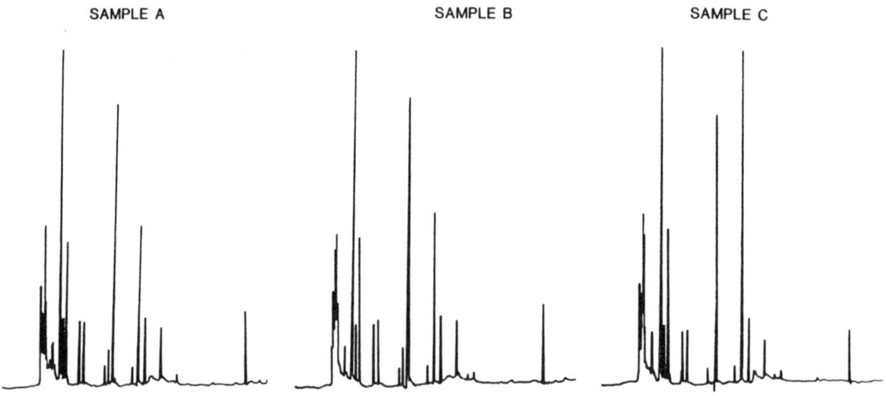

FIGURE 4 Comparison of paint pyrolyses, showing paints A and B matching, C being a different formulation.

with inorganic pigments. Since pyrolysis prepares a volatile organic sample from a polymer or composite, it offers the ability to introduce these organics to an analytical instrument separate from the inorganics, using only a few micrograms of sample. This extends the use of analytical techniques such as mass spectrometry and FT-IR spectroscopy to the investigation of small complex samples. When an opaque paint is pyrolyzed, the organic constituents are volatilized and available for analysis apart from the pigment material. A paint formulation based on methacrylate monomers, for example, will pyrolyze to reveal the methacrylates despite the presence of the pigment, and techniques such as FT-IR, which were previously unable to provide good spectral information, may be applied to the pyrolyzate only. Figure 5 shows the pyrolysis-FT-IR comparison of poly(methyl methacrylate) and poly(ethyl methacrylate). In each case, a 200 μg sample of the solid polymer was pyrolyzed for 5 seconds in a cell fitted directly into the sample compartment of the FT-IR. The cell was positioned so that the FT-IR beam passed directly over the platinum filament of the pyrolyzer. The samples were pyrolyzed, and the pyrolyzate scanned for 10 seconds, producing the spectra

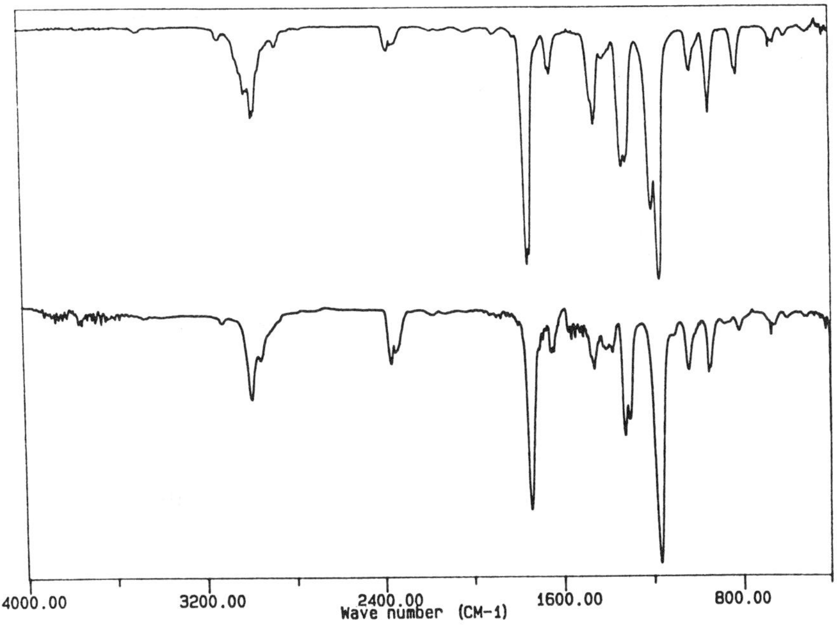

FIGURE 5 Comparison of poly(methyl methacrylate) (top) and poly(ethyl methacrylate) (bottom) by pyrolysis-FT-IR.

shown. This system, details of which are published [12], permits the rapid scanning of polymer-based materials, requiring approximately one minute per sample.

B. Fibers and Textiles

Almost all clothing is made from fibers of natural polymers such a the proteins in silk and wool, cellulose in cotton, synthetic polymers including the various nylons and polyesters, or blends of both natural and synthetic polymers. Since these polymers are all chemically different, the pyrolyzates they generate are all distinctive and provide a ready means of fiber analysis. Significant work has been done in the analysis and comparison of the various nylons by Tsuge et al. [13], among others, as well as acrylate and methacrylate/acrylonitrile copolymer fibers by Saglam [14]. A good overview of fiber analysis by pyrolysis-MS was published by Hughes et al. [15].

Figure 6 shows a pyrogram of silk fibers which are made of the protein fibroin, which is nearly 50% glycene. Figure 7 is a pyrogram of the polyamide nylon 6/12, which is formulated using a diamine containing 6 carbons and a dicarboxylic acid containing 12 carbons. Although both silk and nylon are polyamides, the chemical differences between them make distinctions using pyrolysis gas chromatography relatively simple. The same techniques may be used to differentiate among the various nylon formulations, to distinguish silk from wool, etc.

Polymer blends used in clothing may be analyzed in the same manner. Because the degradation of a specific polymer is largely an intramolecular event, the presence of two different fibers being pyrolyzed simultaneously generally produces a pyrogram resembling the superposition of the pyrograms of the two

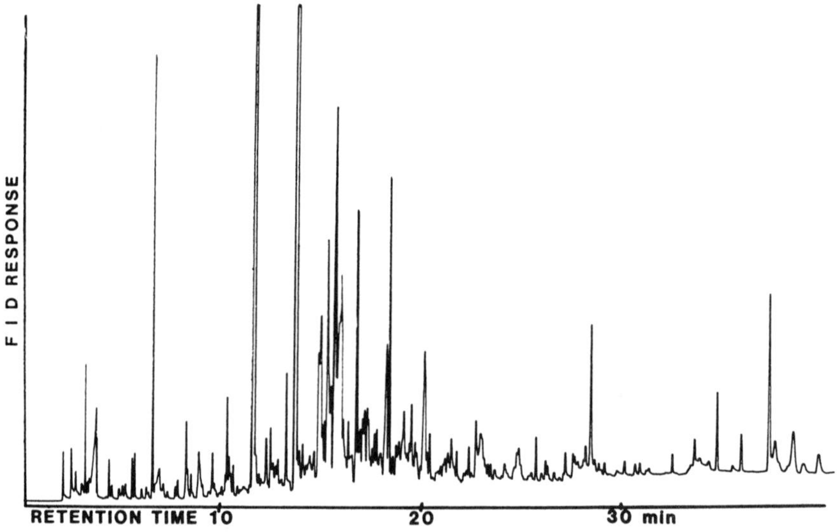

Figure 6 Pyrolysis of silk thread (675°C for 10 s in a glass-lined system).

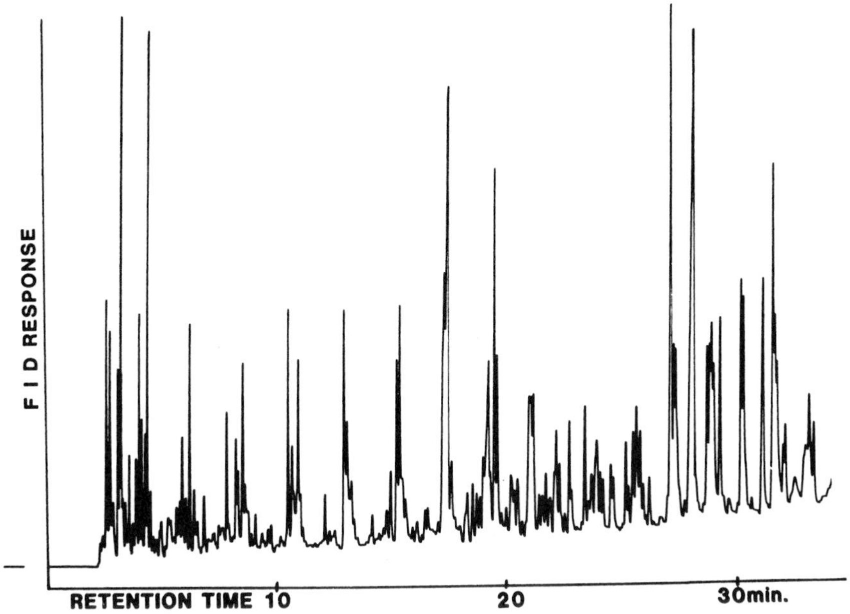

FIGURE 7 Pyrogram of nylon 6/12 (800°C for 10 s).

pure materials. A good example of this is shown in Figure 8, which compares pyrograms of cotton, polyester, and cotton/polyester blend threads. Each sample was a piece of thread 1 cm in length, pyrolyzed at 750°C. The top pyrogram shows just cotton, which produces large amounts of CO_2 and H_2O, as well as some larger molecules, which appear in the chromatogram. The bottom pyrogram is of polyester only, which produces a larger abundance of chromatographic peaks. The center pyrogram is the result of pyrolyzing a 50/50 cotton/polyester blend thread. Even though the cellulose and polyester are copyrolyzed, the patterns of each individually are easily discernible. Specific peaks are marked C for cotton and P for polyester, indicating peaks associated only with the pure material. Washall and Wampler [16] have published further examples of pyrograms of complex, multicomponent systems.

C. Paper, Ink, and Photocopies

Paper, which is primarily cellulose, printing ink, writing ink, and photocopy toners have all been analyzed by pyrolysis, sometimes independently as pure materials, and sometimes intact as a fragment of a document. Zimmerman et al. [17] used pyrolysis GC to extend the specificity of toner identification in photocopies. Ballpoint pen ink, various papers, and photocopy toners have been analyzed using a combination of headspace sampling and pyrolysis by Wampler and Levy [18].

Photocopy toner materials are generally complex formulations of pigments and polymers unique to a specific manufacturer. The formulations of these toners frequently include poly(-styrene), polyesters, acrylic polymers, methacrylic polymers of various side-chain lengths, and various other additives. The formulations vary both from manufacturer to manufacturer and also year to year as improvements are made and new processes introduced. Figure 9 shows a comparison of a Kodak and a Xerox photocopy made in 1985. The pyrograms were prepared by pyrolyzing a single letter punched from each photocopy. The punch, both paper and toner letter, was pyrolyzed at 650°C in a small quartz tube, and analyzed by capillary gas chromatography. The paper, which is essentially cellulose and produces a pyrogram like the one shown for cotton thread in Figure 8, contributes many of the smaller peaks, including the one marked G, which is furfural. The synthetic polymers in the toner materials pyrolyze to generate the larger peaks, including A, methyl methacrylate, and J, which is styrene.

When the same analysis is carried out using a sample of paper which was written upon with ballpoint pen, three sets of peaks are produced. Because the ink was applied as a fluid, there may be traces of the liquid or semiliquid vehicle remaining in the sample. In addition, there are the peaks resulting from actual pyrolysis of nonvolatile ink constituents, and from the paper it-

---

Figure 8 Comparison of pyrograms obtained from cotton (top), polyester (bottom), and cotton/polyester blend (center).

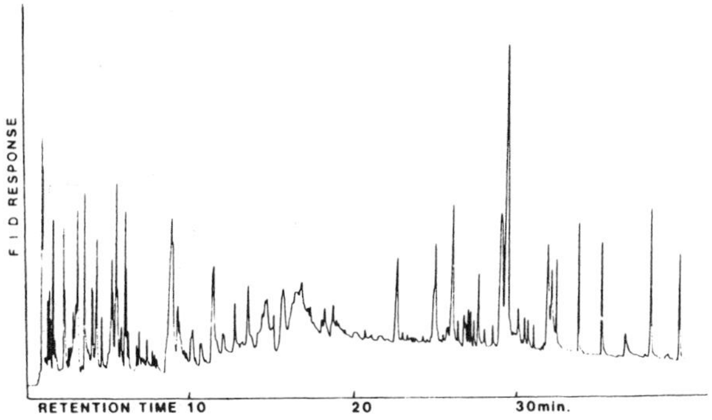
F I D RESPONSE
RETENTION TIME 10
20
30min.

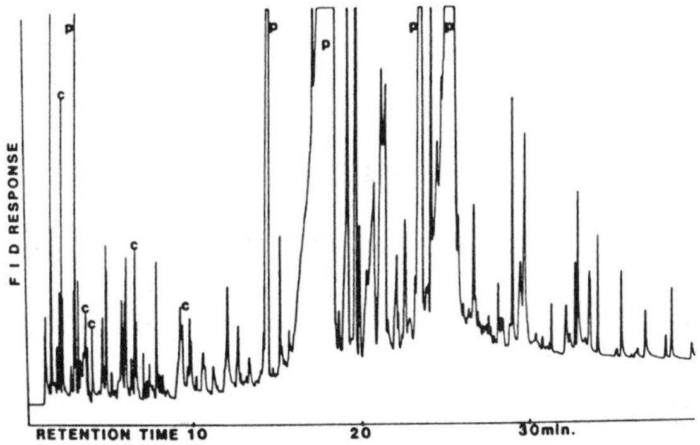
F I D RESPONSE
p
c
RETENTION TIME 10
20
30min.

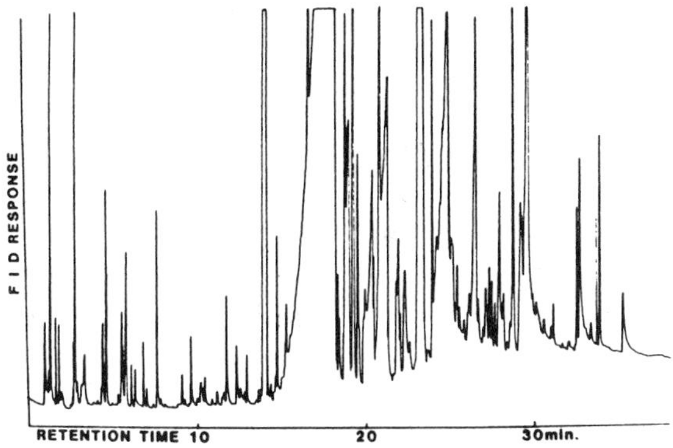
F I D RESPONSE
RETENTION TIME 10
20
30min.

self. Since less ink is applied than in the case for photocopy toner, the peaks from the paper generally make up a larger part of the chromatogram for ink analysis than for toner analysis.

Figure 10 shows the pyrogram resulting from pyrolyzing a 2 mm^2 piece of paper with ballpoint ink on it. All of the peaks labeled with numbers are cellulose pyrolyzates from the paper. Peaks X and Y are from the ink vehicle and will show up in an independent headspace sampling at 200°C, and consequently, they leave the sample before pyrolysis takes place. The peaks labeled A, B, and C are pyrolysis products from the remaining, nonvolatile residue of the ink. Just as with the cotton/polyester blend thread, and the photocopy toner on paper, the presence of several components in the sample causes little effect in the pyrolytic behavior of each individual material, and peaks A, B, and C are seen if the ink is pyrolyzed alone or while it is on the paper.

D. Art Materials and Museum Pieces

Valuable art objects, including paintings, furniture, and archeological artifacts are frequently investigated via pyrolysis of the nonvolatile materials used in their construction. Paints, varnishes, glues, pigments, waxes, and organic binder formulations have been studied from the aspects of both conservancy and authentication. The materials used for varnishes and other protective coatings and glues used in assembling pieces and other polymeric species change from region to region, time to time, and even with specific artists and craftsmen, so that identification of these materials goes far to indicate the authenticity, and even age of a particular object. Natural resins used to formulate varnishes for centuries have been analyzed by Shedrinsky et al. [19], showing good distinctions among dammar, mastic, sandarac, and copals. Wright and Wheals [20] have published results on natural gums, waxes, and resins as they apply to Egyptian artifacts. Frequently art and archeological samples must be investigated in

Figure 9 Comparison of Kodak (top) and Xerox (bottom) photocopy toners on paper.

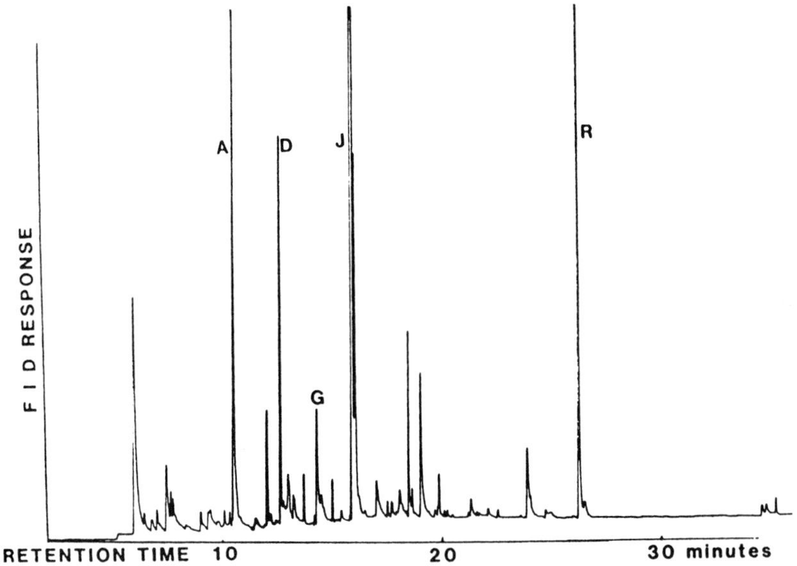
F I D RESPONSE
A
D
J
R
G
RETENTION TIME 10
20
30 minutes

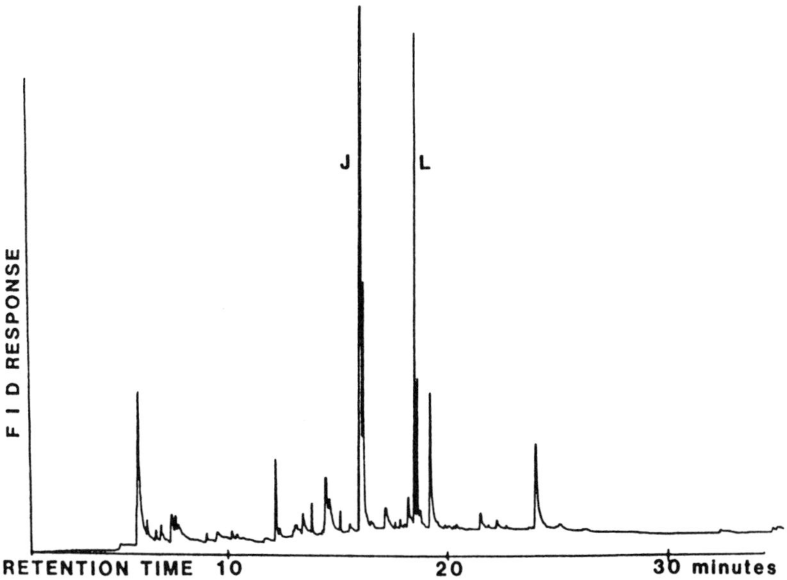
F I D RESPONSE
J
L
RETENTION TIME 10
20
30 minutes

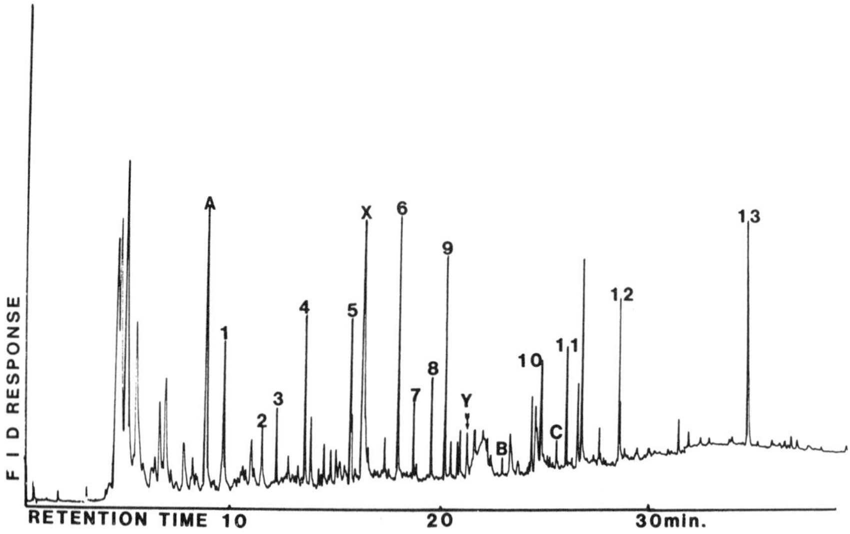

FIGURE 10 Pyrolysis of ballpoint pen black ink (NPC) on paper (650°C for 10 s). Numbered peaks result from paper pyrolysis. Peaks A, B, and C are from pyrolysis of ink nonvolatiles

layers, since protective coatings are generally applied over the original artwork, which itself may have been applied onto a prepared surface. The identification may involve analysis of the binder user in a subsurface, the oil or resin in the artwork, and the natural or synthetic polymer present in the protective coating.

An example of this is shown in Figure 11. The sample came from an Egyptian sarcophagus which was believed to originate from about the fourth century A.D. The object was constructed of wood, which was covered with a white layer (the "ground") of essentially inorganic material, used as a base for decorative paintings. It was decided to investigate the organic binder used in preparing the ground as a measure of the authenticity of the sarcophagus. If an ancient material had been used, the authenticity would be supported, if a modern adhesive was detected, the sarcophagus would be fraudulent. Various natural binder materials were proposed and investigated, including egg, wax, animal

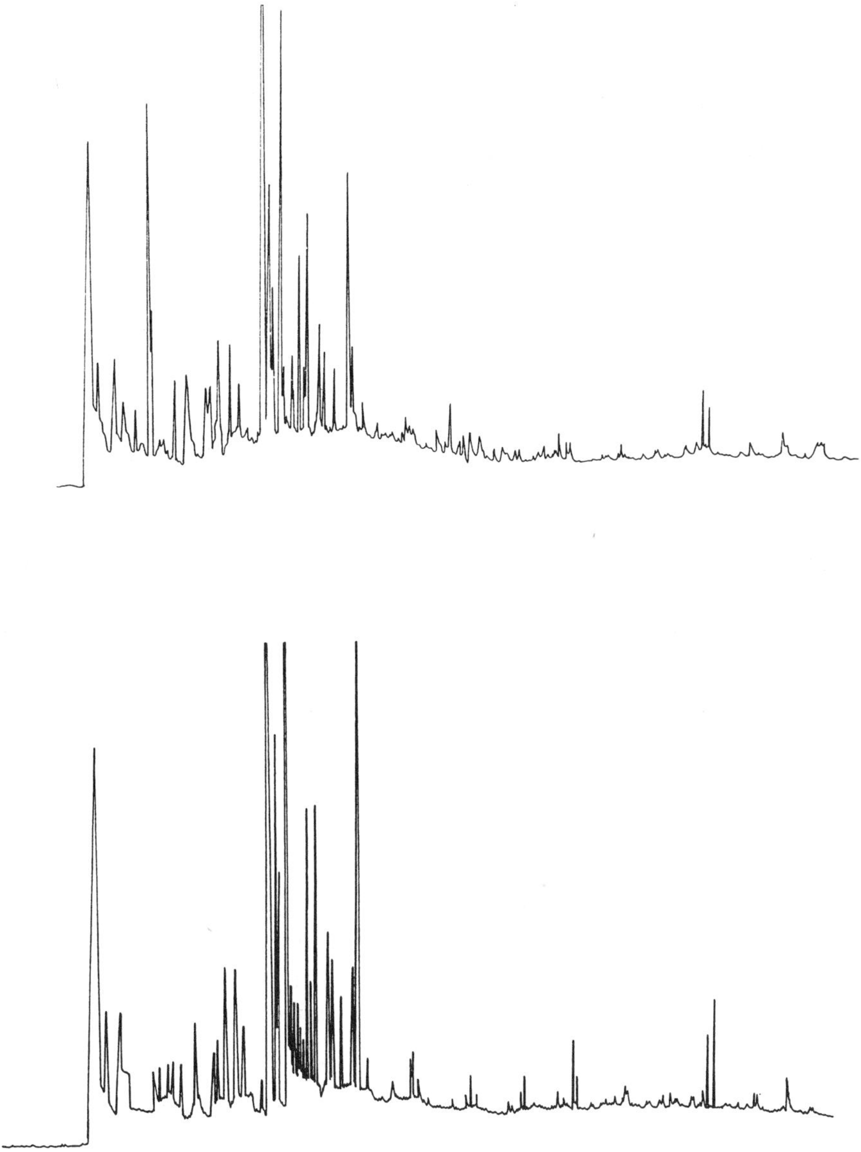

FIGURE 11 Pyrolysis (500°C for 20 s) of Egyptian sarcophagus ground material for binder content (bottom), compared to sample of ancient animal glue (top).

glue, and copal resin. When a sample of the ground layer was pyrolyzed at 500°C, the pyrogram matched that produced from ancient animal glue which had been independently authenticated.

An extensive review of pyrolysis applications in the analysis of artwork and antiquities has been published by Shedrinsky et al. [21], with examples including glues and other adhesives, oil paints, varnishes, natural resins, and reference to the famous van Meegeren case.

E. Synthetic Polymers

Perhaps the widest application of analytical pyrolysis is in the analysis of synthetic polymers, both from the standpoints of product analysis and quality control as well as polymer longevity, degradation dynamics, and thermal stability. Several specific polymers have been discussed in Section II, and subsequent chapters treat specific families of polymers in detail.

That individual polymers may be distinguished from one another fairly readily has been demonstrated. For example, telling poly(ethylene) from poly(propylene) as in Figure 1, or poly(methyl methacrylate) from poly(butyl methacrylate). Since pyrolysis generates fragments that retain molecular structures intact, however, much finer distinctions may be made. Defect structures, branching, head-to-head or tail-to-tail linking, extraneous substitutions, and efficiency of curing have all been investigated. Copolymers may be studied, revealing the monomers involved and even the relative amounts of each monomer present in the final polymer. Even the stereochemistry of polymers may be investigated, since the positioning of side groups in different stereochemical orientations produces pyrolysates of different composition.

Copolymers of methyl methacrylate and ethyl acrylate (EA), varying from 2–32% EA, have been studied by Shen and Woo [22], showing a calibration curve for the determination of monomer ratios in unknowns. Similar work for copolymers of butadiene and acrylonitrile is reported by Weber [23]. When a copolymer is pyrolyzed, the resulting pyrogram may be quite complex, as in the case of polyolefins, or rather simple, if the polymer chain unzips to a monomer. If the monomers are all methacrylates, for example, the pyrogram will show major peaks for each of the corresponding monomers involved. Even for more complex materials, however, the pyrograms are characteristic of the original macromolecular system. Information is present indicating whether the sample is a physical blend of homopolymers, or resulting from the copolymerization of monomers. In the latter case, fragments will be present that incorporate molecules of both monomers. These fragments could not result from the pyrolysis of homopolymers, but rather indicate the position of monomer units relative to each other in the macromolecule. It is possible to identify small fragments resulting from only one or the other of the monomers, and thus study the effect of relative monomer concentration on abundance of specific peaks in the pyrogram. Figure 12 shows the pyrogram of a copolymer of polypropylene and 1-butene, with peaks marked A and B, associated

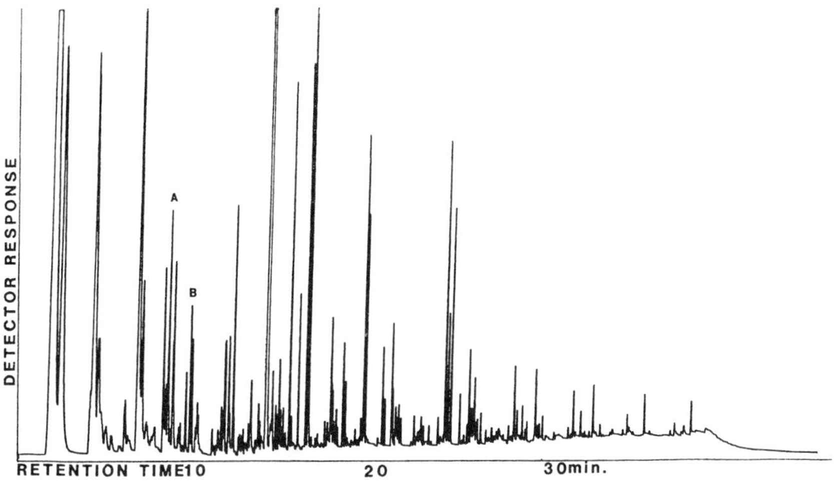

Figure 12 Pyrogram of a copolymer of poly(propylene) and 1-butene, with peaks marked A and B associated with the propylene monomer and 1-butene, respectively.

with the propylene monomer and 1-butene, respectively. Similar work, comparing copolymers and polymer blends of ethylene and propylene, has been published by Tsuge et al. [24].

F. Natural Materials and Biologicals

A rapidly growing area of pyrolytic investigation involves the study of natural materials and biopolymers. Although fuel sources [25,26], kerogens [27,28], and coals [29,30] have been analyzed for decades, more recently analysts are employing pyrolysis techniques in analyzing soil materials [31], microorganisms [32,33], and biomass [34]. Smith et al. [35] describe a chemical marker for the differentiation of group A and group B streptococci, and *Salmonella* strain characterization has been demonstrated by Tas et al. [36]. Environmental samples are being analyzed as well, as in the case of the spruce needle analysis performed by Schulten [37] in conjunction with a study of forest death in Germany.

Although biological samples may be more complex than pure polymers, the degradation of biopolymers must follow the same kinds of chemical processes, producing a volatile distribution which, if complex, is also representative of the original material. Biopolymers include polyamides, as proteins, and polysaccharides, such as cellulose. Biological samples are likely to be complex systems based on such biopolymers with the addition of other, sometimes characteristic materials. Wood, for example, includes the basic macromolecules of cellulose and lignin, and different wood species differ from each other in the presence and amounts of additional substances, including terpenes. Microorganisms, including bacteria and fungi, have been studied in the intact state as well as in isolated parts, such as cell walls.

Figure 13 shows a comparison of two polymers of glucose—cellulose and starch. Since these materials are both comprised of the same monomer, it is understandable that the pyrograms are similar. Cellulose and starch do differ, however, in the orientation of the linkage between the glucose units, and this difference affects the kinds and relative abundances of the pyrol-

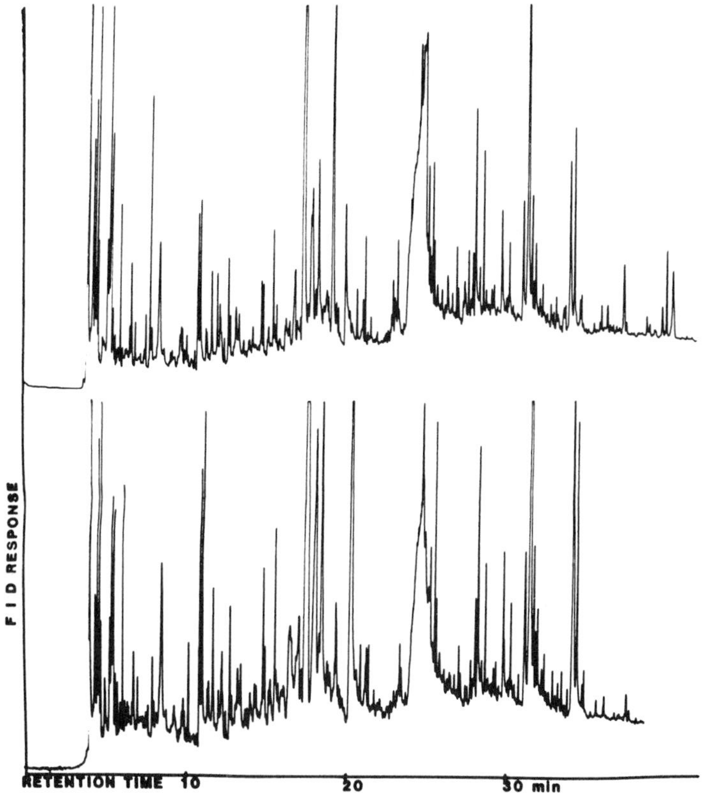

FIGURE 13 Comparison of two polymers of glucose pyrolyzed at 750°C for 10 s—cellulose (top) and starch (bottom).

yzate products formed. Gelatin, hair, and nail are also similar in that they are all protein-based materials, but easily distinguishable in the pyrograms shown in Figure 14.

Polymeric materials then, whether natural, such as cellulose, resins, and proteins; or synthetic, such as polyolefins, nylons, and acrylics, behave in reproducible ways when exposed to pyrolysis temperatures. This permits the use of pyrolysis as a sample preparation technique to allow the analysis of complex materials using routine laboratory instruments. Pyrolytic devices may now be interfaced easily to gas chromatographs, mass spec-

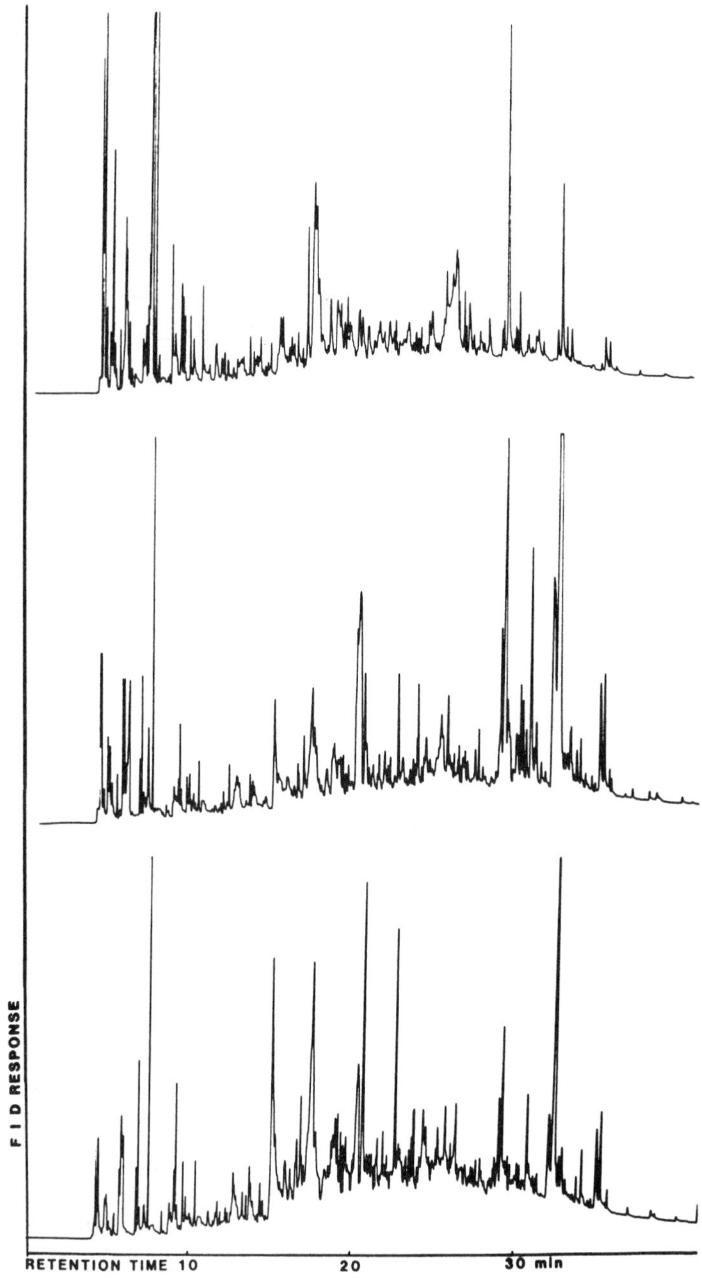
F I D RESPONSE
RETENTION TIME 10
20
30 min

trometers, and FT-IR spectrometers, extending their use to solid, opaque, and multicomponent materials. Forensic laboratories have long made use of pyrolysis for the analysis of paint flakes, textile fibers, and natural and synthetic rubber and adhesives. The list of applications has been expanded to include documents, artwork, biological materials, antiquities, and other complex systems which may be analyzed with or without the separation of various layers and components involved.

REFERENCES

1. W. I. Irwin, *Analytical Pyrolysis: A Comprehensive Guide*, Marcel Dekker, New York (1982).
2. S. A. Liebman and E. J. Levy, *Pyrolysis and GC in Polymer Analysis*, Marcel Dekker, New York (1985).
3. T. P. Wampler, *J. Anal. Appl. Pyrol.*, *16*: 291–322 (1989).
4. S. L. Morgan, A. Fox, L. Larsson, and G. Odham, *Analytical Microbiology Methods, Chromatography and Mass Spectrometry*, Plenum Press, New York (1990).
5. S. Tsuge, Y. Sugimura, T. Nagaya, T. Murata, and T. Takeda, *Macromolecules*, *13*: 928–932 (1980).
6. T. P. Wampler, *J. Anal. Appl. Pyrol.*, *15*: 187–195 (1989).
7. R. P. Lattimer and W. J. Kroenke, *J. Appl. Polym. Sci.*, *25*: 101–110 (1980).
8. R. Saferstein and J. J. Manura, *J. Forens. Sci.*, *22*: 749–756 (1977).
9. B. B. Wheals, *J. Anal. Appl. Pyrol.*, *2*: 277–292 (1981).
10. T. T. Munson, D. G. McMinn, and T. L. Carlson, *J. Forens. Sci.*, *30*: 1064–1073 (1985).
11. K. Fukuda, *Forens. Sci. Inter.*, *29*: 227–236 (1985).
12. J. W. Washall and T. P. Wampler, *Spectroscopy*, *6*, 4: 38–42 (1991).
13. S. Tsuge, H. Ohtani, T. Nagaya, and Y. Sugimura, *J. Anal. Appl. Pyrol.*, *4*: 117–131 (1982).

FIGURE 14 Comparison of three protein materials pyrolyzed at 750°C for 10 s—gelatin (top), human hair (center), and human fingernail (bottom).

14. M. Saglam, *J. Appl. Polymer Sci., 32*: 5717–5726 (1986).
15. C. Hughes, B. B. Wheals, and M. J. Whitehouse, *J. Anal. Appl. Pyrol., 103*: 482–491 (1978).
16. J. W. Washall and T. P. Wampler, *J. Chrom. Sci., 27*: 144–148 (1989).
17. J. Zimmerman, D. Mooney, and M. J. Kimmet, *JFSCA, 31*, 2: 489–493 (1986).
18. T. P. Wampler and E. J. Levy, *LC-GC, 4*, 11: 1112–1116 (1987).
19. A. M. Shedrinsky, T. P. Wampler, and N. S. Baer, The identification of dammar, mastic, sandarac and copals by pyrolysis gas chromatography, *Wiener Berichte uber Naturwissenschaft in der Kunst*, VWGO, Wien (1988).
20. M. M. Wright and B. B. Wheals, *J. Anal. Appl. Pyrol., 11*: 195–212 (1987).
21. A. M. Shedrinsky, T. P. Wampler, N. Indictor, and N. S. Baer, *J. Anal. Appl. Pyrol., 15*: 393–412 (1989).
22. J. J. Shen and E. Woo, *LC-GC, 6*, 11: 1020–1022 (1988).
23. D. Weber, *Int. Lab., 9*: 51–54 (1991).
24. S. Tsuge, Y. Sugimura, and T. Nagaya, *J. Anal. Appl. Pyrol., 1*: 222–229 (1980).
25. R. P. Philp, *Org. Geochem. 6*: 489–501 (1984).
26. H. Bar, R. Ikan, and Z. Aizenshtat, *J. Anal. Appl. Pyrol., 10*: 153–162 (1986).
27. A. J. Barwise, A. L. Mann, G. Eglinton, A. P. Gowear, A. M. Wardroper, and C. S. Gutteridge, *Org. Geochem., 6*: 343–349 (1984).
28. A. K. Burnham, R. L. Braun, H. R. Gregg, A. M. Samoun, *Energy Fuels, 1*: 452–458 (1987).
29. J. J. Delpeuch, D. Nicole, D. Cagniat, P. Cleon, M. C. Foucheres, D. Dumay, J. P. Aune, A. Genard, *Fuel Proc. Technol., 12*: 205–241 (1986).
30. A. M. Harper, H. L. C. Meuzelaar, P. H. Given, *Fuel, 63*: 793–802 (1984).
31. J. M. Bracewell and G. W. Robertson, *Geoderma, 40*: 333–344 (1987).
32. H. Engman, H. T. Mayfield, T. Mar, W. Bertsch, *J. Anal. Appl. Pyrol., 6*: 137–156 (1984).

33. C. Gutteridge, *Meth. Microbiol.*, *19*: 227–272 (1987).
34. T. A. Milne and M. N. Soltys, *J. Anal. Appl. Pyrol.*, *5*: 111–131 (1983).
35. C. A. Smith, S. L. Morgan, C. D. Parks, A. Fox, and D. G. Pritchard, *Anal. Chem.*, *59*: 1410–1413 (1987).
36. A. C. Tas, J. DeWaart, and J. Van der Greef, *J. Anal. Appl. Pyrol.*, *11*: 329–340 (1987).
37. H.-R. Schulten, N. Simmleit, and H. H. Rump, *Int. J. Environ. Anal. Chem.*, *27*: 241–260 (1986).

2

Instrumentation and Analysis

Thomas P. Wampler
CDS Analytical Inc., Oxford, Pennsylvania

I. INTRODUCTION

To perform an analysis by pyrolytic techniques in the laboratory, it is necessary to assemble a system that is capable of heating small samples to pyrolysis temperatures in a reproducible way, interfaced to an instrument capable of analyzing the pyrolysis fragments produced. This chapter will discuss the various methods available for the convenient pyrolysis of laboratory samples and some general considerations regarding the interface of such units to analytical instruments. In addition, concerns about sample preparation, experimental reproducibility, and sources of error will be discussed.

A typical pyrolytic analysis involves sample preparation, pyrolysis, transfer of the pyrolyzate to the analytical instrument, and then analysis. Pyrolysis-gas chromatography (GC) is still the most common technique, but pyrolysis-mass spectrometry (MS) and pyrolysis-Fourier-transform infrared (FT-IR) spectrometry are also common. In any system, the quality of the results will be no better than that permitted by any of its parts, and therefore it is important not only to use a reliable pyrolysis technique, but also to be aware of the effects of sample variations, mechanical and pneumatic connections, and instrument optimization. Although many analysts still design and use their own pyrolysis devices, variations in conditions and design frequently make it difficult to achieve reproducibility. The availability of a variety of pyrolysis instruments commercially has done much to improve the quality of pyrolysis experiments and reduce the frustrations common to early analysts.

Although there is a multitude of ways in which a sample could be heated sufficiently to break bonds, this chapter will treat only those ways that are readily available in the form of laboratory equipment. These instruments may be categorized as furnaces, both isothermal and programmable, inductively heated filaments, and resistively heated filaments. Interesting work has been done using lasers, solar radiation, electric arc, and other nonconventional heating units, but these experiments are frequently one-of-a-kind systems, and are beyond the scope of this book.

II. PYROLYSIS INSTRUMENTS

A. General Considerations

In the typical analysis by pyrolysis, it is essential to heat a small sample to its final temperature as quickly as possible. Samples are generally small because of the sampling capacity of the analytical instrument. For example, most gas chromatographic columns and detectors cannot handle more than a few micrograms of sample. This works to the advantage of an analyst who is using pyrolysis as a sample introduction technique, since a small sample will heat to its endpoint temperature quickly, with less thermal gradient than a large sample. It is important to heat the entire sample to the endpoint quickly, especially if one is studying the effects of pyrolysis temperature on the composition of the pyrolysate. Samples slowly heated will undergo considerable degradation while the pyrolysis instrument is still heating to the final setpoint temperature, and a large sample will degrade according to the temperature distribution across the sample as it is being heated. Consequently, reproducibility may depend heavily on the ability of the pyrolysis instrument to heat the sample uniformly, and to achieve the final temperature before the sample has already begun degradation. Pyrolysis instruments commercially available are capable of heating filaments to temperatures in excess of 800°C in milliseconds, producing rapid degradation of small, thin samples.

There are experiments and conditions, however, under which it is impossible or undesirable to pyrolyze the sample at such a fast rate. Large samples, used because they are nonhomogeneous or low in organic content, may be analyzed by pyrolysis, but the effects of the slower heating rate and thermal gradient through a large sample must be taken into account. In addition, there are experiments and techniques in which slow heating is specifically the point (such as thermogravimetric analysis, or time-resolved spectroscopic analyses) and reproducible, slow-heating profiles or larger sample capacity (or both) are advantages rather than obstacles.

III. ANALYTICAL PYROLYSIS AT RAPID RATES

The three most widespread techniques used to pyrolyze samples rapidly for analysis by GC, MS, or FT-IR are: isothermal furnaces; Curie-point (inductively heated) filaments; and resistively heated filaments. There are specific design advantages to each technique, depending on the sample to be analyzed and the physical requirements of the experiment conducted. Each way is capable of providing reproducible pyrolysis for small samples, and many laboratories use more than one type of instrument. The selection of one technique over another depends frequently on personal preference, experimental requirements, budget, or availability. Since each technique heats in its own unique fashion, it is important to keep the physical differences in mind when comparing pyrolysis results. Having chosen (or inherited) any specific instrument, it is important to understand its heating characteristics to capitalize upon its advantages and minimize the effects of its drawbacks.

A. Furnace Pyrolyzers

In order to pyrolyze samples rapidly for introduction to a gas chromatograph, furnace pyrolyzers are generally held isothermally at the desired pyrolysis temperature, and the samples introduced into the hot volume. Carrier gas is generally routed through the furnace to remove the pyrolyzate quickly from the pyrolysis zone to minimize secondary pyrolysis.

1. Design

Isothermal furnace pyrolyzers are generally designed to be small enough to mount directly onto the inlet of a gas chromatograph. They are normally constructed of a metal or quartz tube which is wrapped with a heating wire, then insulated (Fig. 1). Carrier flow enters the top or front of the furnace, sweeping past a sample inlet or delivery system, then exits directly into the injection port of the chromatograph. They must contain enough mass to stabilize the temperature, which is generally held to within ± 1°C. A temperature sensor (thermocouple or resistance thermometer device) is usually installed between the heater and the furnace tube to indicate the wall temperature.

2. Sample Introduction

Care must be taken to introduce the sample for pyrolysis into the furnace without admitting air, since the pyrolysis zone is already hot and degradation begins immediately. The heating rate of the sample is dependent on the sample material itself, and on the composition of the sample introduction device. In the simplest configuration, a liquid sample is injected into the furnace using a standard syringe, the sample then vaporizes, and is followed by pyrolysis.

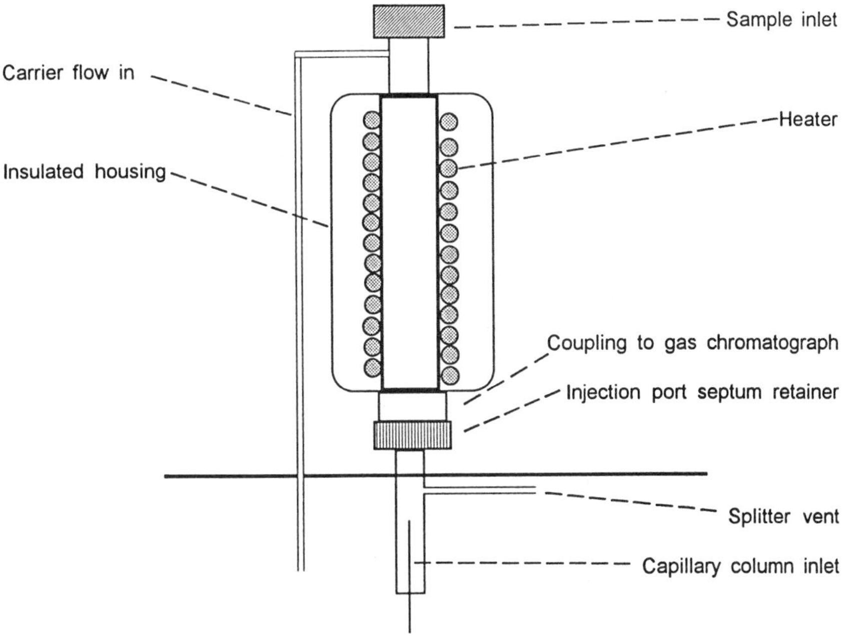

FIGURE 1 Microfurnace pyrolyzer installed on gas chromatograph injection port.

Solid samples present more of a problem, since they cannot be injected using a standard syringe. Some analysts dissolve soluble materials and inject the solution, pyrolyzing both the sample material and the solvent. Solid-injecting syringes have been designed which work on the principle of a needle inside a needle. The inside needle has a groove or slot into which a solid material may be placed. This needle then slides up into the outer needle for injection through the sample port of the furnace. Once inside the furnace, the inner needle is extended beyond the outer sheath, delivering the sample into the pyrolysis area.

Another approach has been developed by Shin Tsuge at the University of Nagoya in Japan [1]. His furnace pyrolyzer includes a cool chamber where samples are loaded into a small crucible above the hot zone. Once the sample is in place, the cup is rapidly lowered into the furnace for pyrolysis.

3. Temperature Control

The heating of a small, isothermal furnace is almost always achieved using a resistive electrical element wound around the central tube of the furnace. The temperature is monitored by a sensor, which feeds back temperature data to the controller, where adjustments are made for deviations from setpoint. It must be stressed that the temperature measured (and displayed) using this sensor is the temperature of the system at that specific location. Depending on the diameter, thickness, and mass of the furnace tube, the temperature experienced by the sample inside the tube may be quite different. It is possible to position a thermocouple inside the furnace to monitor temperature closer to the location of sample introduction. The temperature and rate of heating of the sample will also depend on the sample size and mass, and on the residence time inside the furnace. If the carrier gas is traveling too slowly through the furnace, the sample may degrade, then have time for the pyrolyzate material to interact with the furnace wall, producing secondary pyrolysis products. In general, reasonable results may be duplicated on the same instrument using the same displayed wall temperature, making the assumption that the internal temperature caused by this wall temperature is also reproduced.

4. Advantages of Furnace Pyrolyzers

Because of their simple construction and operation, furnace pyrolyzers are frequently inexpensive and relatively easy to use. Since they are operated isothermally, there are no controls for heating ramp rate or pyrolysis time. The analyst simply sets the desired temperature and when the furnace is at equilibrium, inserts the sample. Although this simplicity may lose its attractiveness as soon as the analyst requires control over heating rate or time, there are some experiments and sample types that capitalize on the design of a furnace. Liquid and especially gaseous samples are pyrolyzed much more easily in a furnace than by a filament-type pyrolyzer. Because filament pyrolyzers depend on applying a cold sample to the filament and then heating, it is very difficult to use them for any liquid which is readily volatile, or for gases. The furnace, however, since it is already hot, may receive an injection of a gas or liquid as easily (or more easily) than a solid sample. Moreover, filament-type pyrolyzers are almost always housed or inserted into a heated zone, which prevents condensation of the pyrolysis products, since the actual pyrolysis filament is heated for only a few seconds. Samples that may be solids at room temperature may still melt or vaporize inside the heated chamber, or may denature in the case of protiens. A furnace with a good sample introduction system may permit the pyrolysis of such samples by thrusting them rapidly into the hot zone of the pyrolyzer, before they have a chance to undergo adverse lower-temperature changes.

5. Disadvantages of Isothermal Furnaces

Perhaps the greatest concern when using an isothermal furnace pyrolysis instrument is the size and construction of the pyrolysis chamber itself. To insure thermal stability, the furnace tube is considerably larger than the sample being inserted into it. This produces a relatively large volume through which the sample must pass before entering the analytical device, with a large hot surface area. In some designs, this pyrolysis tube is quartz, in others, metal. Particularly with metal systems, the possibility exists that the initial pyrolysis will produce smaller organic fragments, which may then encounter the hot surface of the furnace tube and undergo secondary reactions. To counter this, furnaces are almost always operated with a high flow rate through the tube, (e.g., 100 ml/min) generally necessitating split capillary analysis. This high flow rate reduces the residence time for the sample inside the hot zone, and the required high split ratio is generally not a problem unless one is sample limited, or unless the sample analyzed is low in organic content.

The nature of heating and sample introduction in a furnace precludes one from knowing the temperature rise time of the sample, which will depend on the nature of the sample, its size, and geometry. In addition, the mass which gives the furnace its thermal stability also gives it thermal inertia. If one is establishing the effect of pyrolysis temperature on the products, or optimizing the production of fragments of interest, one must allow equilibration time between temperature changes to establish stability at the new temperature.

B. Heated Filament Pyrolyzers

Isothermal furnaces achieve a fairly fast sample heating by keeping the pyrolysis instrument hot and injecting samples into it. Heated filament pyrolyzers take the opposite approach in that the sample is placed directly onto the cold heater, which is then rapidly heated to pyrolysis temperature. Commercially, two heating methods are used: resistance heating, in which a controlled current is passed through the heating filament, and inductive heating, in which the current is induced into the heating filament which is made of a ferromagnetic metal. Each type achieves very fast heating rates by using a small filament, and consequently, the sample size must be limited to an amount compatible with the mass of the filament. This sample size is usually in the low to high microgram range, which is compatible with gas chromatograph column capacities providing the sample is essentially all organic in nature. For larger samples or materials of low organic content, filaments may not be able to heat the sample efficiently, and a furnace style may be more appropriate.

C. Inductively Heated Filaments: The Curie-Point Technique

It is well known that electrical current may be induced into a wire made of ferromagnetic metal by the use of a magnet, and this is the principle used to heat the filaments of a Curie-point pyrolysis system. If one continues to induce a current, the wire will begin to heat, and continue to heat until it reaches a temperature at which it is no longer ferromagnetic. At this point the metal becomes paramagnetic, and no further current may be induced in it. Consequently, the heating of the wire stops—the Curie point of the specific metal alloy being used. This Curie-point temperature is different for each ferromagnetic element, and also for different alloys of these metals. For example, a wire made entirely of iron will become paramagnetic and stop heating at a temperature of 770°C, while an alloy of 40% nickel and 60% cobalt will reach 900°C.

In a Curie-point pyrolyzer, an oscillating current is induced into the pyrolysis filament by means of a high-frequency coil. It is essential that this induction coil be powerful enough to permit heating the wire to its specific Curie-point temperature quickly. In such systems, the filament temperature is said to be self-limiting, since the final or pyrolysis temperature is selected by the composition of the wire itself, and not by some selection made in the electronics of the instrument. Properly powered, a Curie-point system can heat a filament to pyrolysis temperature in milliseconds. Providing that wires of the same alloy composition are used each time, the final temperature is well characterized and reproducible.

1. Design

Curie-point pyrolysis systems must be designed to permit easy insertion of the wire containing the sample into the high-frequency coil, in a chamber which is swept with carrier gas to the analytical instrument. Two approaches may be taken: (1) the pyrolysis chamber, which is surrounded by the coil, is opened and the sample wire is dropped or placed inside; or (2) the sample wire is attached to a probe which is inserted through a septum into the chamber which is surrounded by the coil (Fig. 2). The coil chamber itself may be attached directly to the injection port of the gas chromatograph, or it may be part of a larger module which is isolated from the GC by a valve. In the latter design, the sample may be introduced and removed easily without interrupting the carrier gas flow of the gas chromatograph.

Because the Curie-point filament is heated inductively, no connections are made to the wire. This facilitates autosampling, and permits loading the wires into glass tubes for sampling and insertion into the coil zone. Unlike the isothermal furnace, which

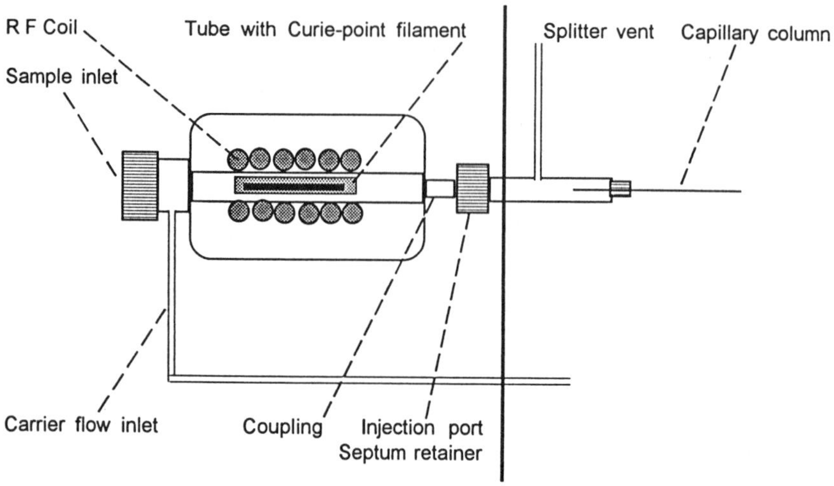

FIGURE 2 Curie-point pyrolyzer installed on horizontal GC injection port.

is on continuously, the Curie-point wire is heated only briefly and is cold the rest of the time. This necessitates heating the pyrolysis chamber separately to prevent immediate condensation of the fragments made during pyrolysis. Therefore, Curie-point pyrolyzers have controls for the parameters of the pyrolysis wire and also temperature selection for the interface chamber housing the wire.

2. Sample Introduction

To capitalize upon the very rapid heating rates possible using the Curie-point technique, the sample and wire should be kept to a low mass. The technique is best suited to the analysis of samples which may be coated onto the filament as a very thin layer. Soluble materials may be dissolved in an appropriate solvent, and the wire dipped into the solution. As the solvent dries, it leaves a thin deposit of the sample material, which will then heat rapidly and uniformly to pyrolysis temperature when the wire is heated. Paints, varnishes, and soluble polymers may be analyzed in this way quite easily.

Pyrolysis samples that are not soluble must be applied to the wires in some other fashion [2]. Finely ground samples may be deposited onto the wire from a suspension, which is then dried to leave a coating of particles on the wire. Another approach is to apply the sample as a melt, which then solidifies onto the wire. To attach materials that will not melt or form a suspension, some analysts flatten the wire, or create a trough in it. Some samples are held in place by bending or crimping the wire around the material, although this introduces concerns about the thermal continuity of a Curie-point material which has been distorted. To encapsulate a sample completely, some systems use a foil [3] of ferromagnetic material instead of a wire. The sample is placed in the center of a piece of metal foil, which is then wrapped around it and dropped into the high-frequency coil chamber, just as a standard wire would be.

3. Temperature Control

Control of the pyrolysis temperature in a Curie-point instrument is achieved solely by the alloy of the ferromagnetic material used. Pyrolysis of a material at a variety of temperatures requires a selection of Curie-point wires of different alloy composition. These wires are generally purchased from the manufacturer of the instrument, and each manufacturer offers a range of alloys covering a fairly wide temperature range. The analyst therefore has no access to temperature settings. Reproducible and accurate temperature control therefore relies on the accuracy of the wire alloy, the power of the coil, and the placement of the wire into the system. One may be reasonably confident that the sample wire achieves the theoretical temperature if the instrument has enough power to induce current sufficient to heat the wire to its limit. Use of wires from one manufacture lot throughout an experiment will increase the likelihood of temperature reproducibility, as will attention to sample loading and placement.

4. Advantages of Curie-Point Systems

The self-limiting temperature of an inductively heated wire and the rapidity with which it heats are major advantages of the

Curie-point system. Using wires of the same manufacture and alloy, one may be confident that the sample is being heated to a specific and known temperature in a rapid and reproducible manner. Since there is no temperature control setting, there is no temperature calibration to perform.

Curie-point systems have some additional advantages from the standpoint of convenience. The sample wires may be coated with sample, then placed into a glass tube for storage so that one may prepare several samples at once and subsequently load them into the pyrolyzer. In addition, since the heating wire need not be connected to a source of power, insertion into the unit is simple, and may be readily automated [4] by placing wires in glass tubes into a feeding magazine, or into a multiposition autosampler. Although the wires may be cleaned and reused, they may also be discarded after one use, eliminating concerns about carryover of sample from one analysis to another.

5. Disadvantages of Curie-Point Systems

Since the temperature of pyrolysis is a function of the Curie-point wire alloy composition, the wire, and consequently the sample, may be heated to that temperature only. If it is desired to evaluate several different pyrolysis temperatures, or to study the behavior of a sample material at different temperatures, it is necessary to use a different Curie-point wire for each run. Therefore, it is not possible with a Curie-point system to optimize the pyrolysis temperature of a sample by placing the material into the instrument and increasing the temperature in a stepwise fashion, observing the pyrolysis products after each heating. The fact that the sample is in contact with a different metal at each different temperature produces concerns about the catalytic effect of the metal during heating for only very small samples, since in larger samples energy is conducted through the sample material itself and pyrolysis takes place without contacting the metal surface. The analyst is, however, limited to the pyrolysis temperatures provided by the Curie-point wires or foils offered by the manufacturer. Curie-point materials are generally provided for a range of temperatures from 350–1000°C, covered by

10–20 specific alloys. Investigations at temperatures between the Curie points of the materials offered are not possible.

Although most pulse pyrolysis analyses benefit from the very rapid temperature rise time experienced by a Curie-point filament, there are investigations in which the analyst wishes to control the heating rate during pyrolysis. This has been attempted by modulating the power supplied to the coil, but for commercial instruments, slowing the heating rate is not possible. For experiments requiring the slow, linear heating of a sample material, a programmable furnace or resisitively heated filament pyrolyzer are required.

D. Resistively Heated Filaments

Like the Curie-point instruments, resistively heated filament pyrolyzers operate by taking a small sample from ambient to pyrolysis temperature in a very short time. The current supplied is connected directly to the filament, however, and not induced. This means that the filament need not be ferromagnetic, but that it must be physically connected to the temperature controller of the instrument. Filaments are generally made of materials of high electrical resistance and wide operating range, and include iron, platinum, and nichrome [5].

1. Design

As with Curie-point systems, the filament of a resistively heated pyrolyzer must be housed in a heated chamber which is interfaced to the analytical device. This interface chamber is generally connected directly to the injection port of a gas chromatograph, with column carrier gas flowing through it. The sample for pyrolysis is placed onto the pyrolysis filament, which is then inserted into the interface housing and sealed to insure flow to the column (Fig. 3). When current is supplied to the filament, it heats rapidly to pyrolysis temperatures, and the pyrolyzate is quickly swept into the analytical instrument.

The pyrolysis filament may be shaped for convenience of sampling, and may be a flat strip, foil, wire, grooved strip, or coil. In the case of the coil, a small sample tube or boat is inserted

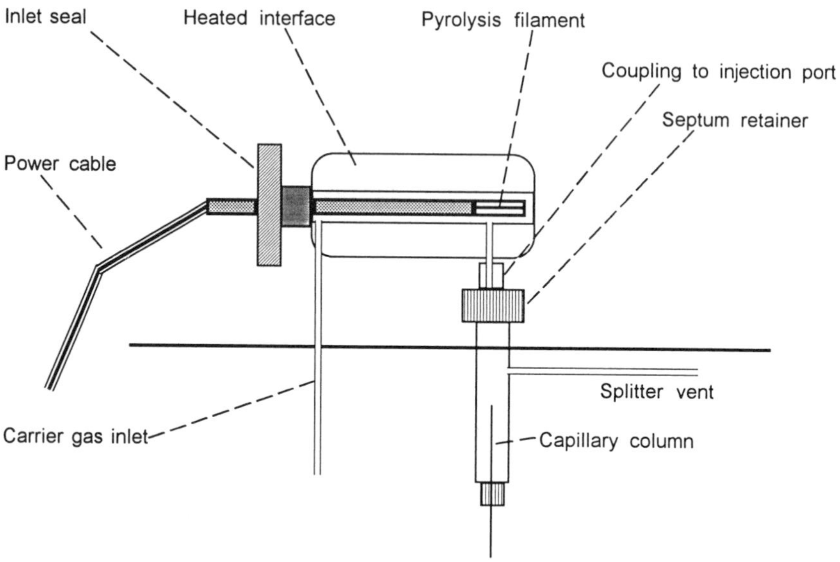

FIGURE 3 Resistive filament pyrolyzer installed on GC injection port.

into the filament so that the sample is not heated directly by the filament, but is in effect inside a very small, rapidly heating furnace. The pyrolysis filament must be connected to a controller capable of supplying enough current to heat the filament rapidly, with some control or limit since the materials used for filaments are not self-limiting. The temperature of the filament may be monitored using the resistance of the material itself, or some external measure such as optical pyrometry [6] or a thermocouple [7].

The filament used may be made to be quite small, and is therefore capable of being inserted into instruments other than gas chromatographs, including the sample cell of infrared spectrometers and the ion source of a mass spectrometer [8].

2. Sample Introduction

Samples may be applied to resistively heated filaments in the same manner used for Curie-point wires. Soluble materials may be deposited from a solvent, which is then dried before pyrolysis.

However, the solution is generally applied to the filament from a syringe, instead of dipping, since the filament is attached to a probe or housing. Insoluble materials may be melted in place to secure them before pyrolysis. Since the filament may be a flat ribbon or contain a grooved surface, placement of some solid materials may be simpler than when using a Curie-point wire.

Since the sample must be placed onto the filament before it is inserted into a heated chamber with carrier flow, placement of fibers and fine powders presents a problem. If these materials may not be melted onto the filament, they may fall off or blow off in the carrier gas stream before pyrolysis. Consequently, these sample types are generally analyzed using a small quartz tube which is inserted into a coiled filament. The sample may be placed into the tube, held in position using plugs of quartz wool, weighed, and then inserted into the coiled element for pyrolysis. It must be remembered that the sample is now insulated by the wall of the quartz tube, so the temperature rise time and final temperature will not be the same as that for a sample pyrolyzed directly on a filament. Use of a quartz tube, however, does extend the use of a filament pyrolyzer to materials such as soils, ground rock samples, textiles, and small fragments of paint.

Viscous liquids, such as heavy oils, may be applied directly to the surface of a filament or may be pyrolyzed while suspended on the surface of a filler material such as quartz wool inside a quartz tube. Lighter liquids, especially anything easily vaporized, will probably be evaporated from the filament by the heat of the interface before pyrolysis, and are probably better studied using a furnace-type pyrolyzer.

3. Interfacing

Filament pyrolysis instruments may be designed small enough to insert directly into the analytical device, especially the injection port of a gas chromatograph or the ion source of a mass spectrometer. In these cases, one need only insure that the filament is positioned so that the pyrolysate may enter the analytical portion of the instrument, and that the filament probe is sealed into the unit so that there are no leaks. For most gas chromatographs equipped with capillary inlets, however, considerable attention has been paid to the elimination of dead volume, and there is not sufficient room to accommodate a pyrolyzer. In these cases, some heated interface must be attached upstream of the injection port to house the pyrolyzer and insure efficient transfer of the pyrolysate onto the column. This interface should have its own heater, independent of the pyrolysis temperature to prevent condensation of pyrolyzate compounds, and be of minimal volume. Flow from the gas chromatograph is brought up into the interface, past the filament and then back into the injection port. In most cases, this means that the chromatograph is open to air when samples are being inserted, so the column oven is cooled before samples are introduced or removed. An alternative is to place a valve between the pyrolyzer and the injection port to isolate the chromatograph flow, permitting removal of the filament during a run [9].

4. Temperature Control

The temperature of a resistively heated filament is related to the current passing through it. Since resistively heated filaments are not self-limiting in the sense that Curie-point filaments are, exacting control of the filament current is essential for temperature accuracy and reproducibility. Early versions of resistively heated pyrolyzers relied on rather approximate control, and, given the difficulty of measuring and calibrating the final temperature, produced some mixed results. It must be remembered that the pyrolysis temperature of the filament depends upon the resistance of the filament, which is affected by both temperature and physical condition. A length of resistive wire could be expected to produce the same temperature each time it is heated using the same current, but relating that temperature to an instrument setpoint, or calibrating the unit after replacing a broken filament, proved problematical.

Current versions of resistively heated pyrolyzers incorporate small computers to control and monitor the filament temperature. These computers may be used to control the voltage used, adjust for changes in resistance as the filament heats, and compensate for differences when broken filaments are replaced. In addition, instruments have been designed which include photodiodes [6], which are used by the computer to measure the actual temperature of the filament during a run. Other instruments include a small thermocouple welded directly to the filament for temperature readout, or use the computer to measure the resistance of the filament itself and make adjustments as needed during a program.

Since the temperature and the rate of heating the filament are completely variable, the instrument may control these parameters independently, as single steps or multiple steps. This gives the analyst the control to select any final pyrolysis temperature, and to heat the sample to this temperature at any desired rate. Instruments are commercially available that heat as slowly as 0.01°C per minute and as rapidly as 30,000°C per second.

5. Advantages of Resistively Heated Filament Pyrolyzers

The central advantage of a resistively heated pyrolyzer is that the filament may be heated to any temperature over its usable range, at a variety of rates. This permits the examination of a sample material over a range of temperatures without the need to change filaments for each temperature. A sample may be placed onto the filament, heated to a setpoint temperature and the products examined, then heated to higher temperatures in a stepwise fashion without removing the sample or filament from the analytical device. This ability also allows pyrolysis at temperatures between the discrete values permitted by Curie-point filaments, and frequently at temperatures higher than those permitted by furnaces.

The ability to control the rate at which the filament, and thus the sample, is being heated extends the use of filament pyrolyzers in two ways. First, it permits the examination of materials and how they are affected by slow-heating, duplicating processes such as thermogravimetric analysis (TGA). Second, it permits the interface of spectroscopic techniques with constant scanning for three-dimensional, time-resolved thermal processing. A sample may be inserted directly into the ion source of a mass spectrometer, or placed in the light path of an FT-IR [10], and the products monitored in real time throughout the heating process.

6. Disadvantages of Resistively Heated Filament Pyrolyzers

Since the filament of a resistively heated pyrolyzer must be physically connected to the controller, two problems arise. First, it is difficult to automate the process since multiple samples must be delivered to the same filament, or multiple filaments must be controlled by the same instrument. Second, the temperature control of a resistively heated filament is based on the resistance of the entire filament loop—the filament as well as its connecting wires. Anything that damages or alters the resistance of any part of the loop will have an effect on the actual temperature produced by the controller.

An additional disadvantage may be produced by the fact that the filament must be housed in a heated zone. Introduction of some samples into a heated chamber before pyrolyzing them may produce volatilization or denaturation, altering the nature of the sample before it is actually degraded.

IV. PYROLYSIS AT SLOW OR PROGRAMMED RATES

Although for most analytical pyrolysis techniques it is advisable to heat the sample to its final temperature as rapidly as possible, there are times when just the opposite is required. To simulate thermal processes, such as TGA on a small scale, or to analyze the degradation products produced from a sample as they are being generated, slower, controlled heating is required. These slow-rate experiments demonstrate, for example, that poly(vinyl chloride) degrades in a two-step process, producing HCl at a relatively low temperature, and aromatics at elevated temperatures.

While a sample material is slowly heated to its degradation temperature, the volatile products may be swept by a carrier gas into an analytical instrument, or collected onto a trap for analysis as a composite. Alternatively, the sample may be pyrolyzed directly in the analytical instrument, while scans are being collected continuously. This produces a time-resolved picture of the production of specific products, as measured by the abundance of specific masses, or absorbance at specific wavelengths.

Slow pyrolysis is generally produced using either a programmable furnace or resistively heated filament pyrolyzer.

A. Programmable Furnaces

The design and interfacing of furnace pyrolyzers has been discussed. The same mass and thermal stability that make typical furnaces useful for isothermal applications may be used to provide reproducible heating profiles at slow rates for large samples. Several programmable furnace pyrolysis instruments are available commercially, with the most common application being the examination of fuel source samples [11]. These materials are generally ground rock from oil well core drillings or shale deposits, and the majority of the sample is inorganic and nonvolatile. Consequently, a rather large sample must be heated to extract the organics, generally performed at a rate of 50–100°C/minute.

Programmable furnaces may be interfaced to a gas chromatograph, generally with some sort of intermediate trapping, but are rarely interfaced directly to spectroscopic techniques. Thermogravimetric analysis coupled with Fourier-transform infrared spectroscopy (TGA-FT-IR) is an important exception, since many thermal units are capable of heating samples to pyrolysis temperatures, with the resulting pyrolysate being swept directly into the cell of the FT-IR [12].

B. Resistively Heated Filaments at Slow Rates

Since the temperature and rate of heating of a resistive filament pyrolyzer are functions of the current, these instruments are frequently used to provide slow programs in addition to pulse-heating pyrolyses. The advantage of the resistive filament is that the sample may be placed directly onto the filament during heating, so that there is no thermal gradient as experienced with a furnace design. In practice, however, the sample is more likely placed into a quartz tube and heated using a coiled filament, creating a small quartz furnace inside of the resistive element.

As with the slowly heated furnace, a collection step is generally required when producing a sample for gas chromatography, since the pyrolyzate may be produced over the course of several minutes. For some applications, collection may be done directly onto the gas chromatograph column, either at ambient or subambient temperatures. For direct mass spectrometry or FT-IR spectrometry in a time-resolved fashion, the pyrolyzer is either inserted into an expansion chamber, which is flushed or leaked into the spectrometer, or the pyrolyzer is inserted directly into the instrument. Specially designed probes permit operation at pyrolysis temperatures directly in the ion source of a mass spectrometer if it was originally equipped with a solids probe. For FT-IR, a heated pyrolysis chamber may be flushed with carrier gas into a light pipe, or a cell used which positions the filament of the pyrolyzer directly below the light path for real-time pyrolytic analysis.

V. OFF-LINE INTERFACING

For many experiments, including the slow rate pyrolyses discussed above, it may be advantageous or necessary to install the pyrolysis device away from the analytical instrument, with a collection or trapping system between. This may be accomplished in a variety of ways, both manually and automatically.

For manual isolation, the pyrolysis device may be flushed with a stream of carrier gas which is routed through a trap, either sorbent or thermal using dry ice or liquid nitrogen. The pyrolysate is swept from the pyrolyzer and through the trap, where the carrier gas is vented but the organics are frozen or sorbed. The trap is then disconnected from the pyrolyzer and taken to the analytical instrument where the pyrolysate is either thermally revaporized or extracted with a solvent and injected. An early method of performing pyrolysis-IR was to allow the pyrolysate to condense onto a window, which was then inserted into the IR and scanned. These techniques also permitted the use of pyrolysis devices to heat samples in atmospheres incompatible with the analytical instrument, or at pressures or flows not permissible in the analyzer. For example, a sample may be heated in oxygen for combustion studies, with the reaction products collected, then transferred to a gas chromatograph [13]. Normal installation of a pyrolyzer would require the use of oxygen as the GC carrier, which is incompatible with column phases.

An analyst performing many off-line pyrolyses, or reactant gas studies, will probably select an automatic means of isolating the pyrolysis end of the experiment from the analytical. This is typically done by employing a valve in a heated oven to place the collection device alternate on line with the pyrolyzer or with the analytical device. Commercial systems are available that incorporate this valve in a programmable controller to automate the process, as well as in manully operated modes. The pyrolysis chamber carrier gas is directed through the valve to the trap, or alternately to vent. During an experiment, the pyrolysis gas, inert or reactant, is routed through a sorbent or cold trap, where the pyrolysate is collected. During this time, the analytical device, generally a gas chromatograph, is being swept with inert carrier as usual. When the valve is rotated, only the trap is inserted into the GC analytical flow, and the collected analytes heated and transferred to the column for analysis. The pyrolyzer itself, and its flow, never touch the analytical instrument. This design permits heating the sample material slowly or for as long as desired, with transfer of the composite-collected pyrolysate for a single GC analysis. Systems have been designed as well with multiple traps for the stepwise collection of multiple fractions from the sample as it is being heated, with subsequent GC analysis of the material from each trap.

VI. SAMPLE HANDLING AND REPRODUCIBILITY

It is not enough to have a pyrolyzer for which the endpoint temperature and heating rate are well characterized to guarantee reproducible results. Sampling, sample handling, introduction, and transfer from the pyrolyzer into the analytical device must be performed with attention to all the inaccuracies that may be introduced. The effects of instrument design and interfacing have been discussed. The most common sources of error in sample manipulation will now be briefly described. The most important areas of concern are sample preparation, including size and shape; homogeneity; and contamination.

Using microsyringes, it is relatively easy to introduce a 1 μl or smaller sample into a gas chromatograph injection port. Preparing and inserting a solid sample that is only a few micrograms, on the other hand, presents some difficulties. This is particularly the case if the sample is an insoluble material of an inconsistent make-up throughout, such as plant material or layers of paint. Even analysts confident in their ability to prepare small slices of a sample are concerned whether or not that piece is representative of the whole.

A. Sample Size and Shape

If a sample material is soluble, microliter-sized portions of the solution may be deposited onto the surface of the pyrolysis instrument using a syringe, which not only regulates the amount of sample material, but also causes the sample to be spread as a thin film, which heats and pyrolyzes readily. One of the advantages of pyrolysis as an analytical technique, however, is its benefits in the analysis of difficult solids, which have limited solubility, such as polymers. In these cases, it is necessary to make samples of the same size and shape consistently for reproducible results. Since analytical pyrolyzers are designed to heat small samples rapidly, it is easy to overload the pyrolyzer with too much sample. This not only affects the rate at which the sample heats, related to the thermal gradient through the thickness of the material, but may also overload the analytical device, causing contamination and carryover into the next analysis. Generally, 10–50 μg of sample is desirable for direct pyrolysis-GC, and about twice that for direct pyrolysis-FT-IR. A thin slice placed on its side is preferable to a cube or sphere, and melting the sample into a film is sometimes helpful if the sample will cooperate. In any event, it is important to use as nearly the same size and shape sample each time to be sure that at a minimum the sample material goes through the same heating process each time.

B. Homogeneity

Insuring that a sample of solid material only a few micrograms large is homogeneous and therefore representative of the material from which it was taken presents a constant problem to analysts. The inability to obtain several samples of identical composition frequently casts suspicion on the technique of analytical pyrolysis, since the results obtained show poor reproducibility. This lack of reproducibility may be interpreted as the result of unexplained or unreliable reactions occurring in the sample during pyrolysis, when in fact it is actually an indication that the samples are different from each other.

Some samples have such large, obvious differences in composition from end to end that nonhomogeneity is self-evident. Materials such as plant leaves, soils and rocks, scrapings, textiles, and laminated products will clearly provide different materials from different physical areas sampled. Since few analysts have the luxury of analyzing pure, homogeneous materials, these concerns must be addressed, and the effects of nonhomogeneous samples on the pyrolysis results established for each analysis.

Analysts sampling materials for which homogeniety is a concern have devised several methods to deal with the problem. If possible, the sample material may be ground to a fine powder from which small portions are taken for analysis. There is sometimes a concern that the grinding process, or the heat produced during it, may alter the sample so that the results are no longer representative. Many materials, however, have been ground successfully under cryogenic conditions, resulting in a powdered sample which has not been heated sufficiently to volatilize any of its constituents. Other analysts have chopped samples finely using a scalpel and then analyzed several of the small fragments together. Again, if the material or materials comprising the sample are soluble, dissolving them and working from a solvent provides an easy way to insure homogeneity and sample size reproducibility.

Some analysts have elected to pyrolyze large samples when concerned that a small sample cannot be made to be representative. A sample of about 1 milligram, if pyrolyzed using an instrument powerful enough to heat it effectively, may produce better reproducibility than a smaller sample which is less representative. In such cases, it is also essential to limit the amount of the pyrolyzate entering the analytical instrument, generally by use of a splitter with a large split ratio, or by passing the pyrolyzate in a carrier gas through a small sample loop attached to a valve which is interfaced to the analytical unit. It is also important to be mindful of the increased amounts of residue and char produced, requiring cleaning to prevent contamination from run to run.

Another way to deal with nonhomogeneous samples is to analyze the individual constituents independently. Since analytical instruments are sensitive enough to respond to very small samples, it is sometimes possible to remove specific portions of an overall sample and investigate them separately. For example, specs of contamination in a polymer melt, discolorations, particles in papers or pulp, different strata in mineral samples, layers of paint, etc. may be studied independently as well as part of the whole sample material. It is important to remember that the individual constituents in a mixture will each pyrolyze in a way consistent with its molecular make-up, and that the entire sample will amount to the sum of its individual parts.

REFERENCES

1. S. Tsuge and H. Matsubara, *J. Anal. Appl. Pyrol.*, *8*: 49–64 (1985).
2. A. Venema and J. Veurlink, *J. Anal. Appl. Pyrol.*, *7*, 3: 207–214 (1985).
3. N. Oguri and P. Kim, *Int. Lab.*, *19*, 4: 59–62 (1989).

4. H.-R. Schulten, W. Fischer, H. J. Wallstab, *J. High Res. Chromatogr. Chromatogr. Commun.*, *10*: 467–469 (1987).
5. E. M. Anderson and I. Ericsson, *J. Anal. Appl. Pyrol.*, *3*: 13–34 (1981).
6. I. Ericsson, *Chromatographia*, *6*: 353–358 (1973).
7. R. S. Lehrle, J. C. Robb, and J. E. Suggate, *Eur. Polym. J.*, *18*: 443–461 (1982).
8. J. B. Pausch, R. P. Lattimer, and H. L. C. Meuzelaar, *Rubber Chem. Technol.*, *56*: 1031–1044 (1983).
9. T. P. Wampler and E. J. Levy, *Am. Biotechnol. Lab.*, *5*: 56–60 (1987).
10. J. W. Washall and T. P. Wampler, *Spectroscopy*, *6*, 4: 38–42 (1990).
11. B. Horsfield, *Geochim. Cosmochim. Acta*, *53*, 4: 891–901 (1989).
12. P. R. Solomon, M. A. Serio, R. M. Carangelo, R. Bassilakis, D. Gravel, M. Baillargeon, F. Baudais, and G. Vail, *Energy Fuels*, *4*, 3: 319–333 (1990).
13. T. P. Wampler and E. J. Levy, *J. Anal. Appl. Pyrol.*, *8*: 153–161 (1985).

3

Pyrolysis Mass Spectrometry: Instrumentation, Techniques, and Applications

JOHN MADDOCK AND T. W. OTTLEY
Horizon Instruments Ltd.
Heathfield, East Sussex, England

I. INTRODUCTION

A. History

Pyrolysis-mass spectrometry has found many applications in microbiology, geochemistry, and soil and polymer sciences. However, its development has been historically hindered by the lack of competitively priced instrumentation, which has restricted the number of laboratories able to use the technique. Furthermore, because of the nature and complexity of the data produced, a heavy reliance upon statistical techniques has been required for data analysis. This specific problem has benefited greatly from the development of powerful personal computers and the increased availability of suitable software.

Probably the first pyrolysis-mass spectrometer was described by Meuzelaar and Kistemaker [1]. This instrument was based on a Riber quadrupole mass spectrometer and used the Curie-point pyrolysis method first described by Giacobbo and Simon [2]. This development eventually produced two commercial instruments—the Extranuclear 5000 (Extranuclear Laboratories, Pittsburgh, PA), effectively a copy of the FOM machine, and the Pyromass 8–80 (VG Gas Analysis, Middlewich, England), based on the same principles but using a small magnetic mass spectrometer.

Both these systems were based on existing instrumentation used for conventional mass spectrometry. This made them prohibitively expensive and only a few have been sold. A significant reduction in the cost of a pyrolysis-mass spectrometer could only be achieved through the development of a dedicated instrument.

Subsequently, a truly dedicated and automated pyrolysis mass spectrometer, the PYMS-200X (Horizon Instruments, Sussex, England) was developed and found its principal use in microbiology. This instrument has been fully described elsewhere [3], together with typical applications [4–6]. Applications were limited because of the mass range offered by this instrument, 12–200 daltons. Further development of the technique led to the production of the RAPyD-400 system (Horizon Instruments) with a mass range of 12–400 daltons, allowing more applications to be addressed.

B. Instrument Design

All the pyrolysis-mass spectrometer systems so far produced have contained common features of loading system, pyrolysis region, expansion chamber, and molecular beam before reaching the mass spectrometer ion source. The FOM Autopyms, the Pyromass 8–80, the Extranuclear 5000, and the PYMS-200X have all been detailed elsewhere [1,3,7]. A brief description of the RAPyD-400 as an example is all that is required here.

The RAPyD-400 is a bench-top, quadrupole-based mass spectrometer utilizing the method of Curie-point heating for sample introduction. The system is capable of being loaded with up to 150 samples which can then be processed unattended.

The three basic elements of the RAPyD-400 can be seen in Figures 1–4: the vacuum system, the inlet system, and the quadrupole analyzer. The vacuum system diagram shows an overview of the complete mass spectrometer system. The quadrupole analyzer, which itself consists of two separate parts, lies inside the main vacuum chamber. The high vacuum is attained using a turbomolecular pump which is backed by a dual-stage rotary pump mounted externally to the main system. The sample inlet system is connected to the ion source of the mass spectrom-

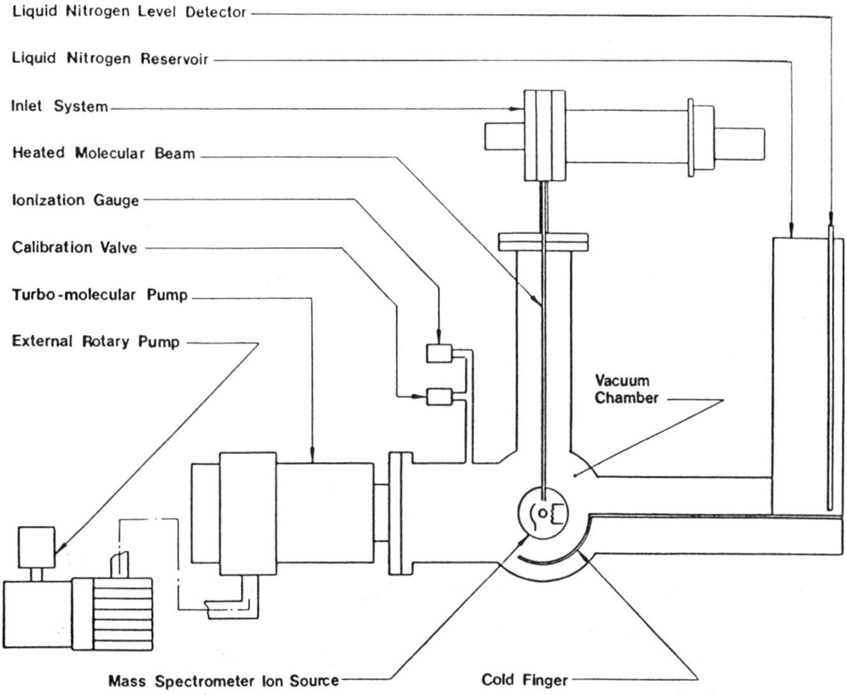

FIGURE 1 Vacuum system schematic for the Horizon Instruments RAPyD-400.

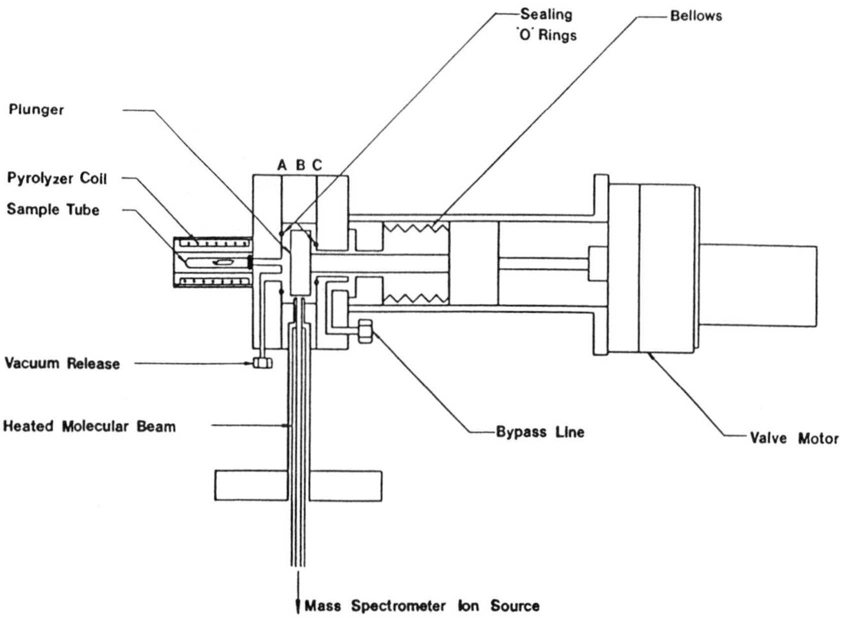

FIGURE 2 Inlet system for the Horizon Instruments RAPyD-400.

eter via a heated molecular beam tube. Around the underside of the ion source is a copper cold finger which is cooled by liquid nitrogen and used as a sample dump to prevent carryover from one sample to another.

Once loaded onto the automation system, each sample tube is taken in turn and presented to the inlet system. The O-ring on the open end of the tube is used to form a seal on the outer face of the inlet system. Once this seal has been made, the inlet system valve plunger can move from position A to position B, where it closes off the molecular beam to the ion source and allows the air inside the glass sample tube to be evacuated via the bypass pumping line. Once this has been achieved, the plunger moves from position B to position C closing the bypass line and opening the molecular beam tube. An expansion chamber has now been created. Once pyrolysis takes place and the resultant gas

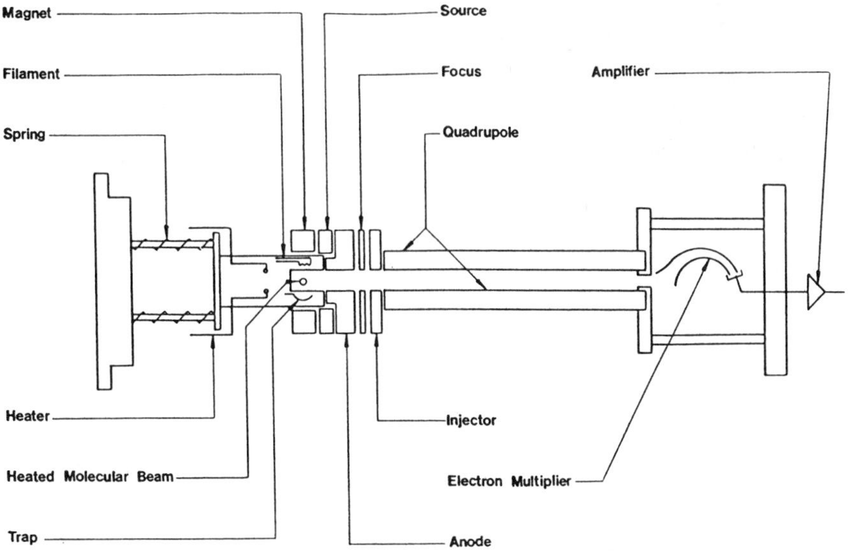

FIGURE 3 Quadrupole mass spectrometer for the Horizon Instruments RAPyD-400.

evolved, the pyrolyzate expands into the vacant space and is then taken to the ion source via the heated molecular beam.

On arrival at the ion source, the pyrolyzate gas is subjected to bombardment by electrons with a nominal energy of 25 eV in classic EI mass spectrometry. The resultant charged molecules produced are transmitted by the quadrupole and a mass spectrum of the pyrolyzate is recorded by an electron multiplier operating in pulse-counting mode.

II. DATA ANALYSIS

A. Multivariate Statistics

Chemometrics is the application of numerical techniques to the identification of, or discrimination between, chemical substances. Expressed more simply, it is the attempt to make rela-

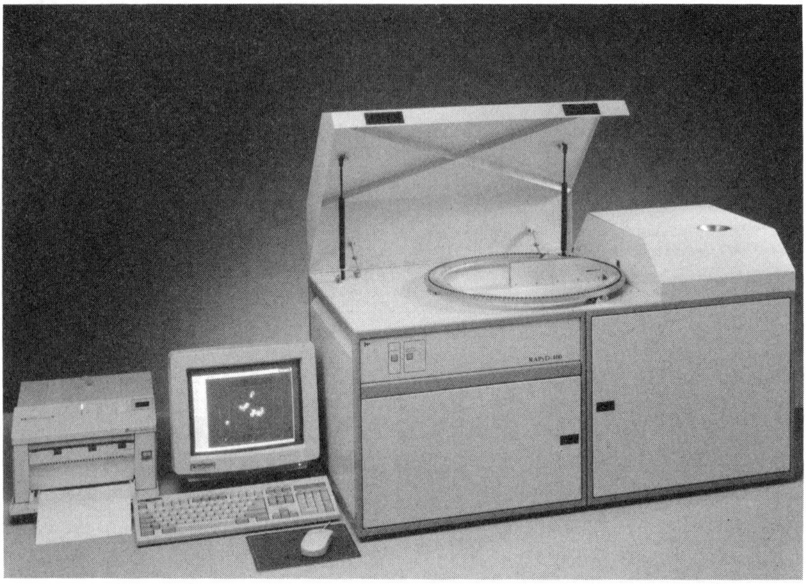

FIGURE 4 Photograph of the RAPyD-400.

tively simple quantitative measurements, and then to use the data obtained in various ways to give the maximum amount of information regarding the differences between samples.

In truth, most analytical methods make use of the concept of chemometrics, at least to some extent, but some techniques are much more suitable than others. Obviously, all forms of spectroscopy can yield a large amount of numerical data, but mass spectroscopy is ideally suited for two reasons.

In a mass spectrum, ions are recorded over a certain mass range, but within that range (at low-to-medium resolution) only integer mass values are possible, the separation between peaks corresponding approximately to the mass of a neutron.

Second, it is possible to make use of single ion counting methods to record an actual numerical result (the number of ions) directly at each mass position. Thus a two-dimensional data matrix is the immediate result.

Multivariate statistical methods as part of the general chemometrics discipline are particularly well suited for the analysis of complex mass spectra. An excellent introduction to these techniques has been written by Manly [8].

One of the main reasons for using multivariate statistics is to reduce a large amount of very complex data down to a form where it can be readily understood. This usually means that any graphical output needs to be in two or three dimensions only. In effect, this means using principal components, canonical variates, and clustering analyses.

Principal components analysis is a well-established multivariate statistical technique which can be used to identify correlations within large data sets and to reduce the number of dimensions required to display the variation within the data. A new set of axes, principal components (PCs) are constructed, each of which accounts for the maximum variation not accounted for by previous principal components. Thus, a plot of the first two PCs displays the best two-dimensional representation of the total variance within the data. With pyrolysis mass spectra, principal components analysis is used essentially as a data reduction technique prior to performing canonical variates analysis, although information obtained from principal components plots can be used to identify atypical samples and/or outliers within the data and as a test for reproducibility.

Canonical variates (CVs) analysis is the multivariate statistical technique that takes into account sample replication or any other a priori structure within the data and attempts to discriminate between the sample groups. CVs are derived in a similar manner to PCs except that their axes are constructed to maximize the ratio of between-group to within-group variance. Thus, a plot of the first two CVs displays the best two-dimensional representation of the sample group discrimination. Examination of the associated mass loadings/weightings allows chemical interpretation. Large mass loadings reflect masses significant to the discrimination in a particular direction. The application of graphical rotation facilitates the construction of factor spectra showing the contribution of masses to the discrimination of a specific sample group. Factor spectra are often representative of pure compounds or classes of components aiding chemical interpretation of the observed discrimination.

The use of hierarchical cluster analysis leading to the construction of dendrograms and minimum spanning trees for data sets where the samples analyzed have a high degree of relatedness is often the most useful way to display the total discrimination within the data because all the variance is displayed in two dimensions.

Although there are other multivariate techniques that can be applied to mass spectral data, the ones we have discussed here tend to be the most widely applied. In reality, there are more checks on validity which are made at different stages. This is to avoid any lack of confidence in the results obtained.

B. Artificial Neural Networks

A related approach to multivariate statistics is the use of artificial neural networks (ANN), which are, by now, a well-known means of uncovering complex, nonlinear relationships in multivariate data. ANNs can be considered as collections of very simple computational units, which can take a numerical input and transform it (usually via a weighted summation) into an output [9–11]. The relevant principle of supervised learning in ANNs is that they take numerical inputs (the training data) and transform them into desired (known, predetermined) outputs. The input and output nodes may be connected to the external world and to other nodes within the network. The way in which each node transforms its input depends on the so called connection weights (or connection strengths), which are modifiable. The output of each node to another node or to the external world then depends on both its weight-strength and on the weighted sum of all its inputs, which are then transformed by a (normally nonlinear) weighting function referred to as its activation function. For the present purposes, the great power of neural networks stems from the fact that it is possible to train them. Training is effected by continually presenting the networks with the known inputs and outputs and modifying the connection weights between the individual nodes, typically according to a back-propagation algorithm [10], until the output nodes of the network match the desired outputs to a stated degree of accuracy. The network, the effectiveness of whose training is usually determined in terms of the root mean square (RMS) error between the actual and the desired outputs averaged over the training set, may then be exposed to unknown inputs and will immediately output the best fit to the outputs. If the outputs from the previously unknown inputs are accurate, the trained ANN is said to have generalized.

The reason this method is so attractive to pyrolysis-mass spectroscopy (Py-MS) data is that it has been shown mathematically [12] that a neural network consisting of only one hidden layer, with an arbitrary large number of nodes, can learn any arbitrary (and hence nonlinear) mapping to an arbitrary degree of accuracy. ANNs are also considered to be robust to noisy data, such as that which may be generated by Py-MS.

A neural network usually consists of three layers, representing inputs, outputs, and a hidden layer, which is used to make the connections (Fig. 5). By training a neural network with known data, it is possible to obtain outputs that can accurately predict such things as polymer concentration mix.

ANNs can be trained using pyrolysis-mass spectra as the inputs and the known concentrations of target analytes as the outputs. For each input (one mass spectrum), there should normally be one output. The trained network can then be tested with the pyrolysis-mass spectra of "unknowns" to accurately predict the concentration or authenticity of the unknowns.

III. APPLICATIONS

A. Microbial Characterization

The rapid accurate discrimination and characterization of microorganisms is the goal of many diagnosticians. Py-MS has proved to be a rapid and inexpensive epidemiological typing technique, applicable to a wide range of bacterial pathogens and it can be used to identify the source of an infectious outbreak as well as

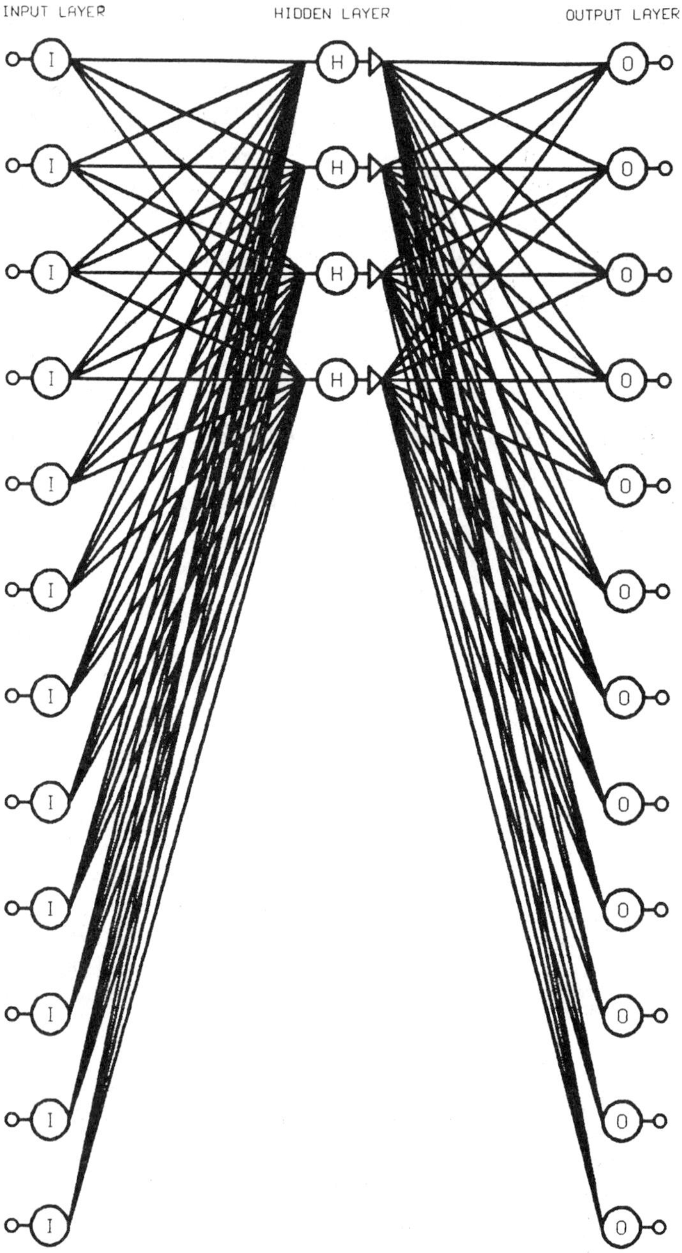

FIGURE 5 Typical artificial neural network structure.

giving an assessment of the relatedness of bacterial isolates [5,13,14].

B. Organic Geochemistry

The characterization of sedimentary organic matter in terms of type and maturity is a prerequisite in the determination of the petroleum-generating potential of sediments. The nonextractable material, the kerogen, consisting of complex high-molecular-weight fragments when subjected to Py-MS and multivariate analysis gives excellent discrimination into the three universally recognized types [15].

C. High-Value Products

Authentication and the detection of adulteration are serious problems within the citrus juice industry. Traditional multicomponent analysis methods are limited by the time required to perform the individual analyses and to construct the data base required. Py-MS rapidly provides fingerprints of the original juice, which facilitates the use of multivariate pattern recognition procedures to detect potentially adulterated samples and to confirm authentication, as well as helping in quality control. More recently, the combination of Py-MS and ANNs has been applied to problems of adulteration within the olive oil industry [16].

D. Forensics

In applications for the characterization of car, house, and fine art paints, Py-MS is valuable in obtaining quick and reproducible fingerprints that can subsequently be matched for identification purposes, whether it be by the forensic scientist or the polymer chemist [17].

E. Polymers

Identification involving the comparison of pyrograms by library matching or multivariate analysis allows an unknown sample to be compared with standard materials in both basic research and quality control. Underlying information concerning backbone structures based on monomer, dimer, trimer ratios, etc. can be elucidated from the pyrograms [18,19].

IV. SAMPLE PREPARATION

The techniques of Curie-point pyrolysis-mass spectrometry and gas chromatography both rely upon a good physical contact between the sample to be analyzed and the supporting ferromagnetic material. Over the years, various shapes of materials have been developed to hold samples prior to pyrolysis. Many of these methods have been described elsewhere [7,20,21].

In all aspects of sample preparation cleanliness is of vital importance. The aim in preparing a sample is to obtain a thin coating of material over the inside surfaces of a foil (which is formed into a "V"), around the surface of a wire or crimped in a ribbon, with between 5 and about 25 μg dry weight. The method for doing this depends on the type of sample.

A. Foils

1. Soluble Solids

Many substances may be dissolved either in water or an organic solvent. The technique is to dissolve sufficient material so that a conveniently small volume of solution will contain the right amount of solid. This concentration is normally a few mg/ml.

As an example, in the case of pectin which dissolves in water, make up a solution by weighing about 4 mg and adding to 1 ml of distilled water. Although this is difficult to do accurately, total accuracy is not required since there is a large degree of latitude in the allowable sample quantities.

Alternatively, however, weigh 40 mg in 10 ml of water. For pectin, agitation for several minutes is required for complete solution. Using a 10 μl syringe, extract about 3 μl of solution and place along the inside fold of the sample foil, which should be protruding about three-fourths of its length from the end of the tube. Ensure that the liquid does not run over the edges of the foil or coat the inside of the glass tube. Even with 5 μl this is not too difficult.

Dry the sample by placing a block holding the tubes in a drying oven at about 70°C, but ensure that the sample does not become cooked. Once dry, push the sample foil into the tube using a depth gauge. The same depth gauge should be used for all samples. By ensuring that all the samples lie in the same position within the induction coil, another potential variable is removed. Finally push an O-ring over the end of each tube, ensuring that the O-ring is positioned approximately 3 mm from the open end of the tube.

2. Insoluble Solids

The method here is similar to that preceding, except that the material must first be ground down to a fine powder, using for example a pestle and mortar, and then kept in suspension while being pipetted onto the foil. A suitable suspension medium is either water or acetone, and ultrasonic agitation will keep the fine particles in suspension. This method is particularly attractive for samples such as glycogen.

In applications where profiling (fingerprinting) is more important than any form of quantification, powdered materials can be mixed into a paste with water or a suitable solvent. The paste can then be smeared onto the foil and the water/solvent evaporated in a drying oven.

Alternatively, for materials such as rubber and other polymers, a small piece of the material may be crimped with a foil using a solid sample preparation jig, as shown in Figure 6. In this device, sample foils are held firmly at each end allowing a small sample to be placed in the open area. The foil can then be crimped with a heavy pair of forceps to hold the sample. Once released, foils can be inserted into glass sample tubes with the same pair of forceps and positioned using a depth gauge.

3. Bacteria and Yeasts

In the case of actively growing colonies on plates, all that is necessary is to pick up a small portion with a disposable loop or flamed wire and smear on both inside faces of the foil. Avoid picking up any of the substrate with the culture. With a little practice it is possible to obtain the correct amount of material

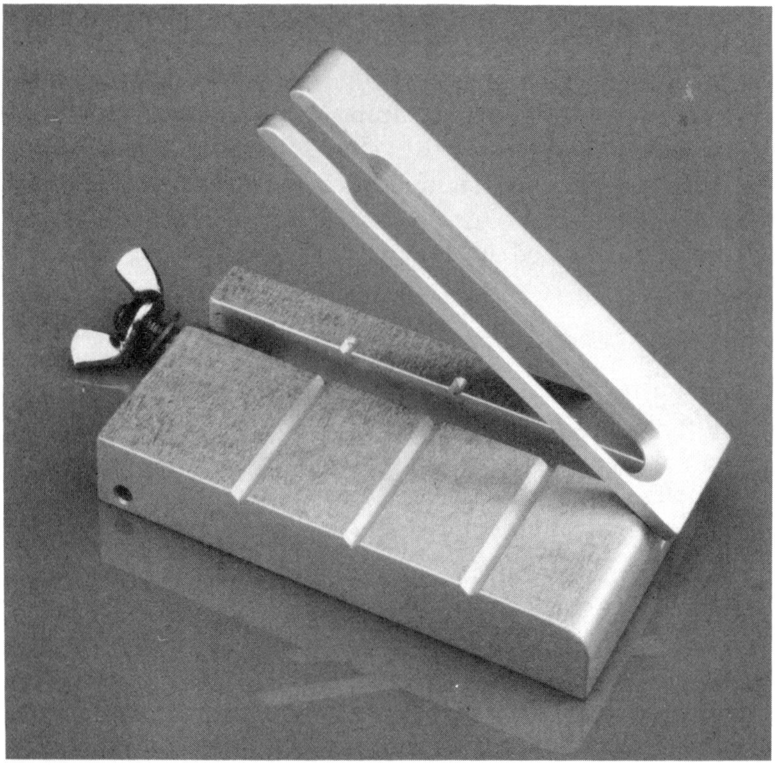

FIGURE 6 Sample preparation jig.

(20 μg) without difficulty. For some bacteria, particularly those in the Actinomycetes family, it is preferential to grow them on a filter placed on top of the medium. This will also prevent the media from being sampled with the colony.

In the preparation of the glass tubes and foils prior to sample loading, it is important that neither are touched by hand. The method described by Sanglier et al. [5] has proved to work well. In this method, pyrolysis foils and tubes are washed in acetone, then dried overnight at 27°C. A single foil is inserted, with flamed forceps, into each pyrolysis tube so as to protrude about 6 mm from the mouth. The tubes may then be stored at 80°C in a clean, dry oven or vacuum dessicator until needed. For each strain, small amounts of biomass (25 μg) are scraped from three different areas of the inoculated plate and smeared onto the protruding foil. The assembled foils are placed in an 80°C oven for 15 minutes to dry the biomass. For analysis, the foils are tamped into the tubes with a flamed stainless steel tool so as to lie 10 mm from the mouth. Viton O-rings were placed on the tubes and then loaded onto a sample carousel.

Liquid cultures can be handled in a way similar to suspended solids in that they can be pipetted onto foils. Cultures presented on slopes can be treated as if they had been plated out.

4. Blood or Blood Cultures

These may be treated in the same way as soluble compounds but the volume of liquid applied to the foil should be much less (1–2 μl). Also, the samples should not be dried in warm air, instead they should be left in a vacuum dessicator for a short time.

5. Liquids

Many liquids requiring analysis will in fact be solutions, but often too dilute to use directly. In the case of whiskey, for example, the concentration must first be increased by a factor of 10 before a sample can be taken. This can be done by blowing a jet of dry nitrogen over the surface of about 10 ml of liquid contained in a beaker, and evaporating down to 1 ml. This takes about 30 minutes. Three microliters of the concentrate may then be applied to the foils and allowed to dry in the vacuum dessicator. Alternatively, the whole 10 ml aliquot can be evaporated to dryness and then reconstituted with 1 ml of the original sample. Either of these two methods should be used in preference to dispensing ten individual 3 μl aliquots on top of each other once the previous aliquot has dried.

In liquid samples, such as fruit juice, olive oil, and milk, the concentration of dissolved solids is sufficient that no preconcentration is required.

B. Ribbons

For some solid samples it may be more convenient to use a ribbon sample carrier. These ribbons take the place of wires in instruments specifically designed to take them.

Taking a ribbon as shown in Figure 7, with the folding lines facing upwards and a flat-ended pair of pliers, fold up the two sides until they are vertical. At this stage it may be advisable to clean the ribbons before further use.

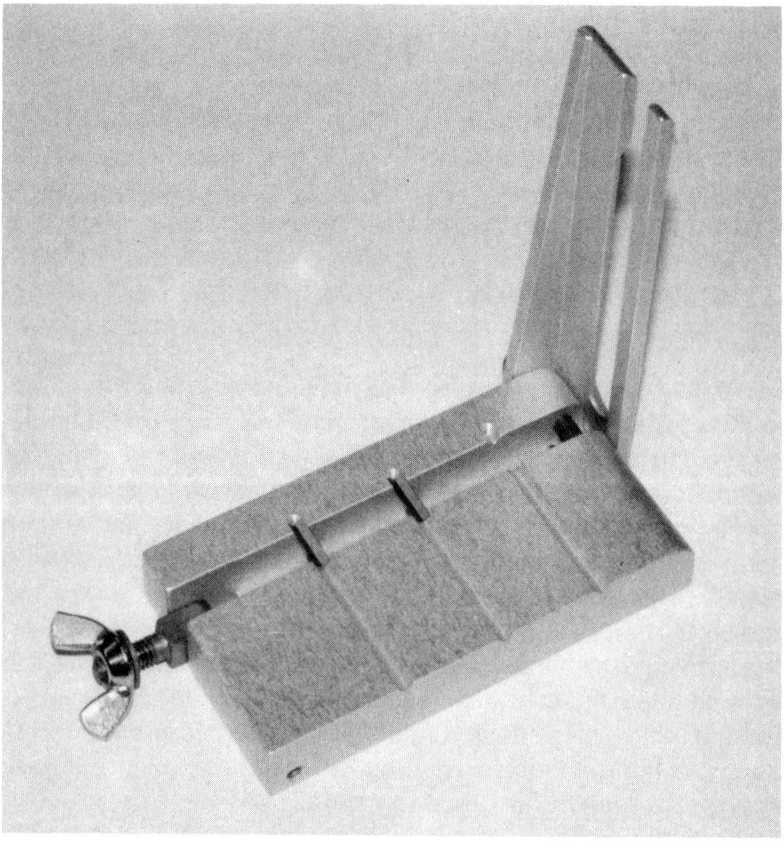

FIGURE 7 Method for holding samples with a ribbon.

Place a sample of material in the ribbon and then fold down both sides to cover the sample, making a tent. Now fold the long end of the ribbon completely over the top of the tent and squeeze gently. The ribbon carrier can now be inserted into the standard glass/quartz tube used by the instrument.

Most of these ribbons are made from pure iron so they need to be protected from water vapor. As supplied, they are normally packed under argon, which should prevent oxidation. After initial folding and cleaning, it may be preferable to store them in a pure solvent—hexane is recommended for this purpose.

C. Wires

For solutions, suspensions and sticky materials (bacteria, resins, etc.), it is quite acceptable to use a straight wire as the sample holder. For solid samples a wire has to be prepared in such a way as to be able to firmly hold the sample in place. It is therefore far better to use foils or ribbons for solid samples, although several methods have been formulated for using wires.

Coating of a wire evenly with a solution or suspension is most desirable to enhance reproducibility. To this end, the use of wire coating units are of great help [1]. This can be achieved by rotating the wires slowly about their axes while applying the liquid medium. As the solvent evaporates, a uniform coating is deposited, which aids reproducibility.

Up to six wires can be prepared at a time. Wires are held by the end 10 mm with tweezers or pliers, and inserted into one of the holes in the guide plate at the front of the wire coater. The wire then pushes into a small socket which will hold it firmly. This part of the wire can be cut off before insertion into the glass sample tube, so slight contamination is not important at this stage.

With the rotation speed set to a mid value, solutions of approximately 3 μl are applied using a small syringe or pipette, one drop at a time to the same position on each wire. After a short time the coating will form uniformly as the solvent evaporates. If necessary, rotation speed can be varied to improve the uniformity of coating.

V. CONCLUSIONS

1. In the case of a solid look for a solvent first, in preference to using a suspension.
2. Try to ensure as even a coating as possible on the foil, wire, or ribbon being used.
3. Ensure tubes and sample carriers are completely clean before using and avoid touching the sample carrier during coating.
4. After preparing, if the samples are not going to be used immediately, store the tubes in a vacuum dessicator until required. This will avoid absorption of water from the air with subsequent deterioration of the sample.
5. Keep complete records of the samples prepared. This information may be transferred to the computer and stored as a name file unique to the samples being analyzed.
6. Do not analyze hundreds of samples without taking the time to understand what the data shows. This assumes that the problem and suitable methods have already been defined.
7. For any one particular data set, the sample sizes used should be constant within an acceptable range. If the data generated is to be analyzed using multivariate statistical packages or artificial neural networks, total mass spectral ion counts over a 51–200 dalton range should be between 1–3 million ions per sample.

REFERENCES

1. H. L. C. Meuzelaar and P. G. Kistemaker, Techniques for fast and reproducible fingerprinting by Pyrolysis Mass Spectrometry, *Anal. Chem*, *45*: 587 (1973).
2. H. Giacobbo and W. Simon, *Pharm. Acta Helvetica*, *39*: 162 (1964).
3. R. E. Aries, C. S. Gutteridge, and T. W. Ottley, Evaluation of a low cost, automated pyrolysis-mass spectrometer, *J. Anal. Appl. Pyrol.*, *9*: 81–98 (1986).

4. K. Orr, F. K. Gould, P. R. Sisson, N. F. Lightfoot, R. Freeman, and D. Burdess, Rapid interstrain comparison by pyrolysis mass spectrometry in nosocomial infection with *Xanthomonus malthophilia*, *J. Hosp. Infect.*, *17*: 187–195 (1991).
5. J. J. Sanglier, D. Whitehead, G. S. Saddler, E. V. Ferguson, and M. Goodfellow, Pyrolysis mass spectrometry as a method for the classification and selection of actinomycetes, *Gene*, *115*: 235–242 (1992).
6. R. Goodacre, D. B. Kell, and G. Bianchi, Olive oil quality control by using pyrolysis mass spectrometry and artificial neural networks, *Olivae*, *47* (1993).
7. W. J. Irwin, *Analytical Pyrolysis: A Comprehensive Guide*, Marcel Dekker, New York (1982).
8. B. J. F. Manly, *Multivariate Statistical Methods*, Chapman and Hall, London (1986).
9. P. D. Waiserman, *Neural Computing: Theory and Practice*, Von Nostrand-Reinhold, New York (1989).
10. D. E. Rumelhart, J. L. McClelland, and the PDP Research Group, Parallel distribution processing, Experiments in the microstructure of cognitation.
11. J. Hertz, A. Krogh, and R. G. Palmer, *Introduction to the Theory of Neural Computation*, Addison-Wesley, CA, (1991).
12. K. Hosnik, M. Stinchcombe, and M. White, *Neural Net.*, *3*: 551–560 (1990).
13. L. A. Shute, C. S. Gutteridge, J. R. Norris, R. C. W. Berkeley, Curie-point pyrolysis mass spectrometry applied to characterisation and identification of selected *Bacillus* species, *J. Med. Microbiol.*, *130*: 343–355 (1984).
14. R. Freeman, F. K. Gould, R. Wilkinson, A. C. Ward, N. F. Lightfoot, and P. R. Sisson, Rapid inter-strain comparison by pyrolysis mass spectrometry of coagulase—Negative staphylococci from persistent CAPD peritonitis, *Epidemiol. Infect.*, *106*: 239–246 (1991).
15. T. I. Eglinton, S. R. Larter, and J. J. Boon, Characterization of kerogens, coals and asphaltenes by quantitative pyrolysis mass spectrometry, *J. Anal. Appl. Pyrol.*, *20* 25–45 (1991).
16. R. Goodacre, D. B. Kell, G. Bianchi, Controllo della qualita

dell'olio di oliva con la pirolisi-spettrometria di massa abbinata ad un sistema di "artificial neural networks. (Olive oil quality control by using pyrolysis mass spectrometry and artificial neural networks), *Olivae*, 47: 36–39 (1993).

17. D. A. Hickman and I. Jane, Reproducibility of PyMS using three different pyrolysis systems, *Analyst*, *104*: 334–347 (1979).
18. C. J. Curry, Pyrolysis mass spectrometry studies of adhesives, *J. Anal. Appl. Pyrol.*, *11*: 213–225 (1987).
19. W. A. Westall, Temperature programmed pyrolysis mass spectrometry, *J. Anal. Appl. Pyrol.*, *11*: 3–14 (1987).
20. *RAPyD-400 Pyrolysis Mass Spectrometer Instruction Manual*, Horizon Instruments, Ltd.
21. *Curie-Point Pyrolyser Instruction Manual*, Horizon Instruments, Ltd.

4

Microstructures of Polyolefins

SHIN TSUGE AND HAJIME OHTANI
Nagoya University, Nagoya, Japan

I. INTRODUCTION

Among the most commonly utilized synthetic polymers are polyolefins, such as polyethylenes (PEs), polypropylenes (PPs), and ethylene-propylene copolymers P(E-co-P)s. Despite their simple elemental compositions, consisting of only carbon and hydrogen, it is well known that their physical properties are quite dependent on the microstructural features, such as short- and long-chain branchings, stereoregularities, chemical inversions of monomer enchainment, sequence distributions, etc. [1].

The structural characterization of polyolefins has been carried out most extensively by means of molecular spectroscopy such as FT-IR, ^{1}H-NMR, and ^{13}C-NMR [1]. On the other hand, high-resolution pyrolysis-gas chromatography (Py-GC), which incorporates the pyrolysis-hydrogenation technique and fused-silica capillary column separation, provides a simple but powerful technique to study the microstructures of polyolefins [2–4]. In this chapter, the instrumental aspects of high-resolution Py-GC are first discussed briefly, and then its applications to the rapid estimation of short-chain branchings in low-density PEs, stereoregularity differences and chemical inversions of the monomer units in PPs, and the study of sequence distributions in P(E-co-P)s are demonstrated.

II. INSTRUMENTATION FOR PYROLYSIS HYDROGENATION-GAS CHROMATOGRAPHY

When saturated polyolefins, such as PE and PP, are exposed to high temperatures under an inert atmosphere, they yield various hydrocarbon fragments that reflect the microstructures of the original polymers. These consist mainly of a series of α-, ω-diolefins, α-olefins, and *n*-alkanes. If some short branches exist in the polymer chain, the resulting degradation products are further complicated by additional diastereomeric, geometrical, and positional isomers. Thus, the number of possible isomers becomes large in those fragments as a function of the containing carbon atoms, with the result that complete chromatographic separation is not an easy task, even when a high-resolution capillary column is extremely useful.

Figure 1 illustrates a flow diagram for a typical pyrolysis-hydrogenation capillary column gas chromatographic system, which was developed in the author's laboratory [4–6]. Between the furnace-type pyrolyzer (A) and the splitter (E), a precut column (B) (3 mm i.d. × 5 cm) containing Diasolid H (80–100 mesh) coated with 5 wt % of OV-101, and a hydrogenation catalyst column (C) (3 mm i.d. × 10 cm) containing Diasolid H (80–100 mesh) coated with 5 wt % of Pt were inserted in series. The former precut column was used to protect the activity of the catalyst and the high efficiency of the capillary column from tarry and/or involatile degradation products which decrease resolution. Both the precut and the catalyst columns were maintained at 200°C. On the other hand, the splitter, with a splitting ratio

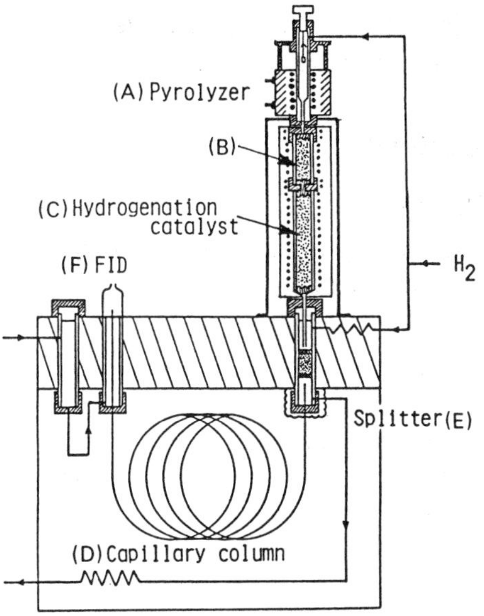

FIGURE 1 Schematic flow diagram for high-resolution pyrolysis-hydrogenation gas chromatographic system.

of about 70:1, was independently held at a high temperature, identical to the maximal temperature of the separation column. About 0.1–0.5 mg of the polymer sample is pyrolyzed, typically at 650°C under a flow of hydrogen carrier gas (50 ml/min), which is also used as the hydrogenation gas. A glass or a fused-silica capillary column (D) is used in a temperature programming mode over a range typically extending from 40 to 280°C at a rate of 2°C/min. A flame ionization detector [FID(F)] is used for the peak detection in the pyrograms, and the peak assignment is mostly carried out by a directly coupled GC/MS system.

Employing this technique, the resulting pyrograms of polyolefins not only become extremely simple, but they are also highly resolved, since those peaks of both α-olefins and α-, ω-diolefin with the same carbon number are combined into the associated alkane peak, and geometrical isomers are extinguished.

This situation is illustrated in Figure 2 for a low density PE (LDPE) before (A) and after (B) hydrogenation.

III. ANALYSIS OF POLYOLEFINS

A. Short-Chain Branching in LDPE

The probable model branching structures for practically utilized PEs are illustrated in Figure 3. Among them, LDPEs, which are

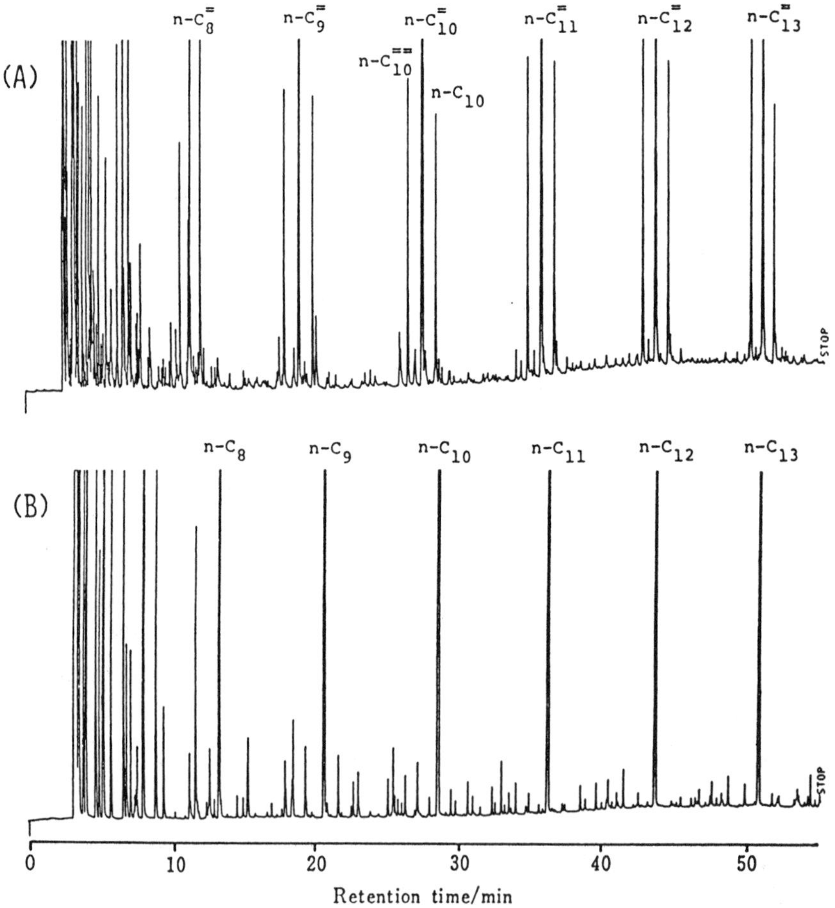

Figure 2 Typical high-resolution pyrograms of low-density PE before (A) and after (B) hydrogenation.

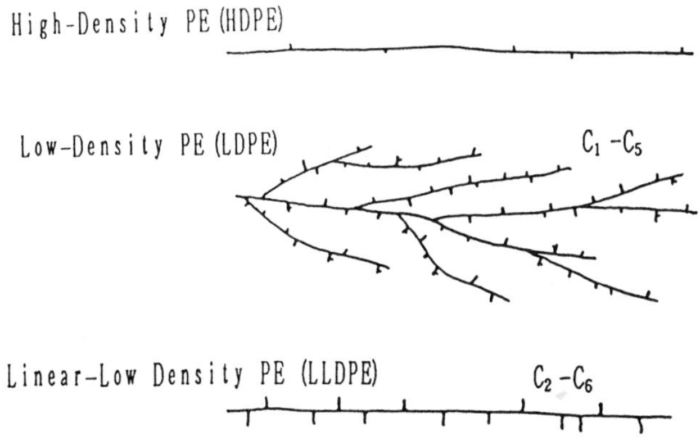

FIGURE 3 Probable branching structures for various PEs.

synthesized in the high-pressure method, are known to have both short-chain branching (SCB) and long-chain branching (LCB). According to the polymerization mechanisms, the former SCB consists mostly of ethyl (C_2), butyl (C_4), and amyl (C_5); and methyl (C_1), propyl (C_3), and larger than hexyl ($>C_6$), if any, are minor components. The type and concentration of SCB in LDPEs, which vary according to the polymerization conditions, affect many properties of these polymers. Therefore, the characterization of SCB plays a very important role in clarifying the structure-property relations for LDPEs.

Although quantitative analyses of SCB have been extensively studied by ^{13}C-NMR spectroscopy recently, such analyses require fairly long measuring times (from hours to days) and relatively large sample sizes (10–100 mg) [7,8]. In contrast, Py-GC requires a minimum sample size (about 0.1 mg) during a simple and rapid (about 1 h) operation.

As shown in Figure 2, the pyrogram of LDPE without hydrogenation (A) consists mainly of serial triplets corresponding to α-, ω-diolefins, α-olefins, and *n*-alkanes with additional weak, but heavily complexed isoalkane, isoalkene, and isoalkadiene peaks between the strong serial peaks. While after hydrogenation (B), these triplets are simplified into singlets of the corresponding *n*-alkanes, and various isoalkanes which are characteristic of the SCB in the LDPE [6,9–11].

Figure 4 shows a typical high-resolution pyrogram of a LDPE measured by use of the in-line hydrogenation technique [12]. As shown in the expanded pyrogram in Figure 4 for the C_{11} components, the isoalkane peaks, which reflect the SCB, are clearly separated between the serial *n*-alkane peaks, which are mostly attributed to the longer methylene sequences in the LDPE. Such isoalkanes are formed through combination of two thermal scissions along the polymer chain containing the SCB:

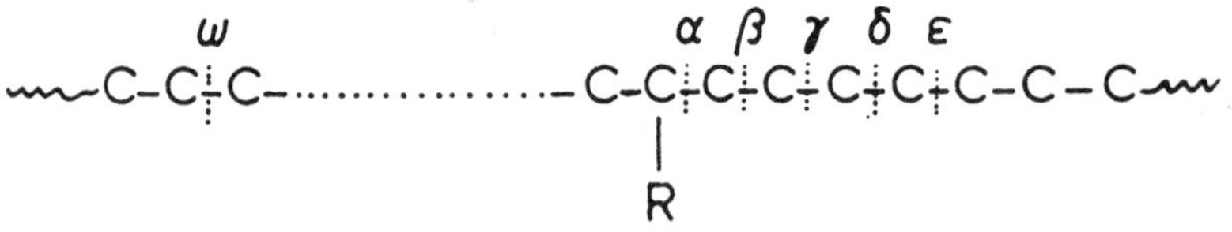

In the case of a methyl branch (R = CH_3) for instance, α and ω, β and ω, γ and ω, δ and ω, and ε and ω scissions followed by hydrogenation would yield *n*-alkane, 2-methyl-, 3-methyl-, 4-methyl-, and 5-methyl-isoalkanes, respectively. Similarly, from the moiety of possible SCBs (C_1–C_6), the corresponding isoalkanes are expected to occur.

Figure 5 shows the typical portions of the expanded pyrograms around C_{11} fragments for a LDPE and those of five model copolymers for methyl, ethyl, butyl, amyl, and hexyl branches [12]. Once the relative peak intensities characteristic of the SCBs are determined using well-defined model polymers with known amounts of possible SCBs, the relative abundance of the SCBs in the LDPE can easily be estimated by the peak simulation of the observed isoalkane peaks in the pyrogram of the LDPE [12–15].

Table 1 summarizes the SCB contents obtained by this method for the four LDPEs, along with those found by ^{13}C-NMR spectroscopy [12]. As a whole, the estimated individual short branch contents and the total values are in fairly good agreement with those obtained by ^{13}C-NMR spectroscopy. Thus, the high-resolution Py-GC combined with in-line hydrogenation is an effective tool for the determination of the SCB with short analysis time and minimal amounts of sample, and can detect SCB even

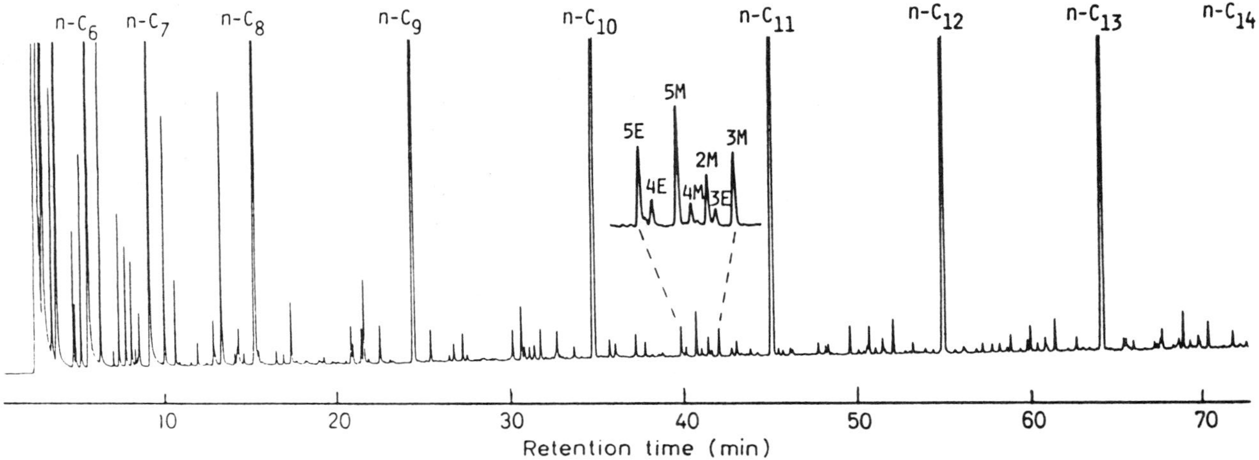

Figure 4 Typical high-resolution hydrogenated pyrogram of LDPE at 650°C. 2M, 2-methyldecane; 3M, 3-methyldecane; 4M, 4-methyldecane; 5M, 5-methyldecane; 3E, 3-ethylnonane; 4E, 4-ethylnonane; 5E, 5-ethylnonane [12].

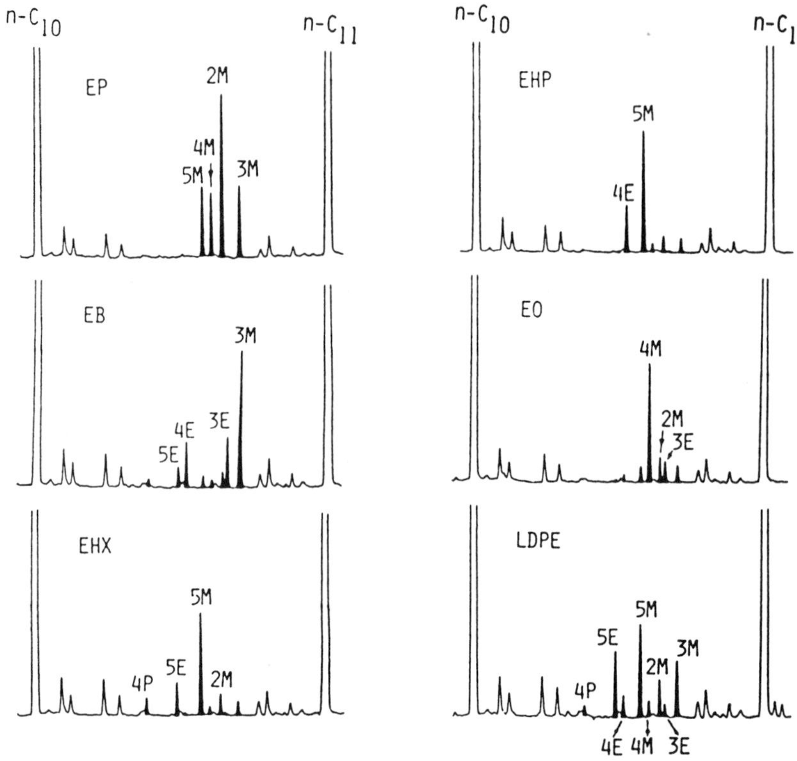

FIGURE 5 Expanded hydrogenated pyrograms of ethylene-propylene (EP), ethylene-1-butene (EB), ethylene-1-hexene (EHX), ethylene-1-heptene (EHP), and ethylene-1-octene (EO) reference copolymers, and LDPE in the C_{11} region. 2M, 3M, 4M, 5M, 3E, 4E, and 5E are the same as those in Fig. 4, and 4P is 4-propyloctane [12].

at concentrations as low as a few per 10,000 carbons [14]. In addition, the existence of pair and/or branched branches in the LDPEs is also suggested [12,16]. This technique was also successfully applied to study the SCB distributions as a function of molecular weight for a given LDPE sample [17].

B. Microstructures of Polypropylenes

Many physical solid and solution properties of PPs are keenly affected not only by the average molecular weight and the molec-

TABLE 1 Estimated SCB Contents in Low-Density Polyethylenes [12].

	Estimated SCB content/1000 C[a]						
Sample	Methyl	Ethyl	Butyl	Amyl	Hexyl	Longer[b]	Total[c]
LDPE-A	1.3	4.9	8.3	2.3	0.2		17.0
	(0.4)	(6.4)	(7.4)	(2.6)		(2.8)	(19.9)
LDPE-B	1.5	7.2	11.2	2.2	0.3		22.4
	(0.5)	(7.2)	(8.5)	(2.7)		(3.5)	(24.9)
LDPE-C	1.3	4.8	8.6	1.9	0.4		17.3
	(0.5)	(5.4)	(6.4)	(2.2)		(2.4)	(17.3)
LDPE-D	1.0	2.1	4.5	0.6	0.3		8.5
	(0.1)	(2.3)	(2.4)	(0.7)		(1.0)	(6.7)

[a] The data in parentheses are obtained by ^{13}C-NMR.
[b] The longer branches than C_7 are not considered in the case of HR (high-resolution) Py-GC.
[c] The values by ^{13}C-NMR involve the propyl branch content between 0.2 and 0.5.

ular weight distribution but also by the configurational characteristics, such as the average stereoregularity, stereospecific sequence length, and the degree of chemical inversions of the monomer units along the polymer chains. The in-line hydrogenation technique combined with highly efficient separation by a capillary column was also successfully applied to the conformational characterization of various PPs with different stereospecificity [18–21].

For example, three kinds of PPs (PP = I, PP = II, and PP = III) synthesized in the presence of different catalysts were treated to prepare the seven samples listed in Table 2. Figure 6 illustrates high-resolution pyrograms of PP-I(A) (iso-PP), PP-I(B) (atact-PP), and PP-II (synd-PP) at 650°C separated by a fused-silica capillary column (0.2 mm i.d. × 50 mm) coated with OV-101 [20,21]. On the pyrogram of iso-PP, a series of triplet peaks with long isotactic sequences are observed up to the undecads (C_{38}–C_{40}), while the intensities of the peaks associated with syndiotactic and atactic sequences become negligibly small at the higher carbon number regions. These data suggest that PP-I(A)

TABLE 2 Polypropylene Samples [21].

Polymer no.	Preparative conditions
1. PP-I(A)	$AlEt_2Cl/TiCl_3$ catalyst, residue of *n*-hexane extraction
2. PP-I(B)	$AlEt_2Cl/TiCl_3$ catalyst, diethyl ether extracts
3. PP-II	$AlEt_2Cl/VCl_4$ catalyst
4. PP-III(A)	$AlEt_2Cl/VCl_4$/anisole catalyst, diethyl ether extracts (extraction I)
5. PP-III(B)	Hexane extracts of the residue in extraction I (extraction II)
6. PP-III(C)	Heptane extracts of the residue in extraction II (extraction III)
7. PP-III(D)	The residue in extraction III

has fairly long isotactic sequences. Similarly, on the pyrogram of PP-II, fairly intense characteristic triplet series associated with long syndiotactic sequences can be seen up to the decads (C_{35}–C_{37}). In this case, however, the triplets rapidly decay, and instead atactic multiplets are observed at the higher carbon number regions. This fact suggests that the stereoregularity of PP-II is fairly lower than that of PP-I(A). On the other hand, on the pyrogram of PP-I(B) (atact-PP), complex atactic multiplets are observed up to the decads (C_{35}–C_{37}).

Before going into further discussion, the possible pentamer components (C_{14}–C_{16}) are illustrated in Figure 7 to explain the complicated diastereoisomers formed from the thermal degradation of PPs followed by hydrogenation. As was predicted by this figure, the observed C_{14}, C_{15}, and C_{16} triads on the pyrogram in Figure 6 are actually composed of a triplet (mm, mr, rr), a quartet (mm, mr, rm, rr), and a triplet (mm, mr, rr), respectively, where m and r are meso and racemic conformations, respectively [19].

From the relative peak intensities among the tetramers and the pentamers, %m and %r (r + m = 100%), and %mm, %mr plus %rm, %rr (mm + mr + rm + rr = 100%) can be estimated.

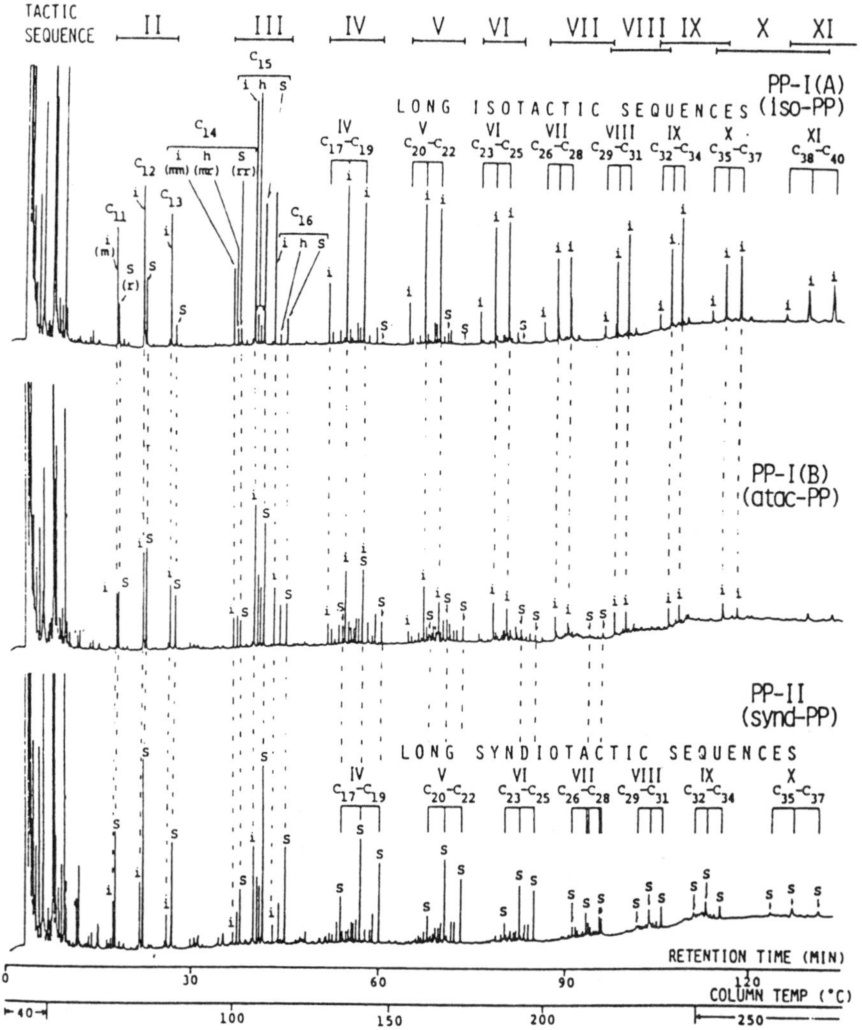

FIGURE 6 Typical high-resolution hydrogenated pyrograms of PP-I(A), PP-I(B), and PP-II. i, s, and h represent isotactic, syndiotactic, and heterotactic products, respectively [21].

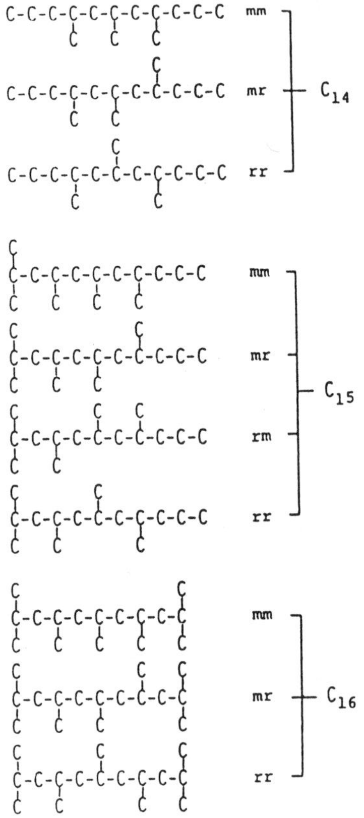

FIGURE 7 Possible diastereomeric isomers for the pentamer clusters of C_{14}, C_{15}, and C_{16}. m and r are meso and racemic conformations, respectively [19].

Then, the average tactic sequence lengths in propylene units can be calculated from the following equation:

$$N_s = 2\frac{\%\mathrm{r}}{\%\mathrm{mr}}, \qquad N_i = 2\frac{\%\mathrm{m}}{\%\mathrm{mr}}$$

where N_s and N_i are the average syndiotactic and isotactic sequence length, respectively [22].

Figure 8 shows two partial pyrograms for PP-I(B) and PP-III(C), both of which are almost comparable (50%) in %m and %r [20,21]. Expectedly, the meso and racemic pair peaks of the tetramer region show almost comparable intensities for both PP samples. However, it is interesting to note that the relative intensities of heterotactic peaks (mr and rm) against those of mm and rr are still comparable for PP-I(B), while the heterotactic peak intensities for PP-III(C) are significantly weaker than those of PP-I(B). These data suggest that PP-I(B) is a typical atactic PP whose N_s and N_i are relatively small, and that PP-III(C) has longer tactic sequences. This result is also supported by the data in Table 3, in which the observed values of N_s and N_i by Py-GC using the data of C_{13} tetramers and C_{16} pentamers are in fairly good agreement with those obtained by ^{13}C-NMR [20,21].

So far in the discussion of the stereoregularity in PPs, only the repeated head-to-tail (H-T) structures were assumed. However, it has been pointed out that the head-to-head (H-H) and tail-to-tail (T-T) structures should exist along the polymer chain to some extent depending on the polymerization conditions [1]. Among the degradation products from PPs, those larger than pentamers ($>C_{11}$) should pertain not only to the positional but also to the diastereomeric isomers. Therefore, C_{10} components, which exclusively reflect the differences in the positions of methyl groups, were carefully separated to determine the chemical inversions of the monomer units [19–21].

Figure 9 shows the expanded pyrogram for the C_{10} region for the highly isotactic PP-I(A) and the highly syndiotactic PP-II. Among these peaks, the main peak representing the successive H-T structures is 2,4,6-trimethylheptane (peak b), while the other minor peaks are associated with either H-H or T-T structures. From the relative peak intensities between these clearly separated C_{10} peaks, the amounts of the chemical inversions in PPs can be estimated.

Figure 10 shows the relationships between the observed %T-T for the various PPs and the %r in the C_{13} tetramers, which is a measure of its syndiotactic nature [20,21]. From these data, it is apparent that the PPs prepared with the vanadium-based

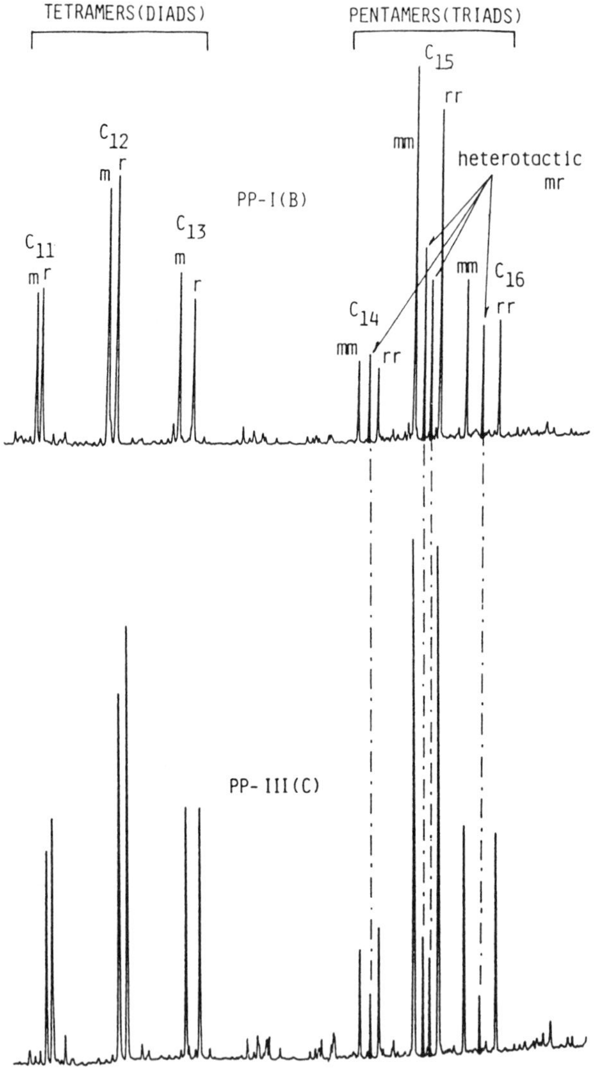

Figure 8 Detailed pyrograms of PP-I(B) and PP-III(C) for tetramer (C_{11}–C_{13}) and pentamer (C_{14}–C_{16}) regions. m and r as defined in Figure 7 [21].

TABLE 3 Estimated Average Syndiotactic Sequence Length (N_s) and Isotactic Sequence Length (N_i) by Py-GC and ^{13}C-NMR [21].

	Py-GC		^{13}C-NMR	
Sample	N_s	N_i	N_s	N_i
PP-I(A)	4.4	32.7	2.5	51.8
PP-I(B)	3.4	3.7		
PP-II	6.5	1.7		
PP-III(A)	6.8	2.8		
PP-III(B)	13.1	4.6	8.3	3.0
PP-III(C)	8.3	8.4	7.2	6.9
PP-III(D)	4.6	15.1		

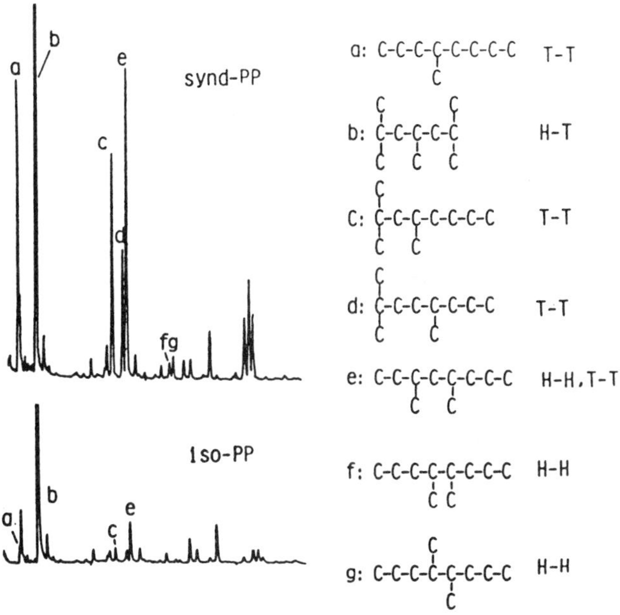

FIGURE 9 Expanded pyrograms for the C_{10} region for the highly isotactic PP-I(A) and the highly syndiotactic PP-II.

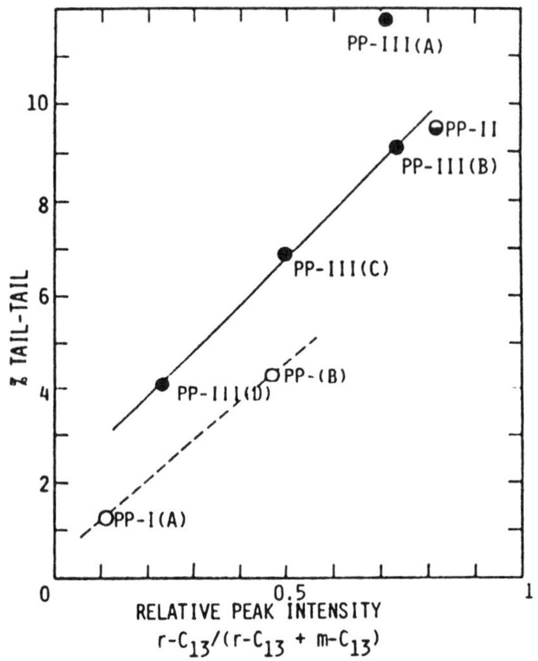

FIGURE 10 Relationships between observed tail-to-tail % and syndiotacticity for various PPs [21].

catalyst system (●, ◒) have many more chemical inversions than those prepared with the titanium-based catalyst system (○), when compared at an equivalent tacticity. In addition, it is very interesting to note that the amounts of chemical inversions increase as the rise of syndiotacticity. This fact suggests that the majority of chemical inversions occurs in the syndiotactic portions, while the catalytic control with isotactic specific propagation yield mostly H-T linkages.

C. Sequence Distribution in Ethylene-Propylene Copolymers

As shown in Figure 11, the pyrogram of P(E-co-P)s generally consist of the peaks associated with those from PE, PP, and hybrid products of ethylene and propylene units [22]. Here it is

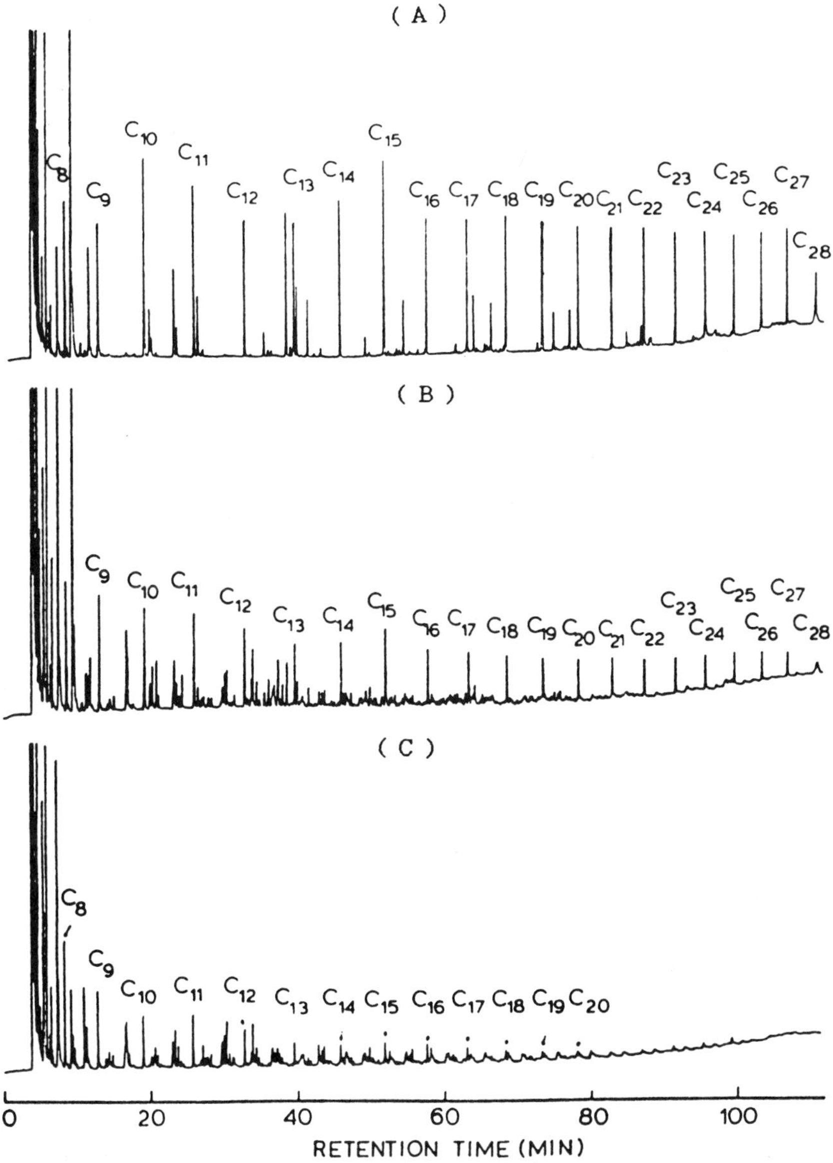

FIGURE 11 High-resolution hydrogenated pyrograms of (A) blend (PE-PP), (B) P(E-co-P) obtained by Ti catalyst, and (C) P(E-co-P) obtained by V catalyst [23].

interesting to note that the P(E-co-P) sample synthesized with the use of a Ti catalyst shows much stronger serial *n*-alkane peaks than the sample synthesized in the presence of a V catalyst despite the fact that the latter has higher ethylene content than the former. This is closely associated with the sequence distribution of ethylene units in the polymer chain.

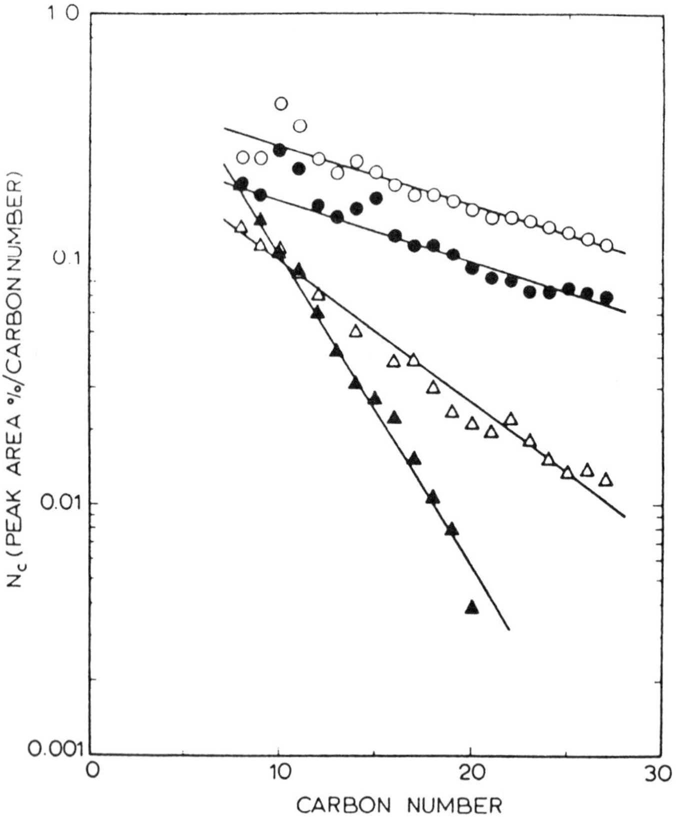

FIGURE 12 Relationships between carbon number and molar yield (N_c) for the series of *n*-alkane peaks on the pyrograms of P(E-co-P)s and the related polymers at 650°C. ○, PE; ●, blend (PE/PP = 50/50 wt %); △, P(E-co-P)(Ti catalyst, E = 40.8 wt %); ▲, P(E-co-P)(V catalyst, E = 52.0 wt %) [23].

In order to illustrate the difference in the ethylene sequences in the P(E-co-P)s, the relationships between carbon number and the relative molar yields (N_c) for the serial *n*-alkane peaks on the pyrogram of the copolymers are shown in Figure 12 together with that for PE and a blend of PE and PP. The intensities of *n*-alkane peaks on the pyrogram of PE is known to decrease as a semilogarithmic function against the carbon number of the *n*-alkanes since random scission is predominating the thermal degradation of PE [24–27]. Naturally, the physical blend of PE and PP shows a gentle slope almost equivalent to PE itself. On the other hand, the steeper the slope, the shorter the average sequence length of ethylene units in the P(E-co-P)s. Hence, this method should be effective for comparing the average sequences in the copolymer system containing ethylene units.

REFERENCES

1. S. van der Ven, *Polypropylene and other Polyolefins; Polymerization and Characterization*, Studies in Polymer Science 7, Elsevier, Amsterdam, The Netherlands (1990).
2. J. van Schotten and J. K. Evenhuis, *Polymer*, *6*: 343 (1965).
3. L. Michajlov, P. Zugenmaier, and H.-J. Cantow, *Polymer*, *12*: 70 (1971).
4. S. Tsuge, *Trends Anal. Chem.*, *1*: 87 (1981).
5. H. Ohtani and S. Tsuge, *Applied Polymer Analysis and Characterization* (J. Mitchell, Jr., ed.), Hanser Publishers, Munich, Vienna, p. 217 (1987).
6. Y. Sugimura and S. Tsuge, *Macromolecules*, *12*: 512 (1979).
7. T. Usami and S. Takayama, *Macromolecules*, *17*: 1756 (1984).
8. J. C. Randall, *J. Macromol. Sci.-Rev. Macromol. Chem. Phys.*, *C29*: 201 (1989).
9. M. Seeger and E. M. Barrall, II, *J. Polym. Sci., Polym. Chem. Ed.*, *13*: 1515 (1975).
10. D. H. Ahlstrom and S. A. Liebman, *J. Polym. Sci., Polym. Chem. Ed.*, *14*: 2478 (1976).
11. O. Mlejnek, *J. Chromatogr.*, *191*: 181 (1980).

12. H. Ohtani, S. Tsuge, and T. Usami, *Macromolecules*, *17*: 2557 (1984).
13. Y. Sugimura, T. Usami, T. Nagaya, and S. Tsuge, *Macromolecules*, *14*: 1787 (1981).
14. S. A. Liebman, D. H. Alstrom, W. H. Starnes, Jr., and F. C. Schilling, *J. Macromol. Sci.-Chem.*, *A17*: 935 (1982).
15. J. Tulisalo, J. Seppala, and K. Hastbacka, *Macromolecules*, *18*: 1144 (1985).
16. M. A. Haney, D. W. Johnston, and B. H. Clampitt, *Macromolecules*, *16*: 1775 (1983).
17. T. Usami, Y. Gotoh, S. Takayama, H. Ohtani, and S. Tsuge, *Macromolecules*, *20*: 1557 (1987).
18. M. Seeger and H.-J. Cantow, *Makromol. Chem.*, *176*: 2059 (1975).
19. Y. Sugimura, T. Nagaya, S. Tsuge, T. Murata, and T. Takeda, *Macromolecules*, *13*: 928 (1980).
20. S. Tsuge and H. Ohtani, *Analytical Pyrolysis, Technique and Applications* (K. J. Voorhees, ed.) Butterworth, London, p. 407. (1984).
21. H. Ohtani, S. Tsuge, T. Ogawa, and H.-G. Elias, *Macromolecules*, *17*: 465 (1984).
22. B. D. Coleman and T. G. Fox, *J. Polym. Sci.*, *Part A*, *1*: 3183 (1963).
23. S. Tsuge, Y. Sugimura, and T. Nagaya, *J. Anal. Appl. Pyrol.*, *1*: 221 (1980).
24. M. Seeger, H.-J. Cantow, and S. Marti, *Z. Anal. Chem.*, *276*: 267 (1975).
25. M. Seeger and H.-J. Cantow, *Makromol. Chem.*, *176*: 1411 (1975).
26. M. Seeger and R. J. Gritter, *J. Polym. Sci.*, *Polym. Chem. Ed.*, *15*: 1393 (1977).
27. M. Seeger and H.-J. Cantow, *Polym. Bull.*, *1*: 347 (1979).

5

Degradation Mechanisms of Condensation Polymers

HAJIME OHTANI AND SHIN TSUGE
Nagoya University, Nagoya, Japan

I. INTRODUCTION

Polymer characterization by analytical pyrolysis often necessitates a detailed understanding of how polymer degradation proceeds by heating in vacuo or in an inert atmosphere. Yet it is not an easy task to elucidate the degradation pathway of condensation polymers with heteroatoms in their backbone chains, because thermal degradation yields a number of complex polar compounds [1]. However, recent developments have provided highly specific pyrolysis devices, advanced gas chromatographs (GC), and mass spectrometers (MS), and specific identification of pyrolysis products by GC-MS systems. Consequently, it has become feasible, to a large extent, to deduce the complex thermal degradation behavior of condensation polymers using analytical pyrolysis methods such as pyrolysis-GC (Py-GC) and pyrolysis-MS (Py-MS). In this chapter, the thermal degradation of two types of typical condensation polymers—polyesters and polyamides—are discussed, focusing on the mechanisms of pyrolysate formation mainly under flash pyrolysis conditions, in which the polymer sample is rapidly exposed to elevated temperatures of 500°C or more.

II. POLYESTERS

Poly(alkylene terephthalate)s such as poly(ethylene terephthalate) (PET) and poly(butylene terephthalate) (PBT) are widely used in the fields of fibers and thermoplastics. Homologous series of poly(alkylene terephthalate)s: PET, poly(trimethylene terephthalate) (PTT), PBT, poly(pentamethylene terephthalate) (PPT), and poly(hexamethylene terephthalate) (PHT) have been studied by Py-GC [2,3]. Figure 1a–1e illustrates the high-resolution pyrograms of PET, PTT, PBT, PPT, and PHT, respectively [3]. The assignment of the characteristic peaks and their relative yields are listed in Table 1. The pyrograms show analogous patterns. The fragments containing two terephthalic acid units (g and h), with high molecular weight (up to 494), provide information about relatively long sequences in the polymer chain. These were observed on the pyrograms by using a thermally stable and chemically inert fused-silica capillary separation column (FSCC).

It has been proposed that the degradation of these polyesters is initiated by random scission of the ester linkages, through a six-membered cyclic transition state, to give an alkenyl and a carboxyl end group [4–6]:

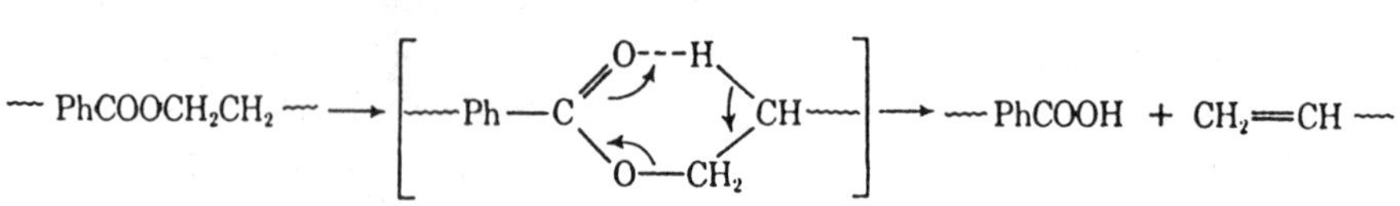

Furthermore, at elevated tempeatures used for flash pyrolysis, the acid-form terminals are often somewhat reduced to the more

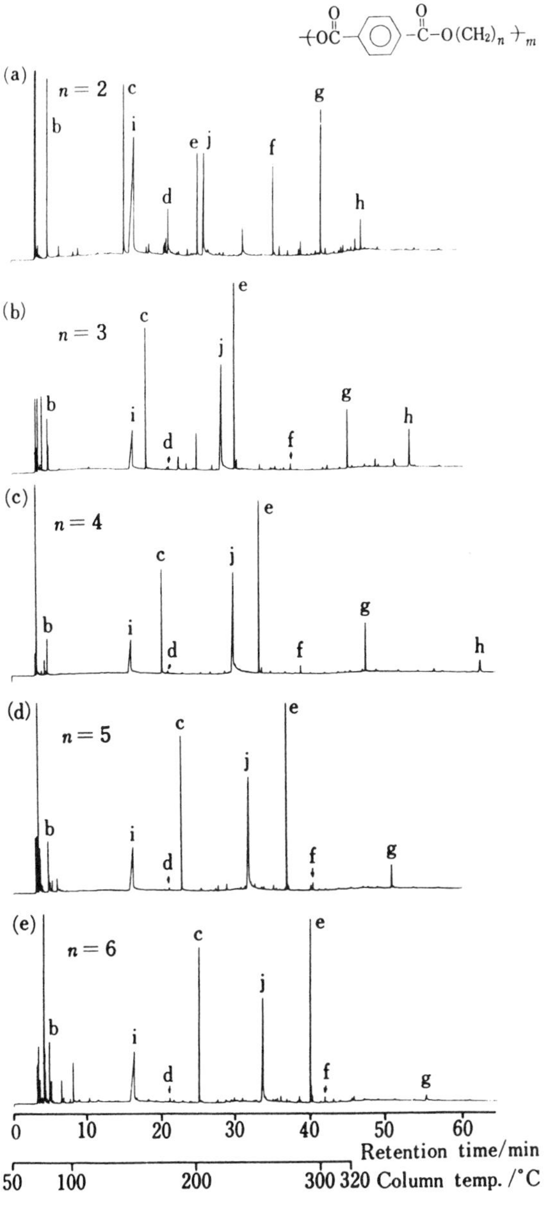

FIGURE 1 Pyrograms of (a) PET, (b) PTT, (c) PBT, (d) PPT, and (e) PHT at 590°C [3]. Peak notations correspond to those in Table 1.

TABLE 1 Assignment and Relative Intensities of the Characteristic Peaks on the Pyrograms of Terephthalate Polyesters (Peak Notations as in Fig. 1) [3]

Peak notations	Structure[a]	Relative peak intensity,[b] % PET	PPrT	PBT	PPeT	PHT
Small products[c]		19.0	9.6	37.6	34.8	38.6
b		5.1	2.3	1.5	1.7	2.7
c	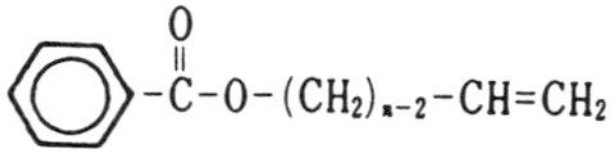	5.9	8.5	6.7	7.5	7.0
i	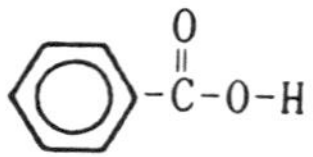	32.0	14.2	10.4	13.3	15.1
d	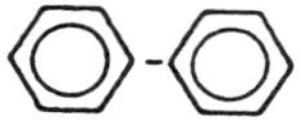	1.8	>0.3	>0.3	>0.3	>0.3

e $CH_2=CH-(CH_2)_{n-2}-O-C(=O)-[C_6H_4]-C(=O)-O-(CH_2)_{n-2}-CH=CH_2$	3.7	21.7	13.2	17.6	11.3
j $H-O-C(=O)-[C_6H_4]-C(=O)-O-(CH_2)_{n-2}-CH=CH_2$	9.2	26.0	21.7	21.0	14.2
f $[C_6H_4]-C(=O)-O-(CH_2)_n-O-C(=O)-$	3.4	0.6	0.7	0.6	>0.5
g $[C_6H_4]-C(=O)-O-(CH_2)_n-O-C(=O)-[C_6H_4]-C(=O)-O-(CH_2)_{n-2}-CH=CH_2$	5.7	4.5	4.1	2.2	0.9
h $CH_2=CH-(CH_2)_{n-2}-O-C(=O)-[C_6H_4]-C(=O)-O-(CH_2)_n-O-C(=O)-[C_6H_4]-C(=O)-O-(CH_2)_{n-2}-CH=CH_2$	1.7	5.6	3.1	—	—

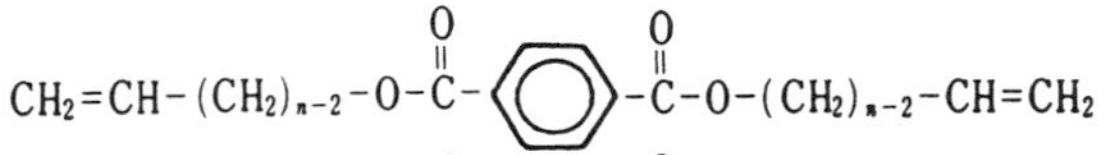

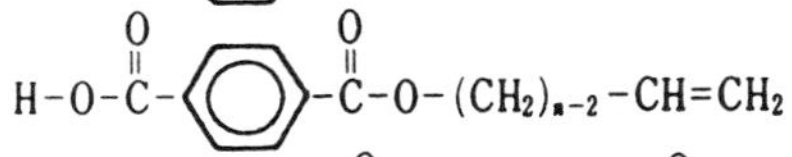

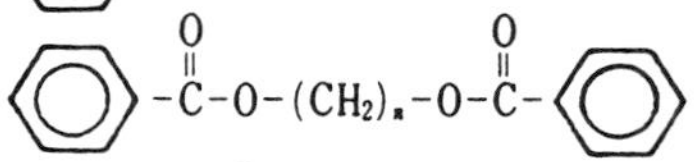

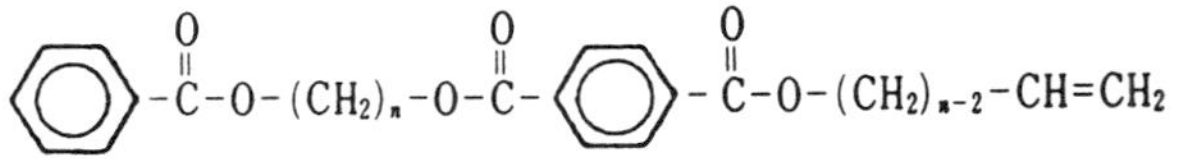

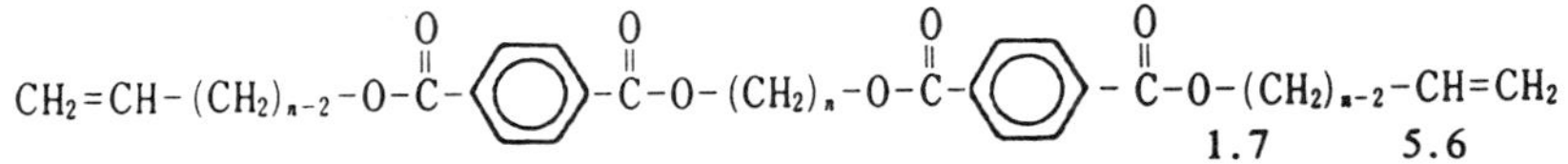

a. $n=2$ (PET), 3 (PPrT), 4 (PBT), 5 (PPeT) and 6 (PHT).
b. Relative peak area (%) among all peaks appearing on the pyrograms.
c. Total intensities of the peaks with shorter retention times than that of benzene.

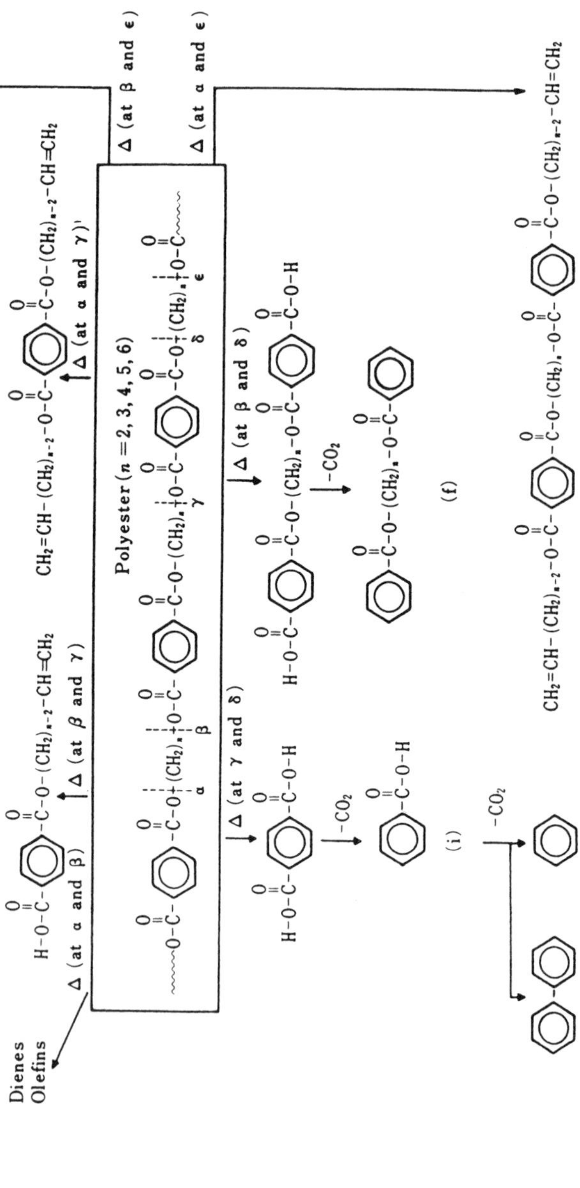
Polyester (n = 2, 3, 4, 5, 6)
Δ (at β and ε)
Δ (at α and ε)
Δ (at α and γ)
Δ (at β and δ)
Δ (at γ and δ)
Δ (at β and γ)
Δ (at α and β)
-CO2
Dienes
Olefins
(a)
(b)
(c)
(d)
(e)
(f)
(g)
(h)
(i)
(j)

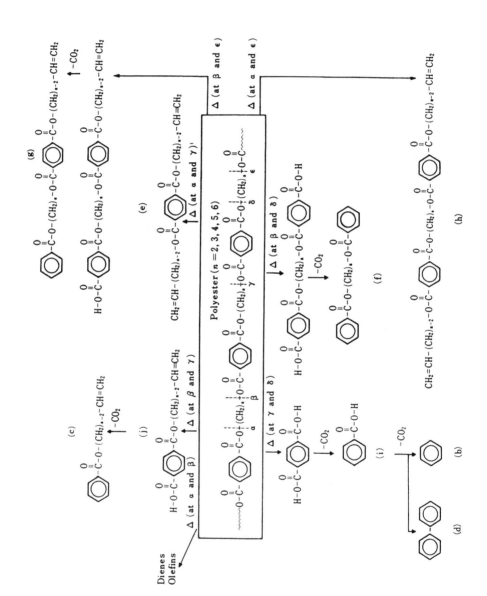

stable phenyl terminals by elimination of carbon dioxide. Thus, the main pyrolysis products are comprised of compounds with alkenyl-, phenyl-, and/or carboxyl end groups reflecting the polymer structure. However, the dibasic acids, such as terephthalic acid, which are also expected to form, were not detected. This is due to their high polarity—even when using FSCC—while the fragments with an acid end group were observed as clearly separated peaks (i and j). On the other hand, α and ω-dienes are most likely formed among the low boiling point products from PBT, PPT, and PHT through the cyclic transition state at two neighboring ester linkages [7]. On the basis of the results described above, the common degradation pathways of poly(alkylene terephthalate) are summarized in Figure 2 [3].

The same basic degradation pathways were reported for PET and/or PBT by other Py-GC [8,9] and Py-MS [10–16] studies. In addition, tetrahydrofuran was observed as one of the major products from PBT [9,13]. Furthermore, another pathway involving the transfer of a hydrogen atom with CO—O bond scission to yield a hydroxyl end group was also indicated by Py-electron impact ionization (EI)-MS as follows [11,12,14]:

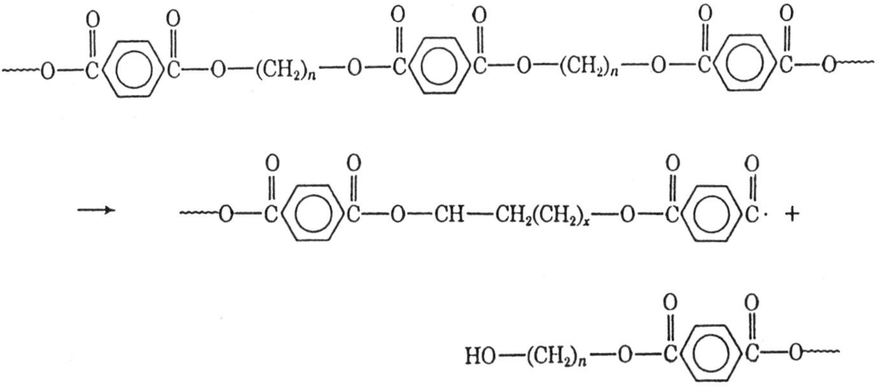

A series of poly(alkylene phthalate)s [17] and some unsatu-

FIGURE 2 Thermal degradation mechanisms of the terephthalate polyesters on the basis of the pyrograms at 590°C [3]: △: thermal cleavage; b–j correspond to the peaks in Figure 1.

rated polyesters [18,19] were also investigated by Py-GC. The major pyrolysis products for these polyesters were the associated diols and anhydrides, such as phthalic anhydride and maleic anhydride, as well as the products formed through the six-membered cyclic transition state. Moreover, various kinds of cyclic ethers and cyclic esters were also observed in considerable quantities. The general scheme of the degradation of poly(alkylene phthalate)s is shown in Figure 3 [17].

Thermal degradation of aliphatic polyesters was investigated by Py-MS [20–27]. Dominant degradation products were cyclic monomer (lactone), cyclic oligomers, or linear oligomers with ketene, hydroxyl, carboxyl, or alkenyl end groups. The relative yields of the characteristic products strongly depended on the chain length and the branching structure between the neighboring ester groups in the polymer chain, as well as on the degradation condition. For example, Py-chemical ionization-MS analysis of several polylactones with various chain lengths indicated that only poly(β-propiolactone) degraded mainly via the six-membered ring transition state to form linear products with a carboxyl and a vinyl end group, whereas the other polylactones predominantly decomposed by intramolecular transesterification reactions to form cyclic oligomers [25].

Fully aromatic polyesters based on *p*-hydroxybenzoic acid (PHB) have been recently recognized as some of the most promising high-performance liquid crystalline polymers (LCPs), especially in fields where ever-increasing thermal stability is required. The thermal degradation mechanisms of typical LCPs prepared from PHB, biphenol (BP) and terephthalic acid (TA) were studied by Py-GC/MS [28,29]. Figure 4 shows a typical pyrogram of an LCP sample (PHB/BP/TA = 2/1/1) at 650°C, with peak identification summarized in Table 2 [29]. The general origin of the characteristic products shown in Figure 5 (estimated by examining the LCPs with different comonomer ratios) and that containing deuterated terephthalate units [29], shows:

1. Phenol is almost exclusively formed form the PHB moiety

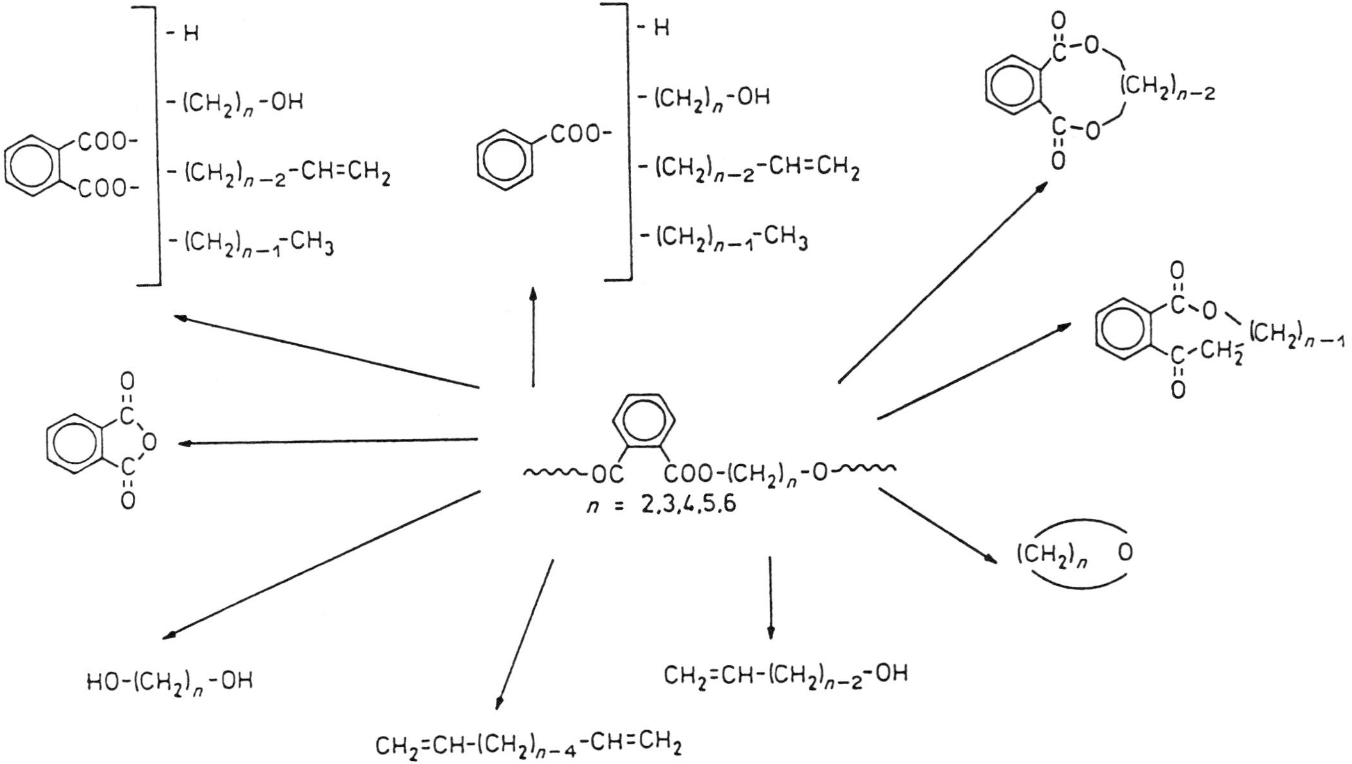

FIGURE 3 General scheme of thermal degradation of poly(alkylene phthalate) [17].

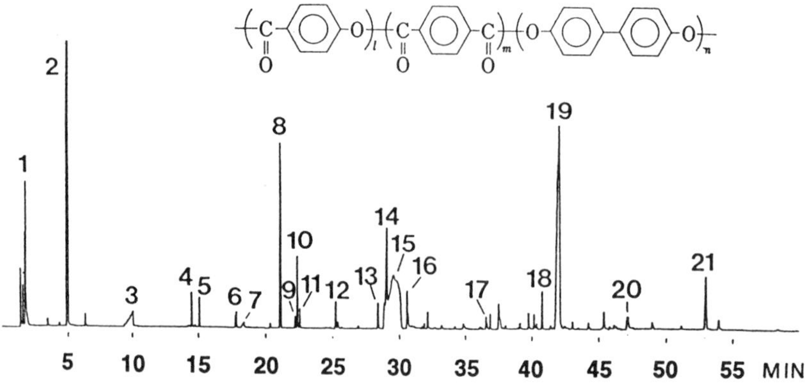

FIGURE 4 Pyrogram of a liquid crystalline fully aromatic polyester consisting of *p*-hydroxybenzoic acid, biphenol, and terephthalic acid at 650°C (PHB/BP/TA = 2/1/1) [29]. Peak numbers correspond to those in Table 2.

2. Benzene is mostly formed from TA units
3. The larger products, such as biphenol, *p*-hydroxyphenylbenzoate, and 4,4′-biphenyldibenzoate are mainly derived from the TA and/or BP moieties
4. Biphenyl is mostly formed from the PHB and TA moieties via recombination rather than directly from the BP moiety
5. Phenyl benzoate is mainly formed by recombination reactions between phenoxy and benzoyl radicals from the PHB and TA moieties, respectively.

These observations suggest that:

1. The C—O bonds between a carbonyl carbon and a phenolic oxygen are preferentially cleaved over those between an aromatic ring and a phenolic oxygen
2. The C—C bonds between an aromatic ring and a carbonyl carbon are preferentially cleaved over the other type of C—C bonds
3. The C—C bonds between aromatic rings are not easily cleaved.

TABLE 2 Assignment of the Characteristic Peaks in the Pyrogram of a Fully Aromatic Polyester [29]

peak no.	products	structure
1	benzene	
2	phenol	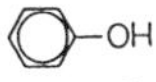
3	benzoic acid	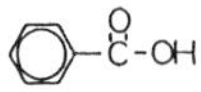
4	biphenyl	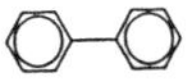
5	diphenyl ether	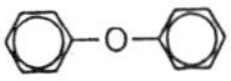
6	benzofuran	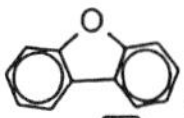
7	p-hydroxybenzoic acid	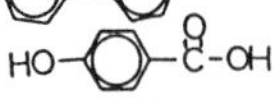
8	phenylbenzoate	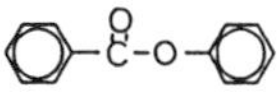
9	m-phenylphenol	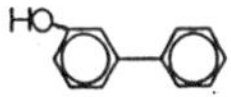
10	p-phenylphenol	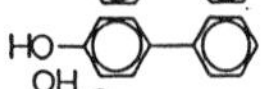
11	o-hydroxybenzofuran	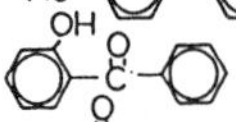
12	xanthone	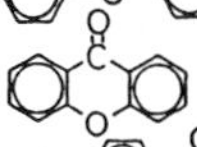
13	p-hydroxybenzophenone	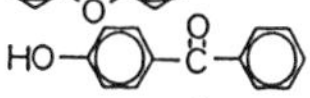
14	phenyl(p-hydroxybenzoate)	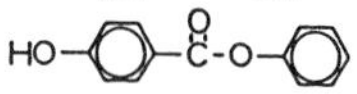
15	biphenol	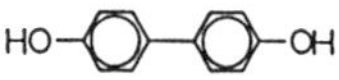
16	4,4'-dihydroxybenzophenone	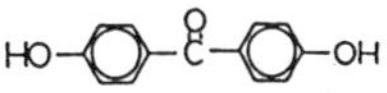
17	biphenylbenzoate	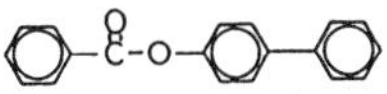
18	acetylbenzoate	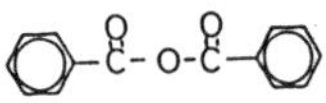
19	p-hydroxybiphenylbenzoate	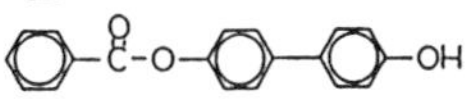
20	p-hydroxybiphenyl(p'-hydroxybenzoate)	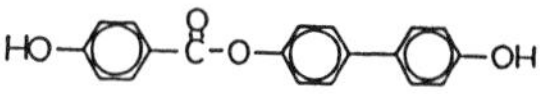
21	4,4'-biphenyldibenzoate	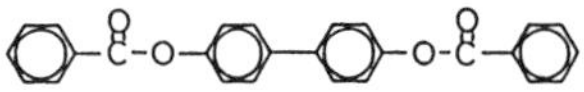

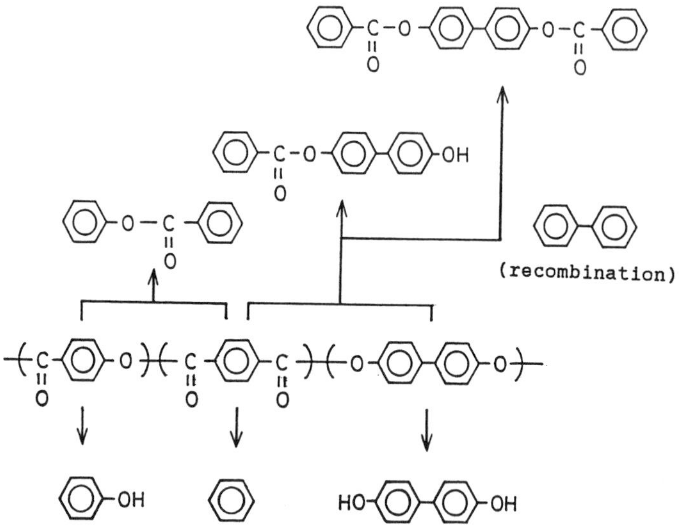

FIGURE 5 Thermal degradation mechanisms of an aromatic polyester [29].

Moreover, inter- and intramolecular ester exchange reactions causing rearranged polyester sequences at the pyrolysis stage were also demonstrated to take place for aliphatics [26] and aliphatic-aromatic polyesters [30] using Py-MS.

III. POLYAMIDES

Synthetic polyamides, or nylons, are widely exploited as fibers, moldings, and films. The thermal degradation of a series of aliphatic polyamides was investigated by high-resolution Py-GC [31]. Both lactam and diamine-dicarboxylic acid types of nylon samples were pyrolyzed at 550°C in a furnace pyrolyzer under a flow of nitrogen carrier gas, and the resulting degradation products were continuously separated by a capillary separation column. Table 3 summarizes the various classes of common characteristic products observed in the resulting pyrograms [31,32]. The

TABLE 3 Characteristic Degradation Products from Aliphatic Polyamides [31,32]

Class of compounds	Abbreviation	Structure
Hydrocarbons	HC	$CH_3-(CH_2)_m-CH_3$ $CH_2=CH-(CH_2)_{m-1}-CH_3$ $CH_2=CH-(CH_2)_{m-2}-CH=CH_2$
Mononitriles	MN	$CH_3-(CH_2)_m-C\equiv N$ $CH_2=CH-(CH_2)_{m-1}-C\equiv N$
Amines	AM	$CH_3-(CH_2)_m-NH_2$ $CH_2=CH-(CH_2)_{m-1}-NH_2$
Lactams	L	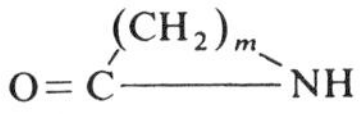
Dinitriles	DN	$N\equiv C-(CH_2)_m-C\equiv N$
Cyclopentanone	CP	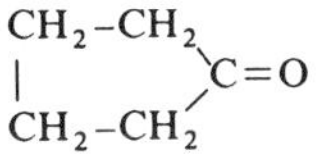
Hydrocarbons containing one amide group	HC(A)	$CH_3-(CH_2)_m-\overset{O}{\overset{\|}{C}}-\overset{H}{\overset{\vert}{N}}-(CH_2)_n-CH_3$ $CH_2=CH-(CH_2)_{m-1}-\overset{O}{\overset{\|}{C}}-\overset{H}{\overset{\vert}{N}}-(CH_2)_n-CH_3$ $CH_2=CH-(CH_2)_{m-1}-\overset{O}{\overset{\|}{C}}-\overset{H}{\overset{\vert}{N}}-(CH_2)_{n-1}-CH=CH_2$
Mononitriles containing one amide group	MN(A)*	$CH_3-(CH_2)_m-\overset{O}{\overset{\|}{C}}-\overset{H}{\overset{\vert}{N}}-(CH_2)_n-C\equiv N$ $CH_2=CH-(CH_2)_{m-1}-\overset{O}{\overset{\|}{C}}-\overset{H}{\overset{\vert}{N}}-(CH_2)_n-C\equiv N$
Mononitriles containing one amide group	MN(A)′**	$CH_3-(CH_2)_m-\overset{H}{\overset{\vert}{N}}-\overset{O}{\overset{\|}{C}}-(CH_2)_n-C\equiv N$ $CH_2=CH-(CH_2)_{m-1}-\overset{H}{\overset{\vert}{N}}-\overset{O}{\overset{\|}{C}}-(CH_2)_n-C\equiv N$

* Formed from ω-aminocarboxylic acid-type nylons.
** Formed from diamine-dicarboxylic acid-type nylons.

degradation products were found to be strongly dependent on the number of methylene groups in the polymer chain units.

Figure 6 shows the pyrogram of (a) nylon 6 and (b) nylon 6/6 observed by using a glass capillary separation column. The main pyrolysis products in (a) are ϵ-caprolactam [4,10,31,33,34] along with small amounts of nitriles (MN and MN(A)). Polylactams consisting of relatively short methylene chains, such as nylon 4 and 6, tend to regenerate the associated monomeric lactame upon heating.

As shown in Figure 6b, the most abundant compound resulting from the thermal degradation of nylon 6/6 is cyclopentanone (CP) [4,31,32,35–39], which is characteristic of adipic acid-based polyamides. It has been demonstrated by a recent study using Py-EI and CI-tandem MS that the formation of CP from nylon 6/6 occurs via a C—H hydrogen transfer reaction to nitrogen to give the primary products bearing amine and ketoamide end groups as follows [39]:

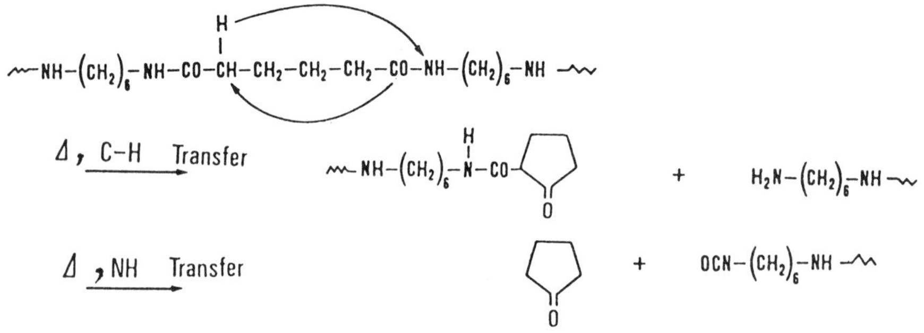

In addition, some nitriles (DN and MN(A)′) and ϵ-caprolactam (L) are observed as minor products. The following mechanism was proposed to accounts for the formation of L from nylon 6/6 [38].

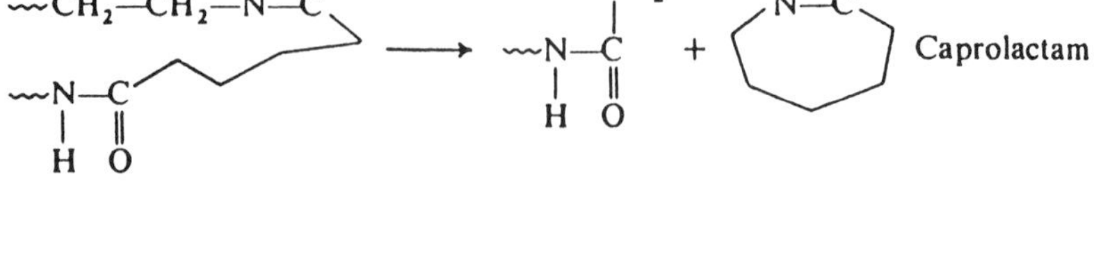

Furthermore, peaks of the associated amines arose from the cleavage of the amide bond are clearly observed on the pyrogram of nylon 6/6 by using a fused-silica capillary separation column [32,40], although they are missing in Figure 6b.

Figure 7 shows the pyrograms of (a) nylon 11, (b) nylon 6/10, and (c) nylon 12/6 observed by using a fused-silica capillary column [32,40]. Fairly strong peaks of mononitriles (MNs) are observed in (a) while that of the associated lactam (L) becomes very small. This suggests that the thermal degradation of nylon 11 is mainly associated with homolytic cleavage of the CH_2—NH bond to form a double bond and an amide or a nitrile group through the six-membered cyclic transition state (*cis* elimination) [4,31,41]. The most intense olefinic MN, where carbon number corresponds to that of the successive methylene groups in the polymer chain plus 1, are formed through the following *cis* eliminations followed by the dehydration reation [31]:

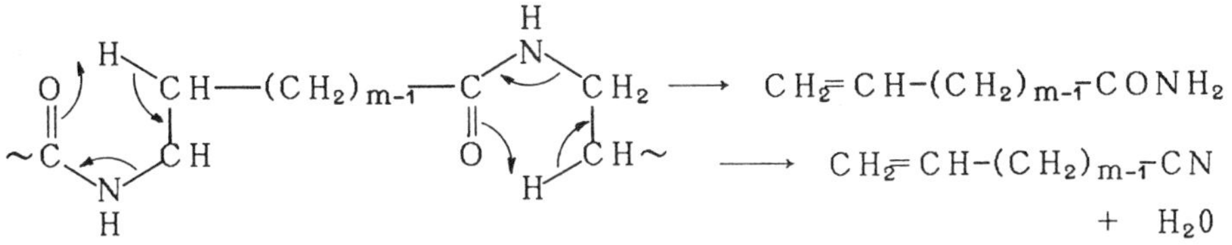

The minor peaks of saturated MNs are always accompanied by those of the olefinic MNs. In addition, the smaller MNs and the associated hydrocarbons (HCs) are formed through the further C—C bond cleavages. The main degradation pathway is basically the same for the polylactams, including relatively longer methylene chains such as nylon 8, 11, and 12 [31].

One of the most characteristic products in Figure 7b is cebaconitrile (DN(m = 8)). Large quantities of dinitriles are formed via the *cis* elimination reactions at the neighboring amide groups across a dicarboxylic acid unit in the diamine-dicarboxylic acid type of polyamide chain except for the adipic acid-based polyamides [31]. Moreover, a series of mononitriles are also observed in the pyrograms. Among these, the most intense peak is that of the second longest unsaturated mononitrile. This fact suggests

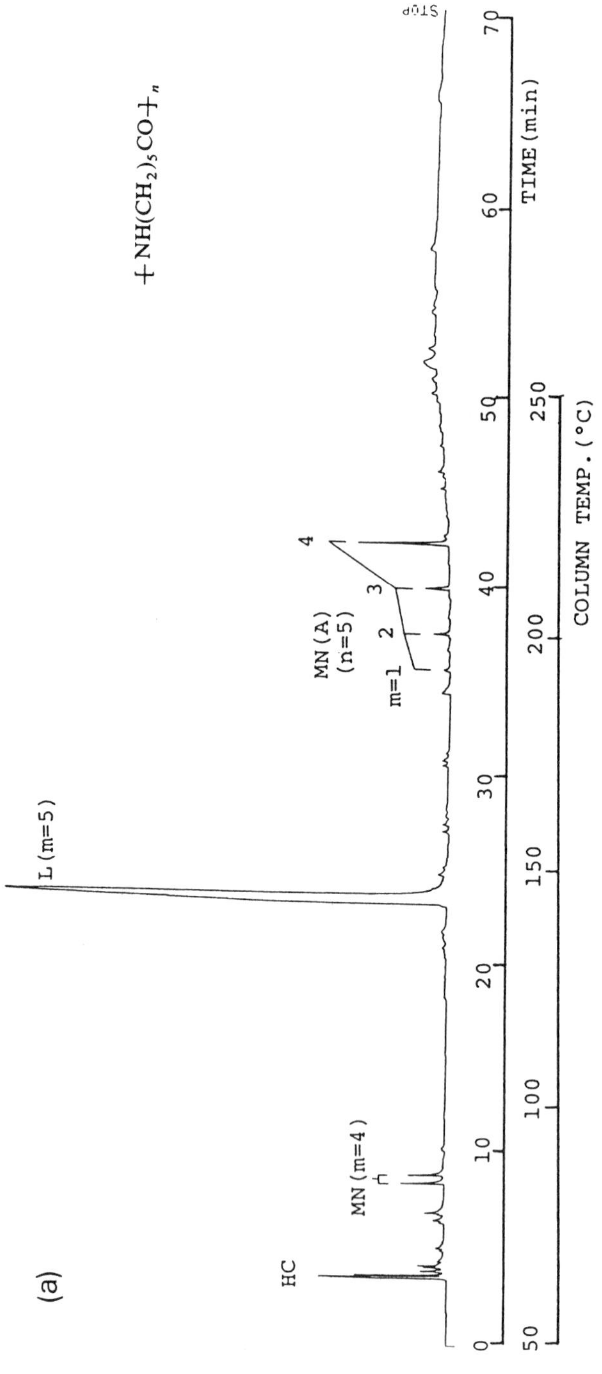

(a)
$+NH(CH_2)_5CO+_n$
HC
MN (m=4)
L (m=5)
MN (A)
(n=5)
m=1
2
3
4
STOP
0
10
20
30
40
50
60
70
TIME (min)
50
100
150
200
250
COLUMN TEMP. (°C)

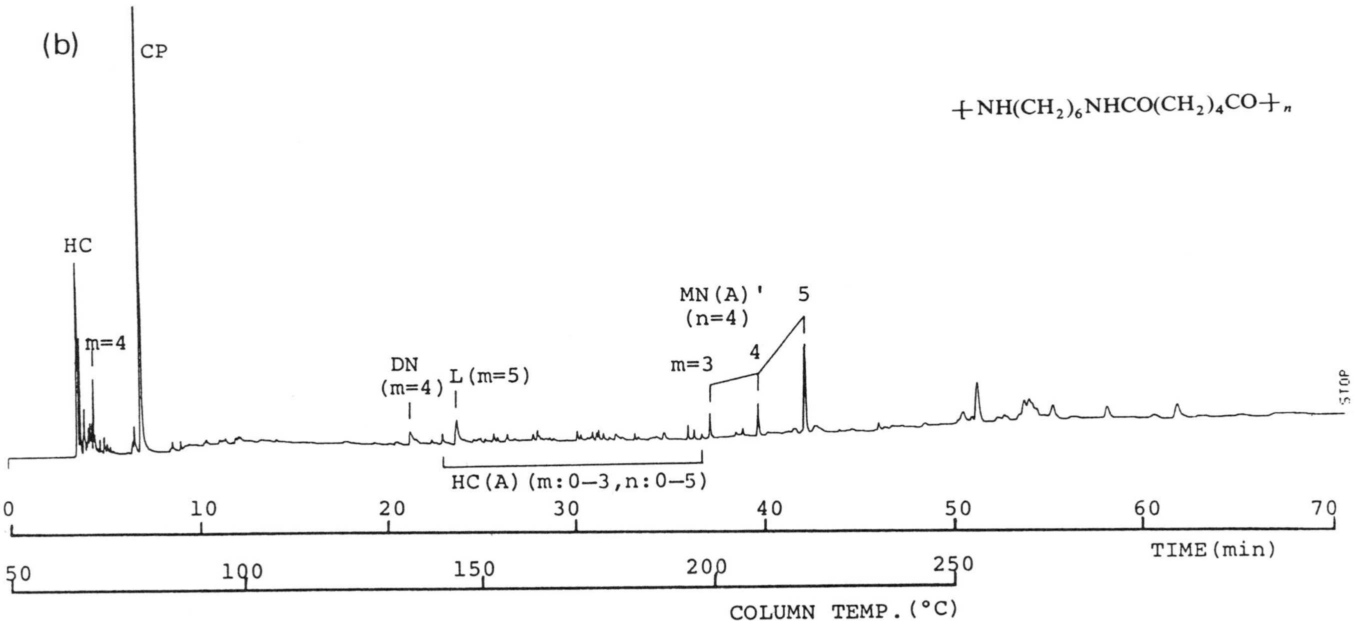

FIGURE 6 Pyrograms of nylons at 550°C observed by a glass capillary column; (a) nylon 6, (b) nylon 6/6 [31].

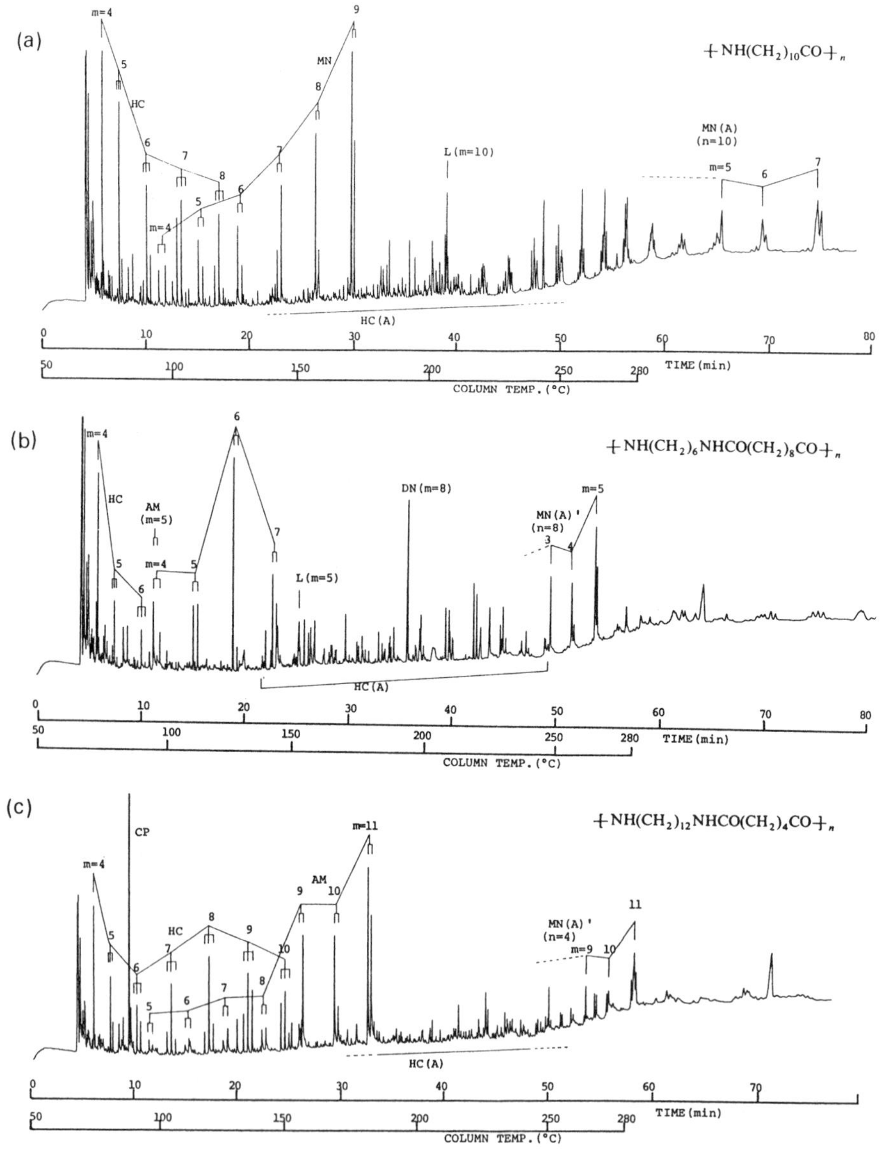

FIGURE 7 Pyrograms of nylons at 550°C observed by a fused-silica capillary column; (a) nylon 11, (b) nylon 6/10, (c) nylon 12/6 [32,40].

that the *cis*-elimination reaction might take place at the opposite site of the amide bond as follows [42]:

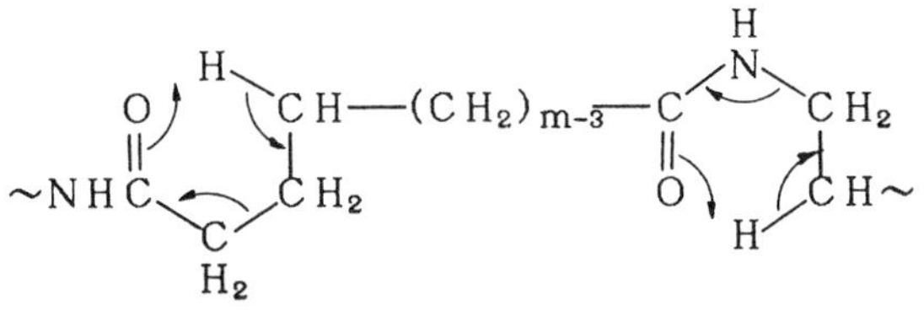

$$\longrightarrow CH_2{=}CH\text{-}(CH_2)_{m-3}\text{-}CONH_2 \longrightarrow CH_2{=}CH\text{-}(CH_2)_{m-3}\text{-}CN + H_2O$$

Similar to nylon 6/6, the most characteristic peak for nylon 12/6 is CP formed from the adipic acid moiety. In addition, a series of HC triplets and those of amine (AM) doublets [32,40] up to C_{12} are formed in considerable amounts through the C—C and amide bond cleavages.

Thermal degradation of various aliphatic polyamides was also studied by Py-EIMS [10,41,43], Py-field ionization (FI), and field desorption (FD)-MS [34,42,44,45]. The Py-EIMS studies demonstrated that the polylactams mainly decomposed to cyclic oligomers whereas the favored decomposition pathways of diamine-dicarboxylic acid-type polyamides were the *cis*-elimination reaction and the cleavage of the amide bond, which also occurred in the polylactams with a large number of methylene groups, such as nylon 12. On the other hand, Py-FIMS of various diamine-dicarboxylic acid-type polyamides illustrated that the main pyrolysis products detected were protonated dinitriles and various protonated nitriles, as well as oligomers up to 1000 daltons, except for the polyamides containing adipic acid subunits, for which protonated amines and diamine were observed in large amounts [42].

Thermal decomposition of aliphatic-aromatic polyamides were mainly investigated by Py-MS. It was found in the early study that the thermal degradation of copolyamides formed from aliphatic and aromatic amino acids occurred almost exclusively through the bond scission in the aliphatic moieties [46]. It was shown by Py-EIMS that thermal degradation of copolyamides of *p*-aminobenzoic acid and some aliphatic amino acids yielded poly-*p*-aminobenzoic acid via the elimination of lactams:

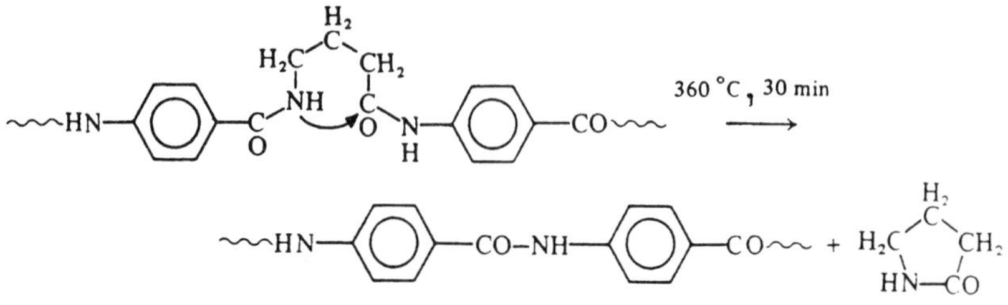

The other favored reactions are the cleavage of the CO—NH bond to form *o*-aminophenyl end groups, the *cis* elimination to form amide and vinyl end groups, and the cleavage of the CH_2—CO bond to form isocyanate end groups [47]:

$$\sim CH_2-CO-NH-Ph\sim \xrightarrow{\Delta} \sim CH_3 + O=C=N-Ph\sim \qquad -e$$

Recently, the thermal decomposition processes of various aliphatic-aromatic polyamides were investigated by Py-GC-MS and by Py-MS using both CI and EI modes [48–50]. The thermal decomposition of the polyamides of aromatic-diamine and aliphatic-dicarboxylic acid was strongly influenced by the structure of the aliphatic subunits [48]. The formation of compounds with succinimide and amine end groups was observed in the pyrolysis of the polyamides containing succinic subunits via an intramolecular exhange and a concomitant N—H hydrogen transfer,

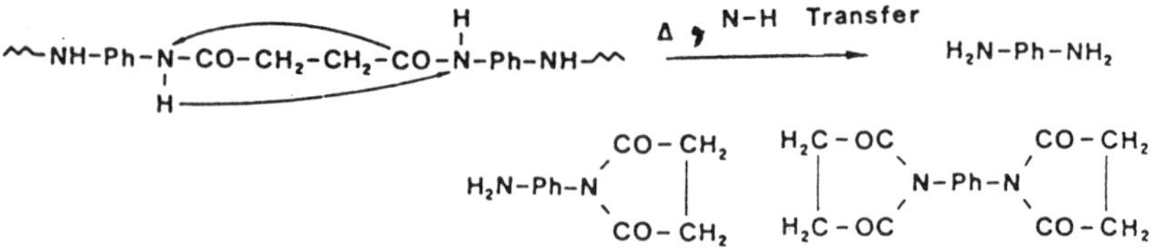

while the primary products were compounds with amine and keto amide end groups in the pyrolysis of the polyamides containing adipic subunits. In addition, the *N*-methyl-substituted polyamide decomposed via a hydrogen transfer process from the methyl group to the nitrogen atom with formation of compounds with amine and/or 2,5-piperidione end groups [49]:

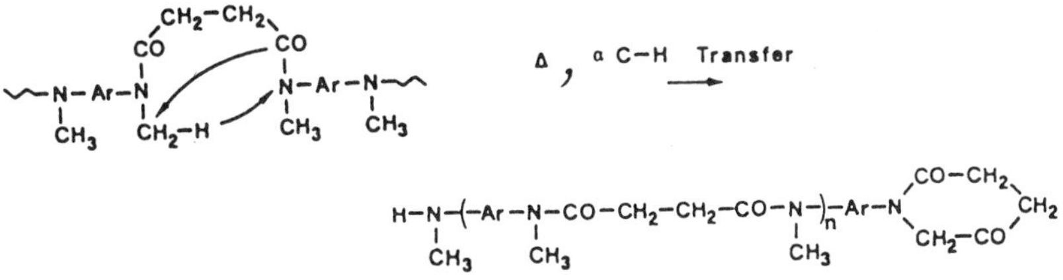

On the other hand, the primary thermal decomposition of polyamides of aliphatic-diamine and aromatic-dicarboxylic acid proceeded via the *cis*-elimination process with formation of the products containing amide and olefin end groups [50]. Nitrile end groups also proved to be formed by dehydration of the amide groups formed in the primary process.

Various investigations of the thermal degradation of wholly aromatic polyamides (aramids) such as poly(1,3-phenylene isophthalamide) (Nomex) and poly(1,4-phenylene terephthalamide) (Kevlar) have also been reported [34,51–55].

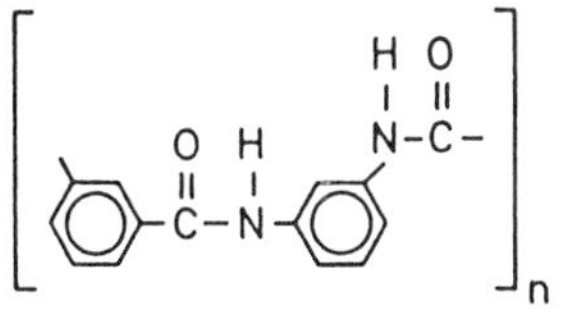

Nomex

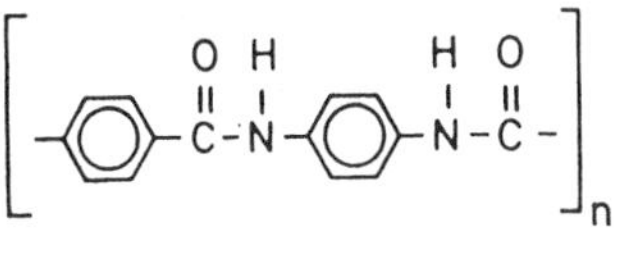

Kevlar

In an early Py-FIMS study [34], benzonitrile and other pyrolysates with amine and/or nitrile end groups were recorded as the main pyrolysis products of Kevlar at 600°C. In another paper on the pyrolysis of Nomex at 550°C, the primary low-boiling volatiles identified by GC-MS were CO, CO_2, H_2O, and benzonitrile [51]. Considerable amounts of benzene, methane, toluene, 1,3-tolunitrile, etc. were also observed. The presence of at least 17 additional degradation products was detected by HPLC in the condensible products, among which the two major components were 1,3-dicyanobenzene and 3-cyanobenzoic acid. These observations supported a mechanism that involved the cleavage of an aromatic—NH bond followed by the loss of H_2O to form aromatic nitriles. Cleavage of the CO—NH bond, hydrolysis, and decarboxylation explained the other major products.

On the other hand, the pyrolysis products of a Nomex-type aramid and its chloro-derivative at 450 and 550°C were identified by GC-FTIR and GC-MS [52]. The end groups of the volatile degradation products reported were amine, nitrile, carboxylic acid, and phenyl both for one-ring and two-ring compounds. The formation of the two-ring compounds occurred preferably at 450°C, whereas pyrolysis at 550°C yielded predominantly the one-ring compounds. Moreover, Nomex and Kevlar were pyrolyzed at several temperatures between 300 and 700°C [53]. At lower temperatures, water was almost exclusively formed as a volatile degradation product, accompanied by traces of carbon dioxide. Hydrolysis products were formed with increasing temperature, followed by the formation of nitriles and products containing a phenyl end group. Derivatives of toluene and biphenyl, HCN and hydrocarbons were observed at high temperatures. The results reported in these two papers [52,53] suggested that the homolytic cleavages involving all the nonaromatic ring bonds as well as hydrolytic reactions took place during the degradation of these aramids. At lower temperatures the hydrolytic mechanism was dominant, whereas at higher temperatures homolytic reactions became increasingly important. Furthermore, it has been reported that for Nomex the cleavage of the N—H bond is both thermodynamically and kinematically favored at the beginning of the thermal degradation over the scission of the other bond in the polymer main chain [54].

Recently, thermal degradation of Nomex and Kevlar was studied by Py-FIMS and Py-GC [55]. In Py-FIMS, the polymer samples were pyrolyzed in the ion source by heating from 50 to 750°C at the rate of 1.2°C/s. The relative abundances of the thermal degradation products are listed in Table 4. Possible intermediate products used to form the observed compounds through the hydrolytic and homolytic decomposition processes are sum-

TABLE 4 Pyrolysis Field Ionization Mass Spectrometry, Tentatively Assigned Pyrolyzates of Aromatic Polyamides [55]

m/z	Thermal degradation products	Relative abundance Kevlar	Relative abundance Nomex
78	C_6H_6	0.5	5
93	$H_2N—C_6H_5$	7	10
103	$C_6H_5—CN$	15	35
108	$H_2N—C_6H_4—NH_2$	29	55
117	$NC—C_6H_4—CH_3$	0.7	3
118	$NC—C_6H_4—NH_2$	12	3
122	$C_6H_5—COOH$	1	13
128	$NC—C_6H_4—CN$	1	5
134	$H_2N—C_6H_4—N=C=O$	7	2
147	$NC—C_6H_4—COOH$	0.5	5
154	$C_6H_5—C_6H_5$	0.3	1
166	$HOOC—C_6H_4—COOH$	1	12
169	$C_6H_5—C_6H_4—NH_2$	1	1.5
179	$C_6H_5—C_6H_4—CN$	1	4
194	$H_2N—C_6H_4—C_6H_4—CN$	6	6
197	$C_6H_5—NH—CO—C_6H_5$	11	2
212	$H_2N—C_6H_4—NH—CO—C_6H_5$	100	70
222	$C_6H_5—NH—CO—C_6H_4—CN$	10	1
237	$H_2N—C_6H_4—NH—CO—C_6H_4—CN$	74	31
238	$O=C=N—C_6H_4—NH—CO—C_6H_5$	23	10
241	$C_6H_5—NH—CO—C_6H_4—COOH$	1	5
256	$H_2N—C_6H_4—NH—CO—C_6H_4—COOH$	4	100
282	$O=C=N—C_6H_4—NH—CO—C_6H_4—COOH$	2	5
316	$C_6H_5—CO—NH—C_6H_4—NH—CO—C_6H_5$ $C_6H_5—NH—CO—C_6H_5—CO—NH—C_6H_5$	55	14
331	$H_2N—C_6H_4—NH—CO—C_6H_4—CO—NH—C_6H_5$	21	25
341	$C_6H_5—CO—NH—C_6H_4—NH—CO—C_6H_4—CN$	37	12
346	$H_2N—C_6H_4—NH—CO—C_6H_4—CO—NH—C_6H_4—NH_2$	18	66
360	$C_6H_5—CO—NH—C_6H_4—NH—CO—C_6H_4—COOH$	2	19
385	monomer + 147	1	3
435	monomer + 197	2	0
450	monomer + 212	11	7
475	monomer + 237	1.5	1.5
476	monomer + 238	4	0.5

marized in Table 5. Kevlar and Nomex mostly gave the same kinds of signals but several showed strong differences in their relative abundance. Higher abundances of the degradation products containing carboxylic acid groups were generally observed for Nomex, whereas Kevlar yields higher abundances of fragments resulting from cleavage of the carbonyl/phenyl bond. In general, nitriles were formed at higher temperatures over carboxylic acids. As the subsequent decarboxylation reactions from carboxylic acids are favored at higher temperatures, the amounts

TABLE 5 Possible Intermediate Products from Aromatic Polyamides Through Hydrolytic and Homolytic Decomposition [55]

Type of decomposition	Possible intermediate products
(A) Hydrolytic decomposition	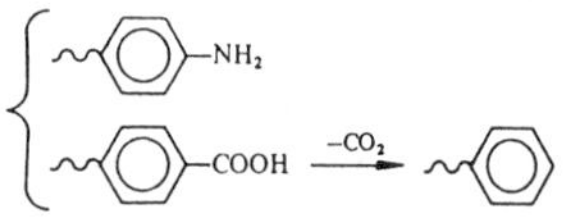
(B) Homolytic decomposition	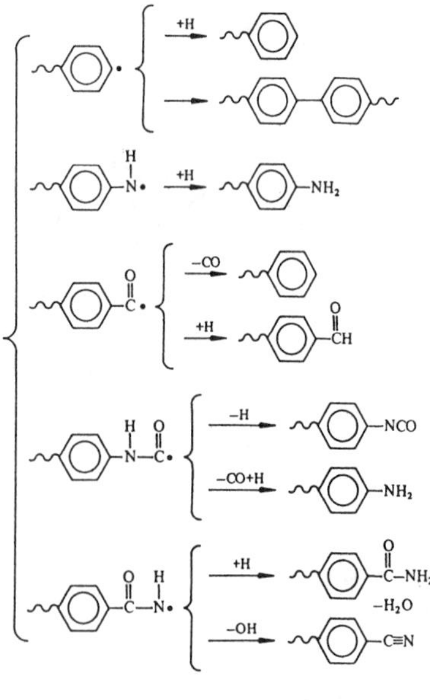

of carboxylic acids from Nomex decrease with increasing pyrolysis temperature, and those from Kevlar are smaller than those from Nomex because of the higher thermal stability of the former.

On the other hand, the pyrograms observed in flash Py-GC at 720°C almost entirely consisted of the fragments formed via homolytic degradations, although many were identical with those observed by Py-FIMS. In addition, the difference between Nomex and Kevlar with Py-GC was much less than those observed in Py-FIMS. Moreover, the formation of secondary products, such as biphenyl derivatives in Py-GC, were much less than that in Py-FIMS. The differences between the results by Py-GC and by Py-FIMS could be attributed to the difference in the final pyrolysis temperature and the heating rate.

REFERENCES

1. G. Montaudo and C. Puglisi, Thermal degradation of condensation polymers, *Comp. Polym. Sci., First Suppl.* (S. L. Aggarwal and S. Russo, eds.), Pergamon Press, Oxford, UK, pp. 227–251 (1992).
2. Y. Sugimura and S. Tsuge, *J. Chromatagr. Sci., 17*: 269 (1979).
3. H. Ohtani, T. Kimura, and S. Tsuge, *Anal. Sci., 2*: 179 (1986).
4. S. A. Liebman and E. J. Levy, eds., *Pyrolysis and GC in Polymer Analysis*, Chromatogr. Sci., Series Vol. 29, Marcel Dekker, New York, (1985); J. H. Flynn and R. E. Florin, Degradation and pyrolysis mechanism, pp. 179–186; D. H. Ahlstrom, Microstructure of synthetic polymers, pp. 256–269; R. Saferstein, Forensic aspect pyrolysis, pp. 360–353.
5. N. Grassie and G. Scott, *Polymer Degradation and Stabilization*, Cambridge Univ. Press, Cambridge, UK, pp. 33–41 (1985).
6. I. C. McNeil, Thermal degradation, *Comprehensive Polym. Sci., Vol. 6, Polymer Reactions* (G. C. Eastmond, A. Ledwith, S. Russo, and P. Sigwald, eds.), Pergamon Press, Oxford, UK, pp. 490–495 (1989).
7. V. Passalacqua, F. Pilati, V. Zamboni, B. Fortunato, and P. Manaresi, *Polymer, 17*: 1044 (1976).

8. M. E. Bednas, M. Day, K. Ho, R. Sander, and D. M. Wiles, *J. Appl. Polym. Sci.*, *26*: 277 (1981).
9. C. T. Vijayakumar and J. K. Fink, *Thermochim. Acta*, *59*: 51 (1982).
10. A. Zeman, *Angew. Makromol. Chem.*, *31*: 1 (1973).
11. I. Luederwald and H. Urrutia, *Makromol. Chem.*, *177*: 2079 (1976).
12. I. Luederwald and H. Urrutia, Direct pyrolysis of aromatic and aliphatic polyesters in the mass spectrometer, *Analytical Pyrolysis* (C. E. B. Jones and C. A. Cramers, eds.), Proc. 3rd Int. Symp. Anal. Pyrolysis, Amsterdam, 1976, Elsevier, Amsterdam, The Netherlands, pp. 139–148 (1977).
13. R. M. Rum, *J. Polym. Sci., Polym. Chem. Ed.*, *17*: 203 (1979).
14. R. E. Adams, *J. Polym. Sci., Polym. Chem. Ed.*, *20*: 119 (1982).
15. D. C. Conway and R. Marak, *J. Polym. Sci., Polym. Chem. Ed.*, *20*: 1765 (1982).
16. I. Luederwald, *Pure Appl. Chem.*, *54*: 255 (1982).
17. C. T. Vijayakumar, J. K. Fink, and K. Lederer, *Eur. Polym. J.*, *23*: 861 (1987).
18. G. H. Irzl, C. T. Vijayakumar, J. K. Fink, and K. Lederer, *Polym. Degrad. Stab.*, *16*: 53 (1986).
19. C. T. Vijayakumar and K. Lederer, *Makromol. Chem.*, *189*: 2559 (1988).
20. I. Luederwald and H. Urrutia, *Makromol. Chem.*, *177*: 2093 (1976).
21. I. Luederwald, *Makromol. Chem.*, *178*: 2603 (1977).
22. H. R. Kricheldorf and I. Luederwald, *Makromol. Chem.*, *179*: 421 (1978).
23. E. Jacobi, I. Luederwald, and R. C. Schultz, *Makromol. Chem.*, *179*: 429 (1978).
24. M. Doerr, I. Luederwald, and H.-R. Schulten, *Fresenius Z. Anal. Chem.*, *318*: 339 (1984).
25. D. Garazzo, M. Gluffrida, and G. Montaudo, *Macromolecules*, *19*: 1643 (1986).
26. B. Plage and H.-R. Schulten, *J. Anal. Appl. Pyrol.*, *15*: 197 (1989).
27. B. Plage and H.-R. Schulten, *Macromolecules*, *23*: 2649 (1990).

28. B. Crossland, G. J. Knight, and W. W. Wright, *Br. Polym. J.*, *18*: 371 (1986).
29. K. Sueoka, M. Nagata, H. Ohtani, N. Nagai, and S. Tsuge, *J. Polym. Sci, Part A*, *29*: 1903 (1991).
30. M. Giuffrida, P. Marvigna, G. Montaudo, and E. Chiellini, *J. Polym. Sci, Part A*, *24*: 1643 (1986).
31. H. Ohtani, T. Nagaya, Y. Sugimura, and S. Tsuge, *J. Anal. Appl. Pyrol.*, *4*: 117 (1982).
32. S. Tsuge, *Chromatogr. Forum*, *1*: 44 (1986).
33. H. Senoo, S. Tsuge, and Takeuchi, *J. Polym. Sci.*, *9*: 315 (1971).
34. H.-J. Duessel, H. Rosen, and O. Hummel, *Makromol. Chem.*, *177*: 2434 (1976).
35. L. J. Peebles, Jr. and M. W. Huffma, *J. Polym. Sci, Part A*, *9*: 1807 (1971).
36. F. Wiloth, *Makromol. Chem.*, *144*: 263 (1971).
37. C. David, Thermal degradation of polymers, *Comprehensive Chemical Kinetics*, *Vol. 14*, *Degradation of Polymers* (C. H. Bamford and C. F. H. Tipper, eds.), Elsevier, Amsterdam, The Netherlands, pp. 104–121, 130–153 (1975).
38. D. M. MacKerron and R. P. Gordon, *Polym. Degrad. Stab.*, *12*: 277 (1985).
39. A. Ballistreri, D. Garozzo, M. Giuffrida, and G. Montaudo, *Macromolecules*, *20*: 2991 (1987).
40. S. Tsuge, H. Ohtani, H. Matsubara, and M. Ohsawa, *J. Anal. Appl. Pyrol.*, *17*: 181 (1987).
41. I. Luederwald and F. Merz, *Angew. Makromol. Chem.*, *74*: 165 (1978).
42. H.-R. Schulten and B. Plage, *J. Polym. Sci, Part A*, *26*: 2381 (1988).
43. I. Luederwald, F. Merz, and M. Rothe, *Angew. Makromol. Chem.*, *67*: 193 (1978).
44. U. Bahr, I. Luederwald, R. Mueller, and H.-R. Schulten, *Angew. Makromol. Chem.*, *120*: 163 (1984).
45. B. Plage and H.-R. Schulten, *J. Appl. Polym. Sci*, *38*: 123 (1989).
46. H. R. Kricheldorf and E. Leppert *Makromol. Chem.*, *175*: 1731 (1974).

47. I. Luederwald and H.-R. Kricheldorf, *Angew. Makromol. Chem.*, *56*: 173 (1976).
48. A. Ballistreri, D. Garozzo, M. Giuffrida, P. Maravigna, and G. Montaudo, *Macromolecules*, *19*: 2963 (1983).
49. A. Ballistreri, D. Garozzo, G. Montaudo, and M. Giuffrida, *J. Polym. Sci, Part A*, *25*: 2531 (1987).
50. A. Ballistreri, D. Garozzo, P. Maravigna, G. Montaudo, and M. Giuffrida, *J. Polym. Sci, Part A*, *25*: 1049 (1987).
51. D. A. Chatfield, I. N. Einhorn, R. W. Mickelson, and J. H. Futrell, *J. Polym. Sci., Polym. Chem. Ed.*, *17*: 1367 (1979).
52. Y. P. Khanna, E. M. Pearce, J. S. Smith, D. T. Burkitt, H. Njuguna, D. M. Hindenlang, and B. D. Forman, *J. Polym. Sci., Polym. Chem. Ed.*, *19*: 2817 (1981).
53. J. R. Brown and A. J. Power, *Polym. Degrad. Stab.*, *4*: 179 (1989).
54. A. L. Bhuiyan, *Eur. Polym. J.*, *19*: 195 (1983).
55. H.-R. Schulten, B. Plage, H. Ohtani, and S. Tsuge, *Angew. Makromol. Chem.*, *155*: 1 (1987).

6

The Application of Analytical Pyrolysis to the Study of Cultural Materials

Alexander M. Shedrinsky
Long Island University, Brooklyn, New York and Conservation Center, New York University, New York, New York

Norbert S. Baer
Conservation Center, New York University, New York, New York

I. INTRODUCTION

The application of analytical pyrolysis in any of its several variations (Py-GC, Py-MS, Py-GC/MS, Py-GC/FTIR) has proved to be one of the most useful approaches to analyzing materials of art and archaeology. Since our previous review [1] was published in 1989, four major conservation laboratories (Metropolitan Museum of Art, Getty Conservation Institute, Musée du Louvre, and the National Gallery of Art (U.S.)) have become involved in the application of this analytical technique [2–7] to the analysis of organic media such as gums, waxes, natural resins, and synthetic polymers.

There are many reasons for the increasing popularity of this method in museum laboratories. Primary is the circumstance that many art materials—such as amber, ivory, paper, and wood—are both nonvolatile and insoluble. Other materials (e.g., drying oils, lacquers, and some synthetic polymers) become insoluble and nonvolatile on aging. This alone makes them unsuitable for conventional analysis requiring solubility (e.g., HPLC) or high volatility (e.g., GC). Another attractive feature of analytical pyrolysis is the very small sample size required (10–100 μg), which makes analytical pyrolysis a virtually "nondestructive" technique. A further most attractive reason is the absence in most applications of any required preliminary chemical treatment of the sample prior to the analysis, coupled with the simplicity of the procedure and the relatively low-cost equipment (in the case of Py-GC). Of course, a high degree of sophistication for the mass spectrometry in Py-MS or Py-GC/MS can bring the instrumentation costs into the high-cost range. The absence of chemical treatment is especially important because very often such pretreatment can cause rearrangements in labile molecules of analyzed materials.

At the same time, there are certain limitations and difficulties in the interpretation of the resulting pyrograms for cultural materials which deserve serious consideration. In spite of the growing interest in the pyrolysis technique, one should not overstate its apparent superficial simplicity.

One may note that some researchers have simply purchased a set of reference materials from reputable chemical producers, pyrolyzed them, and published the results, claiming that the problem of distinguishing among them has been solved. Unfortunately, this is not the case. A thorough investigation of pyrolysis patterns of fresh and aged materials must be conducted. The aged materials of definite provenance can often be obtained from museum collections or as a result of intensive artificial aging procedures, though one must again exercise caution since the end products of natural and artificial aging may differ.

Comparison of the results of analysis of fresh and aged materials can lead to very different conclusions. For example, in the case of dammar and mastic, it is very easy to distinguish between fresh samples but after extensive artificial aging, dammar will retain its characteristic profile while mastic can give a completely unrecognizable one [8]. One must emphasize that analytical pyrolysis is a very valuable tool in analysis of artistic and archaeological materials only as a result of very thorough and systematic research programs rather than through sporadic efforts.

There are two areas of complication related to analytical pyrolysis. One has to do with sampling and another with instrumentation. The first is very specific for this particular area of research, while the second will affect any application of analytical pyrolysis. From the point of view of sampling, it is quite obvious that as with every comparative technique, analytical pyrolysis requires an extensive reference library of pyrograms. This condition can easily be met in the case of modern materials but becomes almost insurmountable when looking for samples of material ranging from a few hundred to a few thousand years in age.

Another potential complication is that artists seldom use absolutely pure materials, but rather prefer to use complicated mixtures. Considering old varnish recipes [9–11] or oil media [12,13], one can immediately recognize the complexity and the resulting uncertainty in the composition of these mixtures: it is easy to overlook minor peaks that are overshadowed by the signals of the major components.

Another trap is that very often the art object has been restored using materials superficially similar, but quite different from the original (e.g., beeswax substituted by paraffin). Identification of this later addition could lead one to misleading conclusions. A further stumbling block is the understandable reluctance of curators to provide samples of adequate size. In practical terms this means that it is very difficult to undertake controlled or replicate experiments to confirm or reject ambiguous results. Similarly, the extremely small size of the sample leads one to ask how representative is this sample? This is especially the case when artists use mixed media (or vary composition throughout the artifact). In addition to the difficulties arising from the samples of art and archaeological materials, there are those posed by the pyrolysis equipment itself.

II. INSTRUMENTATION CONSIDERATIONS

Though the detailed "pros" and "cons" of different types of pyrolysis apparatus are discussed elsewhere [14], we feel obliged to share our own experience, working for six years with the CDS Pyroprobe 120. This system is well known and in wide use. It produces a highly predictable temperature–time profile for the filament and also provides a means of varying the heating rate linearly over the initial temperature rise period (ramp control).

Among the definite advantages of the Pyroprobe over Curie-point pyrolyzers one should mention the absence of solvent or grinding for sample introduction, ease in weighing the sample, and freedom of temperature choice. It is also hard to overestimate the ability of the CDS Pyroprobe to carry out so-called sequential pyrolyses [15], i.e., the pyrolysis temperature and time is chosen in a way that each pyrolysis affords only fractional decomposition of the sample. This additional capability of this instrumentation was successfully used to identify the provenance of an amber artifact from Hasanlu [16].

At the same time, a CDS Pyroprobe combined with a GC instrument through an interface can cause certain problems.* Because of the relatively large surface area of the interface between the GC instrument and the pyrolyzer, a "memory effect" is often obtained where high-molecular-weight compounds from previous analyses still produce GC traces in subsequent runs. Especially notorious in this regard are paraffins, waxes, and various oils. For example, it is very difficulty to get rid of traces of palmitic acid. Even heating of the interface to 300°C between runs does not always permit elimination of these "shadow" traces. We strongly recommend a blank run (sometimes two)

* There is no interface in instruments with packed columns.

between the actual analyses because these shadow traces can interfere with current pyrograms.

Similar complications were observed for Curie-point pyrolyzers. In a recent review [17], the author complained that

> A well-known trapping effect in Curie point pyrolysis devices is condensation of tar on the glass liners which are generally used to hold the ferromagnetic probes. This condensation can be reduced by heating of the glass liners and a very close interfacing of GC column and pyrolysis chamber . . ., but drop out of higher molecular weight fractions is unavoidable.

Another danger associated with the instrument is the possibility that during the actual run the needle, which introduces the gaseous product into a GC column, can become clogged. It is unavoidable that after hundreds of analyses some of the less volatile components will be deposited inside the needle. Sooner or later they will clog it, but unfortunately it is very difficult to predict this precisely because the clogging depends not only on the number of runs but also on the nature of the materials analyzed. For example, it is rare to have problems with "unzipping" synthetic materials, which under pyrolysis conditions produce mainly monomers, dimers, and trimers. However, significant problems can occur when analyzing old, heavily oxidized oil samples, ambers, and similar materials.

A serious consideration when comparing results from different laboratories is the temperature of pyrolysis. What must be stated clearly is that the temperature of pyrolysis differs dramatically depending on the filament used, e.g., coil filament with a quartz interior or a platinum ribbon. In the second case, one needs 8 ms to reach 600°C [13] in a heating rate that depends on the pyrolysis system. The CDS Pyroprobe uses the resistive filament simultaneously as the heating element and as the temperature sensor. This means that when the temperature of the coil is 650°C, the temperature inside the quartz tube will be substantially lower. Comparing published results on acrylic polymers [19] and our own observations, we can say that this difference is about 150°C; i.e., to obtain the same pyrogram with Paraloid B-72, which was obtained at 500°C on the Pt ribbon, we had to heat our sample in the quartz tube to 650°C.

A word of caution on comparative studies where researchers have used both types of filaments: Perhaps as a result of Pt catalytic effects or because of differences in temperature rise profiles, it is quite difficult to obtain identical pyrograms on different filaments—even when the difference in temperature is taken into consideration.

Finally, we note another "limitation." The CDC Pyroprobe coil filaments are initially evenly spaced; however, after intensive use, the individual loops very often touch each other. This has a far-reaching and quite important consequence: the resistance of the coil will increase and as a result the real temperature will be much higher than the nominal reading.

In this chapter we will critically review those publications that deal directly with the application of analytical pyrolysis to the examination of art or archaeological objects. Nevertheless, it should be indicated that a substantial literature devoted to analysis of such materials as wood, polysaccharides, textiles, proteins, and numerous synthetic polymers can be found in other chapters of this book.

In this chapter we will only briefly note work done using "homemade" equipment or old-type packed columns, even if the reported results were quite significant. For a more extended discussion of these studies consult our previous review [1].

III. ANALYSIS OF MATERIALS

For the convenience of specialists working in this field, we will discuss all the data according to material.

A. Waxes

Analytical pyrolysis proved itself a most suitable technique for analysis of waxes because they are not readily soluble in organic solvents and, considering the molecular size of their major components (from 30 to 60 carbon atoms), are not very volatile.

One of the first publications dealing with a great variety of waxes [20] used Py-MS for their analysis. There is some confusion about the results of this investigation. On the one hand the author claims that, "Py-MS was found to provide a rapid and unambiguous method of characterizing the waxes in the reference collection but discrimination within the class was rather poor," while the same publication contains a table (no. 4) that represents "major ions in the zone and their relative abundance," where all the samples but three show the same very uncharacteristic ion mass numbers 43, 57, and 111; and some (no. 18, spermaceti wax) fragments that are very difficult to imagine (for example, no. 229). Also, it is difficult to accept the explanation that "Fig. 1 is typical of 15 of the reference waxes," and includes such chemically different materials as beeswax, carnauba wax, spermaceti, shellac wax, rice wax, earth wax, and paraffin wax. It appears that the high ionization energy (70 eV) caused such an excessive fragmentation that, as a result, very uncharacteristic ions were obtained. At the same time, the relatively low temperature of the MS jet separator (200°C) prevented heavy fragments from reaching the mass spectrometer. Also surprising is the choice of the mass range used for "Fig. 1". It covers the area between 1 and 250 mass units, in spite of the fact that the most characteristic and informative ions should be found in much higher mass numbers. For example, Py-MS analysis of the notorious "Flora" bust in the Berlin museum [21] revealed the presence of spermaceti wax mixed with beeswax through the presence of the undeniably strongest ions with mass 480 (cetyl palmitate ester, the major component (95.4%) of spermaceti max, MW = 480) and also a few very strong signals at 224, which represent the alkene that results from the dehydration of cetyl alcohol, and at 257, which could be accounted for as a protonated molecule of palmitic acid (MW = 256). This analysis was performed using a very gentle ionization energy (12 eV) and choosing a very wide range of mass, from 1 to 1000.

Interesting results were also published by Puchinger and Stachelberger [22], applying Py-GC and Py-GC/MS to the study of two synthetic waxes and beeswax, and finding very clear differences in their fingerprints. A few technical details raise some questions: very short time (2 s), which makes it difficult to pyrolyze material inside the quartz tube; very high temperature (800°C), which can cause extensive fragmentation; and high carrier gas rate (25 ml/min), which can cause substantial cooling and decrease the degree of product separation in the GC column.

As an illustration of the practical application of their method, the same authors, using Py-GC, analyzed a few samples from the surface of the Renaissance sculpture *Jüngling von Magdalensberg* and found that all contained beeswax [23]. The analytical results of this investigation are quite convincing, but it is hard not to raise an objection from an art historical point of view. There is no evidence that this beeswax represents original material. It is well known that beeswax can be used as a protective coating for bronzes and it could have been applied quite recently, for example, in the eighteenth, nineteenth, or even twentieth century. As was demonstrated earlier [23], the fingerprints of beeswax are practically unaffected even by very substantial intervals of time.

Our own results [24], using Py-GC for more than 40 analyzed wax samples, produced quite distinctive fingerprints even within the same class (for example, plant waxes). Also, as was demonstrated by the comparison of beeswax from fresh samples to some approximately 4000 years old, the fingerprints of this material are very stable and are easily recognized in spite of minor differences in the middle portion of the pyrogram (see Fig. 1). It is only in beeswax, in the last portion of the pyrogram that represents undecomposed hydrocarbons, that one can easily spot differences in the amount of even- and odd-carbon substances (in beeswax, odd-carbon hydrocarbons will always dominate).

As a practical application of these results, we note that the "oldest wax sculpture in the world" [25], in the collections of the Metropolitan Museum of Art, was found to be made of modern paraffin [26] (see Fig. 2). This finding definitely casts doubt on the authenticity of this object because paraffin is a product of the nineteenth century. For comparison, Figs. 3–6 show pyrograms of spermaceti, ceresin 101, candelilla, and carnauba waxes, respectively.

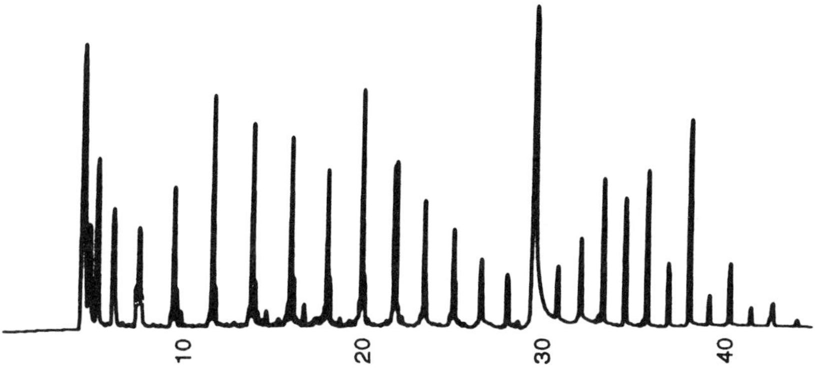

FIGURE 1 Pyrogram of natural, unbleached beeswax.

Another interesting approach to wax analysis, "simultaneous pyrolysis-alkylation-gas chromatography," was also reported [27] by Challinor who used TMAH (tetramethylammonium hydroxide) to alkylate acids of different waxes, making them more volatile. He concluded that the different waxes may

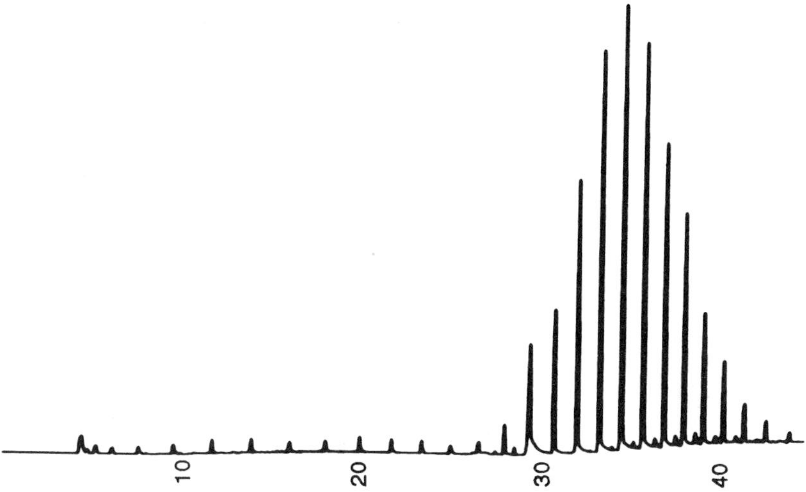

FIGURE 2 Pyrogram of paraffin wax.

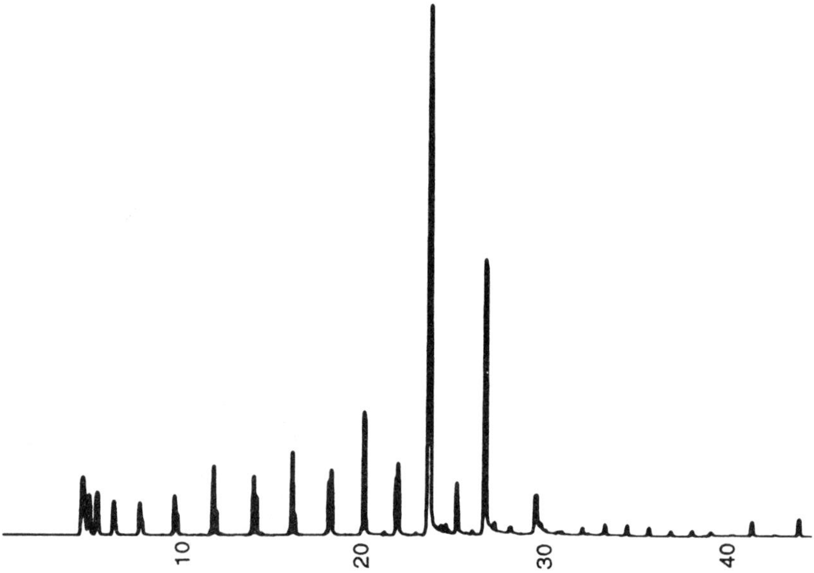

FIGURE 3 Pyrogram of spermaceti wax (USP).

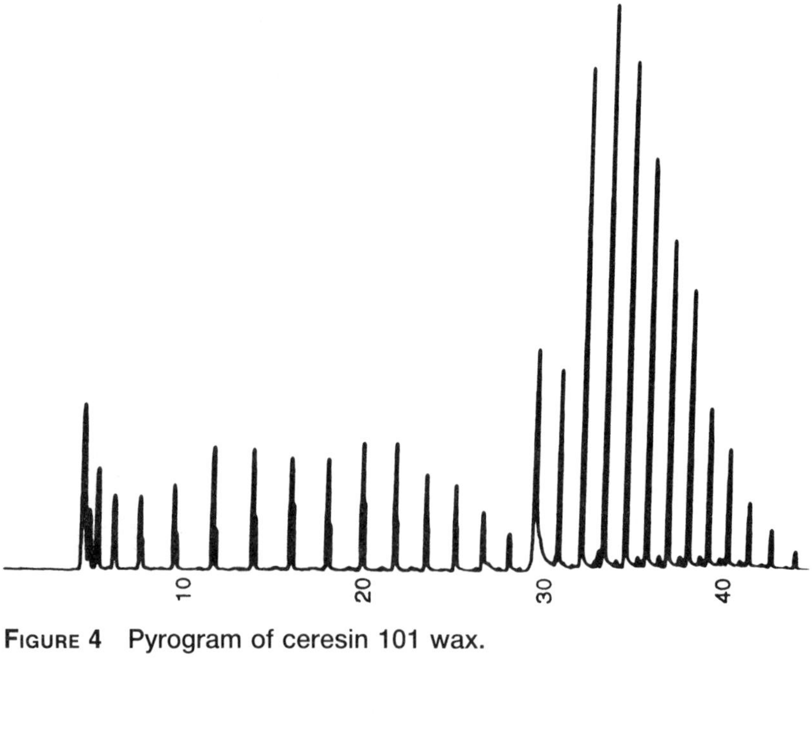

FIGURE 4 Pyrogram of ceresin 101 wax.

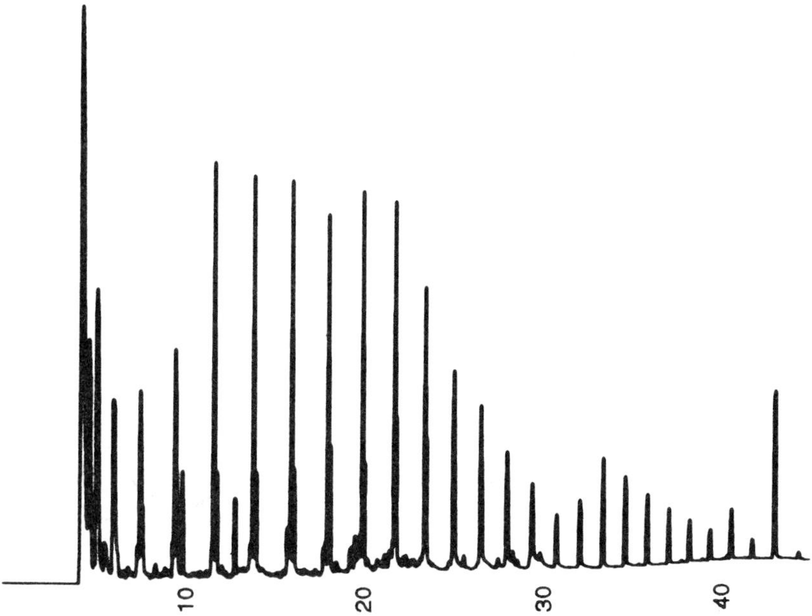

FIGURE 5 Pyrogram of crude candelilla wax.

be readily distinguished by the distribution of fatty acid methyl esters and fatty alcohols.

B. Natural Resins

One of the most interesting classes of compounds from the point of view of art and archaeology is natural resins. The past five years brought some remarkable successes in this area of analysis and at the same time revealed some very fundamental difficulties which are not easily overcome.

From initial, sometimes quite naive attempts to characterize natural resins on the basis of low-molecular-weight uncharacteristic products [28], research in this particular area moved in very sophisticated directions: Py-GC, Py-GC/MS, in-source Py-MS, and simultaneous pyrolysis methylation (SPM) [29–37].

As a result of dramatic improvement in GC equipment (e.g., introduction of fused silica capillary columns) and a choice of

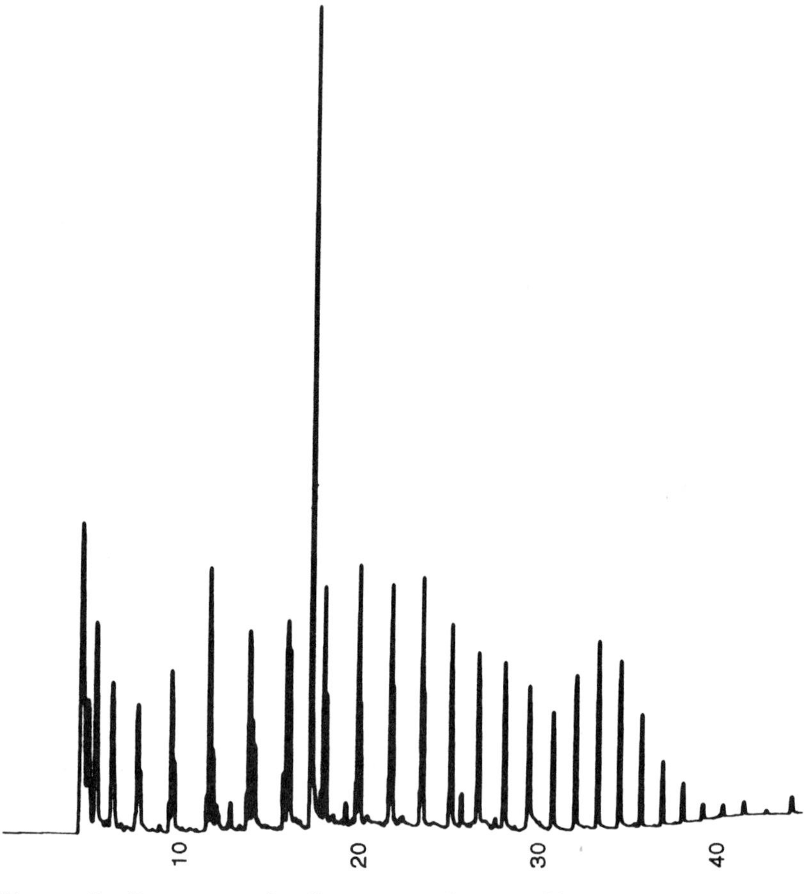

FIGURE 6 Pyrogram of yellow carnauba wax No. 1.

correct GC conditions, even a regular Py-GC was able to solve some problems, which other analytical methods were unable to solve. For example, IR analyses were unable to differentiate between dammar and mastic: two major natural resins used for varnishes [38]. This problem was settled by using Py-GC [8,29], and as it was proved by this investigation, dammar has an extremely characteristic fingerprint (see Figs. 7 and 8 for pyrograms), which is practically independent of age or source of

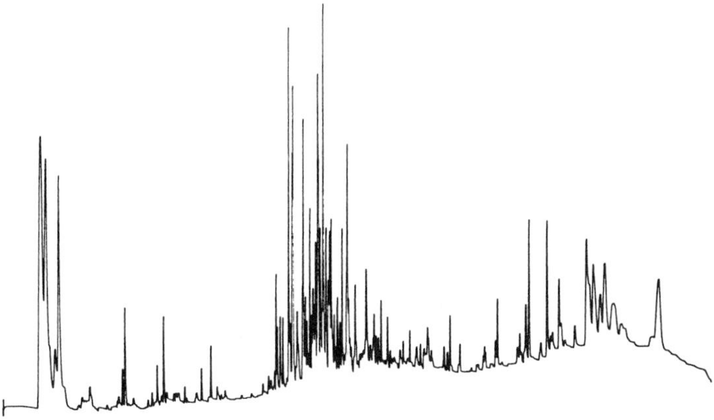

FIGURE 7 Representative pyrogram of typical dammar sample.

resin. The same paper also deals with sandarac and copals, but the very uncertainty of the term "copal" creates difficulties in distinguishing among them. The same authors examined a wide range of different copals in their other work [39], and this result will be discussed with amber analysis later in this chapter.

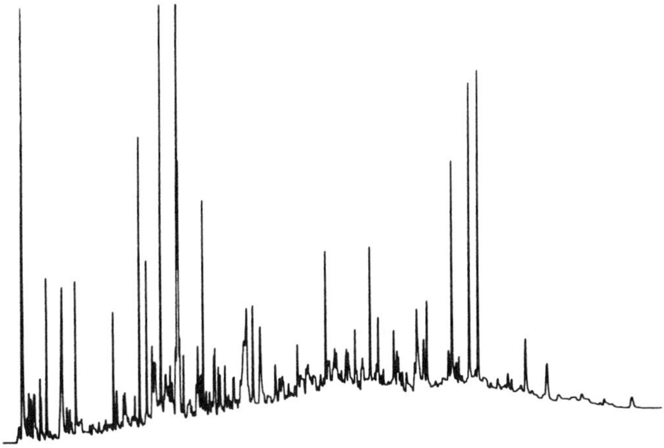

FIGURE 8 Pyrogram of typical mastic sample.

A most impressive study in natural resins analysis was done by Van Aarssen, de Leeuw, and Horsfield [40] and was devoted to characteristics of a biopolymer isolated from fossil and extended dammar resin (this fraction is known to conservators or conservation scientists as β-resin fraction in dammar). The most interesting result of this study was not just the finding that this fraction consists of a linear polymeric cadinene (the presence of the cadinane carbon skeleton in dammar was suggested much earlier [41]), but the whole approach; that is, three different pyrolysis methods were used: flash pyrolysis, open system isothermal furnace pyrolysis, and closed system isothermal pyrolysis. These three different pyrolysis methods gave complementary results and yielded information about the structure of β dammar. It was concluded from flash pyrolysis data that this polymer is "a linear polysesquiterpene in which the monomers are linked by one carbon bond. From the open system isothermal furnace pyrolysis it became clear that the basic structural unit is a cadin-5, 6-ene. The closed system isothermal pyrolysis revealed that bicadinanes can be formed from the polymer, implying the presence of a bond between monomers at position C_1-C_{13}."

The practical importance of this finding is that even after extensive aging by UV irradiation and heat, dammar still possesses these very characteristic peaks in the middle of the pyrogram caused by the stereoisomer of C_{15} compounds with a cadinane carbon skeleton [8].

C. Amber

In light of the recently developed interest in ambers, we discuss these fossilized natural resins as a separate topic. We also include information about fake ambers (mostly synthetic polymers) in this discussion.

Ambers have long proved difficult to examine. IR and FTIR spectroscopy proved useful tools in the determination of amber origins, but in a comprehensive review of the subject, Beck stated:

> I regret to say that we have not yet found the way to distinguish with any degree of confidence between the

> various kinds of non-Baltic amber. All we can do with a rather high level of certainty, is to say whether a given artifact is of Baltic amber or not. [42]

When ambers were analyzed by Py-GC [39], Py-MS [31], and Py-GC/MS [30–37], this situation changed dramatically.

The very first attempt in 1985 to analyze amber by Py-MS provided "a general picture of the chemical structure of amber and plant resins which after close comparison by multivariate (statistical) analysis, can be used for differentiation classification and identification purposes" [31]. At the same time, the authors admitted that it is difficult to give a detailed interpretation of the mass peaks and it is almost impossible to trace back the various fragments to the original molecular entities. Another limitation of this study was the modest set of samples available. Further, several samples were of unclear origin. In this study, the authors used a Curie-point pyrolyzer (Fe/Ni alloy) at 510°C. The volatile pyrolysis products passed through a heated expansion chamber (150°C) into an open cross-beam type ionizer operating at 14 eV. Processing of spectral data was carried out on CDC Cyber 7600 system.

The next interesting study was done by Boon's group at the FOM Institute for Atomic and Molecular Physics (Amsterdam). They reported on data obtained by Py-GC and Py-GC/MS of recent and fossil copals and ambers from six continents.* As the authors of this project stated:

> It is our intention to make a mass spectrometric atlas of geological and archaeological ambers and other resins. This atlas will contain a high resolution pyrolysis gas chromatogram, the EI and CI pyrolysis mass spectra and relevant PyGC-MS data on the specimens.

This ambitious and very useful undertaking will require the collective efforts of scientists all over the world to overcome the

* Though this study has not yet been published, after its original presentation at the 8th International Symposium "Pyrolysis 88," it became a part of the Amber Profiling Project at FOM.

main problem plaguing researchers in this area, i.e., the unreliability of samples.

In the meantime, Grimaldi, Beck, and Boon published the results of an investigation entitled, "Occurrence, Chemical Characterization and Paleontology of Fossil Resins from New Jersey" [43]. Using FTIR and Py-GC to examine fossil resins containing insect inclusions, the authors found that Cretaceous fossil resin is a true amber (it has terpenoid origin) but the unique tertiary fossil resin is composed mostly of polystyrene (a product of the genus *Liquidamber* from the Hamamelidaceae family).

In our own early work [44], we analyzed a modest group of ambers and copals to demonstrate the ability of Py-GC to differentiate among specimens of different origins.

Recently, as a confirmation of our long-standing interest in natural resins in general and ambers in particular, we used Py-GC to analyze a wide variety of different ambers and copals from the collection of the American Museum of Natural History and other sources. In the course of this investigation, which included more than 40 samples from areas around the world, we were able for the first time to analyze by Py-GC samples of Japanese amber from the Fuji area. In an interesting finding, we established that one fossil resin sample from Montana was almost pure natural polystyrene [39]. On the basis of data obtained, attribution of some doubtful samples was done in a "blind" experiment and a number of fake amber imitation materials were uncovered. As a result of this work, we reached the conclusion that Py-GC can be a valuable addition to the analytical methods routinely used in the study of amber, giving in many cases much more convincing "fingerprints" than other methods, e.g., FTIR.

Much richer information about ambers can be obtained by using the combined approach of Py-MS and Py-GC/MS. The most interesting work to date was presented at the conference "Amber in Archaeology" and published in its proceedings [30].

This paper is a preliminary report on the potential of Py-MS and Py-GCMS to characterize ambers, copals, and other resins on a microgram scale. A wide range of analyzed samples makes this publication especially attractive. The Py-MS methods employed by these authors, as compared to the earlier work [31] of Poinar and Haverkamp, permitted discrimination of thermally desorbable compounds from fragments of the macromolecular skeleton resulting from pyrolytic dissociation. Using Py-GC/MS, the researchers overcame an important disadvantage of the Py-MS approach, i.e., the lack of isomer information, which is a particular drawback in the case of terpenoids.

The combined efforts of an international group of scientists [45] also permitted the full characterization of fossil polystyrene which was found initially in Germany (1883) and later in the United States (Montana and New Jersey). Curiously enough, this material is atactic as proved by Py-GC and confirmed by H-NMR spectra.

A systematic study on the nature and fate of natural resins in the geosphere was recently published by K. Anderson et al.[32–37]. It includes major amber and copals and very rare material, such as Tymir amber, which had not been systematically studied from a chemical point of view.* The authors suggested a very broad classification scheme for resinites. They suggest that most resinites be classified on the basis of structural characteristics into four classes [35]. In a later publication, they added a fifth category [36]. One can question the validity of this classification or some of the particular conclusions regarding the chemical structure of some of the described resins, but in general it is a valuable study, providing much useful material for fruitful discussion.

A recent study was devoted to amber forgeries [46]. It was shown that analytical pyrolysis (Py-GC, Py-GC/MS) can provide undeniable proof that materials under investigation were amber forgeries and also quite precisely characterize the nature of these substitutes. Early Bakelites, modern phenolic resins, polystyrene, epoxy resins, and a wide range of unsaturated polyesters were identified as the most often used materials.

* The authors used Py-GC/MS and alternated direct pyrolysis with pyrolytic methylation, which was achived by copyrolysis with tetramethylammonium hydroxide.

With the wide commercial availability of unsaturated polyesters of compounds and sophisticated "artwork," convincing imitations of large transparent amber pieces with a wide variety of "inclusions" (ants, bees, lizards, mosquitoes, etc.) have been prepared [47]. Py-GC provides a simple test for unmasking such fakes, and Py-GC/MS can identify quite precisely the structures of the materials used for these purposes.

D. Oils and Fats

Identification of oils and fats is crucial for art and archaeology. In art, after the first third of the fifteenth century, drying oils became the most often used paint media, and may have been used even earlier. Archaeological findings of oil-containing jars are common and can range from 400 to a few thousand years in age [48–53].

In our previous review [1], we discussed all earlier attempts to analyze these materials using pyrolysis techniques [43,53–58]. What has become quite clear is that fingerprinting alone with Py-GC is not enough to obtain a sufficient amount of information about the nature of the oil under examination.

Many previous studies dealt with fresh oil samples (or relatively young samples just a few years old), but oxidation processes take place under different conditions. For example, direct intensive sunlight in many cathedrals, constant candle heat in front of altarpieces, numerous restorations using unknown and sometimes very strong solvents—all these factors can dramatically affect the condition of the oil layer under examination. Even more, assume that as a result of our analysis we obtained a very accurate description of the modern condition of the old paint layer or archaeological oil sample. It does not mean that we can easily reconstruct the formula of the original material. All fats and oils (with rare exceptions) have more similarities than differences and the later ones most often are represented by unsaturated acids, which suffer the most in the case of oxidation. If one would add the deep hydrolysis of oils, which is always the case in archaeological samples, the problem of tracing the origin of this fatty material becomes almost insurmountable.

As a rare illustration of the successful application of analytical pyrolysis for oil analysis, J. Boon demonstrated [59] that high-temperature in-source Py/MS can deal simultaneously with the analysis of the oily compound of an artist's paint and the analysis of inorganic components derived from the pigments. This work deals with "yellow ochre" oil paint (Talen's van Gogh series), but even in the case of fresh material the author refrained from jumping to any conclusion about the nature of the oil media (linseed, walnut oil, poppyseed, etc.).

In light of this statement, it is easy to appreciate the restraint of the authors in their publication dealing with molecular archaeology [60]. They determined, using Py-MS and Py-GC/MS, the remains of lipids in prehistoric pottery and residues but stated that "the identification of original foodstuffs based on the relative distribution of free fatty acids in an archaeological sample is a difficult process." Also, they admitted that mono-, di-, or triacylglycerides could not have been detected with the pyrolysis technique utilized.

E. Synthetic Polymers

The application of synthetic materials in art and conservation increased dramatically over the past 50 years. Among the reasons for this phenomenon are the loss of availability of some traditional materials and the improved longevity, strength, resistance to yellowing, etc., offered by synthetic substitutes. A further consideration is cost.

Synthetic adhesives and acrylic paints are but two widely used examples of synthetic polymers used in art and conservation. However, these products generally are not specifically developed for conservation purposes. Most manufacturers, citing "trade secrets," refuse to reveal the precise contents of their products. This is especially true of such additives as antioxidants, flocculants, antifoamers, emulsifiers, etc. Often, companies change the contents of products, keeping the same name and not informing customers about these changes. A well-known example is that of ketone-resin varnishes [61]. An extensive review of the thermal analysis of polymers is given in Chapters 4 and 5 of this book. Here we note a few of the applications of pyrolysis in the area of art and conservation.

In an early study, now primarily of historical interest, W. Noble et al. [62] dealt with the characterization of adhesives by pyrolysis gas chromatography and IR spectroscopy. The scope of this study, in which 179 commercial products were analyzed and two complementary analytical methods were employed, provides a useful model though the applicability of the results is limited by the obsolescence of the instrumentation used.

In a much more comprehensive study [63] from the same laboratory, the authors studied 94 adhesives available in the U.K. Some are of considerable importance in art and conservation: poly(vinyl acetate) (PVA), PVA copolymers, acrylics, silicones, carbohydrates, and epoxy-based adhesives. It was demonstrated that Py-MS is capable not only of determining the main products of pyrolysis and producing reliable "fingerprints," but also of detecting some of the additives present in small amounts, e.g., tackifiers, antioxidants, and plasticizers. This is very important in light of growing concern among conservation scientists over the long-term effects of additives on conservation and art materials. Except for the widely known and well-described work of E. DeWitte [64,65] et al., who used analytical pyrolysis to identify a number of synthetic resins of importance in a very systematic study, there is the work published by the Louvre conservation laboratories [4–6]. Sonoda and co-workers investigated ketone resins by using Py-GCMS, and made a quite thorough study on the identification of synthetic materials in modern paintings, including polymeric varnishes and binders and synthetic organic pigments and media using Py-GC and x-ray diffraction (in the case of pigments). Among the many interesting conclusions, the authors observed that analytical pyrolysis is a very convenient tool for analysis of new synthetic materials and could be "indispensable to determine or to confirm the painting technique or the study of its alteration."

Furthermore, they note that

> The dates of the first synthesis or of the first industrial and commercial production being generally known, the

> identification of these materials gives a time reference which may permit one to study the authenticity of the painting or to specify the nature and the age of the repainting. [4]

In the second part of their investigation [5], the authors provided extensive historical data on the nature and the methods generally used for the analysis of organic synthetic pigments, and also demonstrated that two techniques, Py-GC and x-ray diffraction, could be complementary and very useful in identifying these pigments. Moreover, Py-GC was also able to identify and differentiate the synthetic binding medium used in some paintings, without any further analysis. Figure 9 shows an example of an acrylic painting medium with its pigment, along with the medium and the pigment material pyrolyzed separately. What is also of practical importance is that several examples were given showing the relevance of these methods to the investigation of modern paintings: identification of the technique and diagnosis of alterations, not to mention a characterization of retouchings or restoration materials.

F. Siloxanes

The thermal decomposition of linear and branched poly-dimethyl siloxanes and their derivatives [65–73] has been extensively studied. These papers were published outside the conservation literature, but all are relevant since in the last 6–7 years great interest in the use of these polymers as stone consolidants has developed. Some of these publications despite the sometimes antiquated equipment used, could be a very useful tool in the analysis of siloxanes.

G. Proteins

This field should be as important to art and conservation as natural resins and waxes. It includes such diverse materials as gelatin, egg white, egg yolk, casein, and vegetable proteins. In practical terms, this field covers quite a range of very different substances and includes adhesives, binding media, horn, ivory, tortoise shell, wool, silk, etc.

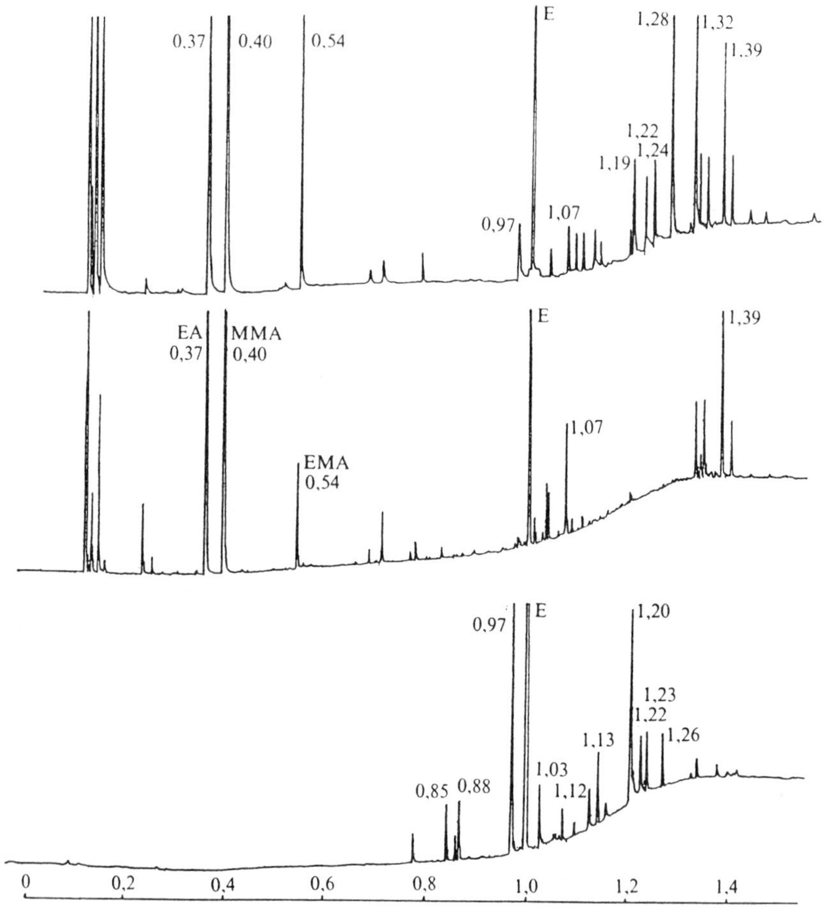

FIGURE 9 Pyrogram of Hooker Liquitex Green (top), Liquitex Medium (center), and the green pigment. Retention times are relative to the internal standard dodecane (marked E). EA = ethyl acrylate, MMA = methyl methacrylate, EMA = Ethyl methacrylate. (Adapted from Ref. 5.)

However, as J. Mills and R. White state,

> . . . As the proteins of museum objects are concerned there are no unique amino acids for particular proteins whose presence serves to identify them, but an identification is often possible by means of a quantitative assay of the amino acids. [74]

One can argue that, for example, hydroxyproline is quite noticeable in collagen (12.8%) and completely absent in keratin (wool), or fibroin (silk), or that it could be found in gelatin (7.4%) but has a zero content in egg white, egg yolk, and casein—materials very often used as adhesives and binding media. Another example is cystine, used in some structural proteins. It can be as high as 12.8% (wool) or 8.2% (feather), but is completely absent in collagen and fibroin (silk).

In theory analytical pyrolysis, with its absence of sample pretreatment, should be very useful in light of the fact that "some epimerization of the amino acids at the common asymmetric center can take place when they are liberated by base hydrolysis of proteins. Also loss of amino acids during hydrolysis occurs through formation of humins. These are dark brown substances formed by condensation of amino groups and the indole nucleus of tryptophan with aldehydes formed *in situ*" [74].

The other complication was mentioned in the thorough research done by Halpine [75], who used HPLC for protein identification, observing that tryptophan was not detectable because it was completely destroyed during derivatization.

In an effort to avoid problems of derivatization, Chiavari et al. recently published two articles [76,77] dealing with the subject of amino acids and proteins analyzed by Py-GC/MS and its application to the identification of ancient painting media.

According to their results, in contrast to HPLC, the presence of tryptophan is "neatly revealed through the generation of a series of indoles by Py-GC/MS." At the same time, they found that "hydroxyproline gave a dehydration product similar to that of proline." Glycine, alanine, and threonine yielded pyrograms of little diagnostic significance. For unknown reasons, histidine also did not produce any meaning for pyrolysis products. But 15 to 19 analyzed amino acids did provide some very interesting mass spectra.

As usual, some questions arose when the authors moved to the analysis of real paint media. They stated that "the main pyrolysis fragment of rabbit glue was pyrrole which was absent or of very low intensity in all the other media." They also detected the presence of pyrocoll ($2C_{10}H_{22}O_3N_3$, MW-186) and 3,6-(2-methylpropyl)-2,5-diketopiperazine ($C_{12}H_{22}O_2N_2$, MW-226). After presenting these data, the authors state that "these compounds can be ascribed to the thermal decomposition of hydroxyproline . . ., which has been shown to generate such fragments" (in their previous investigation of amino acids [77]). There is some problem with this statement, since pyrrole was generated also by glutamic acid, which is plentiful not just in animal glue (10%), but also in egg yolk (15%), egg white (14%), and casein (20%). As for the other two compounds, they are absent in the fragmentation product of hydroxyproline according to Table 1 of their earlier publication [74].

At the same time, there is a very interesting observation about indole and methylindole revealed by Py-GC/MS as products of thermal decomposition of tryptophan in egg white, because this amino acid is absent in animal glue and usually lost in the acid hydrolysis work-up common to conventional GC and HPLC. Also, there is a detailed comparison between conventional techniques (like GC and HPLC) and analytical pyrolysis methods at the end of the article.

What is quite obvious is that much more work should be done in this particular area to arrive at reliable qualitative and quantitative results. It is hard to argue with the pessimistic assessment given in one of the most interesting articles devoted to this subject [78] in 1987:

> The composition of pyrolysates of proteins is still obscure despite the effort of several researchers. This is partly due to the different structures of the amino acids, the different stabilities of the various peptide bonds and the overall heterogeneity of the peptide. On the other hand

there has been very little uniformity in the methods used for pyrolysis and hence different parts of the pyrolysates are always studied.

Finally and for the sake of completeness, three other papers dealing with proteinaceous materials [27,79,80] or identification of substances applied during chemical processing of wool should be consulted. Otherwise, this field is not very well explored.

Acknowledgments: The authors thank Jaap Boon for his careful reading of the manuscript and helpful suggestions. One of the authors (AMS) acknowledges the generous support of a Long Island University Faculty Released Time Grant.

REFERENCES

1. Shedrinsky, A. M., T. P. Wampler, N. Indictor, and N. S. Baer, Application of analytical pyrolysis to problems in art and archaeology: A review, *J. Anal. Appl. Pyrol.*, *15*: 393–412 (1989).
2. Shedrinsky, A. M., R. E. Stone, and N. S. Baer, Pyrolysis gas chromatographic studies on Egyptian archaeological specimens: Organic patinas on the "Three Princesses" gold vessels, *J. Anal. Appl. Pyrol.*, *20*: 229–238 (1991).
3. Derrick, M. R. and D. C. Stulik, Identification of natural gums in works of art using pyrolysis-gas chromatography, Proceedings of the ICOM Conference for Conservation, 1990, Dresden, Vol. 1, pp 9–13.
4. Sonoda, N. and J. P. Rioux, Identification des Matériaux Synthétiques dans les Peintures Modernes. I. Vernis et Liants Polymères, *Stud. Conserv.*, *35*: 189–204 (1991).
5. Sonoda, N., J. P. Rioux, and A. R. Duval, Identification des Matériaux Synthétiques dans les Peintures Modernes. II. Pigments Organiques et Matiere Picturale, *Stud. Conserv.*, *38*: 99–127 (1993).
6. Mestdagh H., C. Rolando, M. Sablier, and J. P. Rioux, Characterization of ketone resins by pyrolysis/gas chromatography/mass spectrometry, *Anal. Chem.*, *64*: 2221–2226 (1992).
7. National Gallery of Art, Washington D.C., 1992 Annual Report, pp 43–44.

8. Shedrinsky, A. M. and N. S. Baer, Py-GC analysis of aged damar and mastic, unpublished results.
9. Hurst, H. G., *A Manual of Painters, Colours, Oils and Varnishes*, 5th ed., revised by N. Heaton, London (1913).
10. Theophilus, *An Essay upon Various Arts*, trans. with notes by R. Hendrie, London (1847).
11. Mills, J. S. and R. White, *The Organic Chemistry of Museum Objects*, Butterworth, London (1987), p. 99.
12. White R., A review with illustrations of resin/oil varnish mixtures, methods applicable to the analysis of resin/oil varnish mixtures, ICOM Committee for Conservation, Report 2/16/81, 6th Triennial Meeting, Ottawa (1981).
13. Havel, M., Un element nouveau dans la couche picturale de la peinture moderne: le vernis de retoucher. Un exemple: le vernis de retoucher Vibert, ICOM Committee for Conservation, Report 19/72/10 Triennial Meeting, Madrid (1972).
14. Irwin, W. J., *Analytical Pyrolysis*, Marcel Dekker, New York, (1982).
15. Ericson, I., *J. Chromatogr. Sci.*, *16*: 340 (1978).
16. Shedrinsky, A., C. Beck, and N. S. Baer, Analysis of amber artifacts from Hasanlu, unpublished results.
17. Boon, J. J., Analytical pyrolysis mass spectrametry: New vistas opened by temperature-resolved in-source PYMS, *Int. J. Mass Spectr. Ion Proc.*, *118/119*: 755–787 (1992).
18. Boon, J. J., A. D. Pouwels, and G. B. Eijkel, *Biochem. Soc. Trans.*, *125*: 170–174 (1987).
19. DeWitte, E. and A. Terfve, The use of Py-GC/MS technique for the analysis of synthetic resins, *Science and Technology* (N. S. Brommelle and G. Thomson, eds.), International Institute for Conservation, London (1982), pp 16–18.
20. Wright, M. M. and B. B. Wheals, Pyrolysis-mass spectrometry of natural gums, resins and waxes and its use for detecting such materials in ancient egyptian mummy cases (Cartonajes), *J. Anal. Appl. Pyrol.*, *11*: 195–211 (1987).
21. Boon, J., private communication.
22. Puchinger, L. and H. Stachelberger, Pyrolyse-gaschromatographische Prüfung organischer Rohstoffe am Beispiel der Wachse, *Fat. Sci. Technol.*, *92/6*: 243–248.

23. Puchinger, L., H. Stachelberger, and G. Banik, Identifizierung Organischer Materialien on Objekten der bildenden Kunst, *Wiener Berichte über Naturwissenschaft in der Kunst, Band 6/7/8* (1989/90/91).
24. Shedrinsky, A. M., T. P. Wampler, and N. S. Baer, Pyrolysis gas chromatography (PyGC) applied to the study of natural waxes in art and archaeology, *Science, Technology and European Cultural Heritage* (N. S. Baer, C. Sabbioni and A. I. Sors, eds.), Butterworth-Heinemann Oxford, (1991), pp 553–558.
25. Von Büll, P. R., Vom Wachs, *Hoechster Beiträge zur Kenntnis der Wachse,*" *Band I*, Beitrag 1, Farbwerke Hoechst A. G., Frankfurt (M) -Hoechst (1963), p 36.
26. Shedrinsky, A. M., J. Boon, R. E. Stone, and N. S. Baer, The oldest artistic wax sculpture in the world revisited, *Anal. Chem.* (in press).
27. Challinor, J. M., The scope of pyrolysis methylation reactions, *J. Anal. Appl. Pyrol., 20*: 15–24 (1991).
28. Takiura K., A. Yamaji, and H. Yuki, Analysis of natural resins by pyrolysis gas chromatography. I. Classification of natural resins by pyrograms, (Y. Zasshi), *J. Pharm. Soc. Jpn., 93*(6): 769–775 (1973).
29. Shedrinsky, A. M., T. P. Wampler, and N. S. Baer, The identification of dammar, mastic, dandarac and copals by pyrolysis gas chromatography, *Wiener Berichte über Naturwissenschaft in der Kunst, Doppelband, 4/5* (1987/88), pp 12–25.
30. Boon J. J., A. Tom, and J. Pureveen, Microgram scale pyrolysis mass spectrometric and pyrolysis gas chromatographic characterisation of geological and archaeological amber and resin samples, Amber in Archaeology: Proceedings of the Second International Conference on Amber in Archaeology (edited by Curt W. Beck and Jan Bouzek in collaboration with Dagmar Dresleová), Liblice, Tsjechoslovakia, Oct. 30–Nov. 2, 1990 Institute of Archaeology, Prague (1993), pp. 9–27.
31. Poinar, G. O. and J. Haverkamp, Pyrolysis mass spectrometry in the identification of amber samples, *J. Baltic Stud., 16*(3): 210–220 (1985).
32. Anderson, K. B., R. E. Botto, G. R. Dyrkacz, R. Hayatsu, and R. E. Winans, Analysis and comparison of two victorian brown

coal resinites, *ACS Div. Fuel Chem. Preprints*, *34*(3): 752–758 (1989).

33. Anderson, K. B. and Randall E. Winans, Structure and structural diversity in resinites as determined by pyrolysis-gas chromatography-mass spectrometry, *ACS Div. Fuel Chem. Preprints*, *36*(2): 765–773 (1991).
34. Anderson, K. B. and R. E. Winans. The nature and fate of natural resins in the geosphere. I. Evaluation of pyrolysis-gas chromatography-mass spectrometry for the analysis of natural resins and resinites, *Anal. Chem.*, *63*: 2901–2908 (1991).
35. Anderson, K. B., R. E. Winans, and R. E. Botto, The nature and fate of natural resins in the geosphere. II. Identification, classification and nomenclature of resinites, *Org. Geochem.*, *18*(6): 829–841 (1992).
36. Anderson, K. B. and R. E. Botto, The nature and fate of natural resins in the geosphere. III. Re-evaluation of the structure and composition of highgate copalite and glessite organic geochemistry. *Org. Geochem.*, *20*(7): 1027–1038 (1993).
37. Anderson, K. B., The nature and fate of natural resins in the geosphere. IV. Middle and upper Cretaceous amber from the Taimyr Peninsula, Siberia—Evidence for a new structural subclass of resinite. *Org. Geochem.*, *21*(2): 209–212 (1994).
38. Feller, K. L., Dammar and mastic infrared analysis, *Science*, 120: 1069–1070 (1954).
39. Shedrinsky, A. M., D. Grimaldi, T. P. Wampler, and N. S. Baer, Amber and copal: Pyrolysis gas chromatographic (PyGC) studies of prevenance, *Wiener Berichter über Naturwissenschaft in der Kunst*, *6/7/8* (1989/90/91), pp 37–62.
40. Van Aarssen, B. G. K., J. W. de Leeuw, and Horsfield, A comparative study of three different pyrolysis methods used to characterize a biopolymer isolated from fossil and extant dammar resins, *J. Anal. Appl. Pyrol.*, *20*: 125–139 (1991).
41. Shigemoto, T., Y. Ohtani, A. Okagawa, and M. Summoto, *Cell. Chem. Technol.*, *21*: 249 (1987).
42. Beck, C. W., *J. Baltic Stud.*, Special issue: Studies in Baltic amber, Vol. XVI, No. 3 (1985).
43. Grimaldi, D., C. Beck, and J. Boon, Occurrence, Chemical Char-

acterization and Paleontology of the Fossil Resins from New Jersey, *Novitaties*, AMNH, No. 2948 (1989), pp 1–27.

44. Shedrinsky, A., T. Wampler, N. Indictor, N. S. Baer, The use of pyrolysis gas chromatography in the identification of oils and resins found in art and archaeology, *Conserv. Cult. Prop. India*, *21*: 35–41 (1988).
45. Shedrinsky, A., H. Ohtani, Y. Freidzon, J. Boon, and N. S. Baer, Fossil polystyrene. A missing chapter, *Euro. Polym. J.*, (1994) (in press).
46. Shedrinsky, A. M., D. A. Grimaldi, J. J. Boon, and N. S. Baer, Application of Py-GC and Py-GC/MS for unmasking of faked ambers, *J. Anal. Appl. Pyrol.*, *25*: 77–95 (1993).
47. Grimaldi, D., A. Shedrinsky, A. Ross, and N. S. Baer, Forgeries in amber: History, identification, and case studies, *Curator*, AMNH (in press), 1994.
48. Basch, A., Analysis of oil from two Roman glass bottles, *Israel Explor. J.*, *22*: 27–32 (1972).
49. Jaky, M., J. Peredi, and L. Palosl, Untersuchungen eines aus römischen Zeiten stammenden Fettproduktes, *Fette, Seifen, Anstrichmmitte*, *66*: 1012–1017 (1964).
50. Condamin, J., F. Formenti, M. O. Metais, M. Michel, and P. Blond, The application of gas chromatography to the tracing of oil in ancient amphorae, *Archaeometry*, *18*: 195–201 (1976).
51. Condamin, J. and F. Formenti, Detection du contenu d'amphores Antiques (Huile, Vin). Etude méthodologique, *Rev. d'Archéometrie*, 2: (1978).
52. Seher, A., H. Schiller, M. Krohn, and G. Werner, Untersuchungen von Ölproben aus archäologischen Funden, *Fette, Seifen, Anstrichmittel*, *82*: 395–399 (1980).
53. Jain, N. C., C. R. Fontan, and P. L. Kirk, Identification of paints by pyrolysis-gas chromatography, *J. Forens. Sci. Soc.*, *5*: 102–109 (1965).
54. Stolow and G. deW. Rogers, Further studies by gas chromatography and pyrolysis techniques to establish ageing characteristics of works of art, *Application of Science in Examination of Works of Art* (W. G. Young, ed.), Museum of Fine Arts, Boston, 1973, pp 213–228.

55. Rogers, G. deW., An improved pyrolytic technique for the quantitative characterization of the media of works of art, *Conservation and Restoration of Pictorial Art*, (N. Brommelle and P. Smith, eds.), Butterworth, London, 1976, pp 93–100.
56. Breek, R. and W. Froentjes, Application of pyrolysis gas chromatography on some of Van Meegeren's faked Vermeers and Pieter de Hooghs, *Stud. Conserv.*, *20*: 183–189 (1975).
57. Wampler, T. P., S. A. Liebman, E. J. Levy, and Z. Barov, Authentication of archaeological specimens by pyrolysis capillary gas chromatography, *Chromatography as a Quality Control Tool* (R. B. Huntoon, ed.), Marcel Dekker, New York (in press).
58. Nazer, J., C. Young, and R. Giesbrecht, Pyrolysis-GC analysis as an identification method of fats and oils, *J. Food Sci.*, *50*: 1095–1100 (1985).
59. Boon, J. J., Analytical pyrolysis mass spectrometry: New vistas opened by temperature-resolved in-source PYMS, *Int. J. of Mass Spect. Ion Proc.*, *118/119*: 758–762 (1992).
60. Oudemans, T. and J. Boon, Molecular archaeology: Analysis of charred (food) remains from prehistoric pottery by pyrolysis-gas chromatography/mass spectrometry, *J. Anal. Appl. Pyrol.*, *20*: 197–227 (1991).
61. de la Rie, E. R. and A. M. Shedrinsky, The chemistry of ketone resins and the synthesis of a derivative with increased stability and flexibility, *Stud. Conserv.*, *34*: 9–19 (1989).
62. Noble, W., B. Wheals, and M. J. Whitehouse, The characterization of adhesives by pyrolysis gas chromatography and infrared spectroscopy, *Forens. Sci.*, *3*: 163–174 (1974).
63. Curry, C. J., Pyrolysis-mass spectrometry studies of adhesives, *J. Anal. Appl. Pyrol.*, *11*: (1987).
64. De Witte, E., M. Goessens-Landrie, E. Goethals, and R. Simonds, The structure of "old" and "new" Paraloid B-72, in ICOM Committee for Conservation, Report 78/16/4, 5th Triennial Meeting, Zagreb (1978).
65. de Witte, E. and A. Terfve, The use of a Py-GC/MS technique for the analysis of synthetic resins, *Science and Technology*, (N. S. Brommelle and G. Thomson, (eds.), International Institute for Conservation, London (1982), pp 16–18.

66. Thomas, T. H. and T. C. Kendrick, *J. Polym. Sci*, A-2 *7*: 537 (1969).
67. Blazsó, M., G. Garzo, and T. Székely, *Chromatographie*, *5*: 485 (1972).
68. N. Grassie and K. F. Francey, *Polym. Degr. Stab.*, *2*: 53 (1980).
69. Kleinert, J. C. and C. J. Weschler, *Anal. Chem.*, *52*: 1245 (1980).
70. G. Garzo, J. Tamas, T. Székely, and K. Ujszaszi, *J. Organometal. Chem.*, *165*: 273 (1979).
71. Blazsó, M., G. Garzo, K. Andrianov, N. Makarova, A. Chernavski, and I. Petrov, *J. Organometal. Chem.*, *165*: 273 (1979).
72. Blazsó, M., E. Gal, and N. Makarova, *Polyhedron*, *2*: 455 (1983).
73. Blazsó, M., T. Szekely, and N. Makarova, Thermal decomposition of cyclolinear and cyclic methylsiloxane polymer, *J. Polym. Sci.*, *23*: 2589–2599 (1985).
74. Mills, J. S. and R. White, *The Organic Chemistry of Museum Objects*, Butterworth, London (1987).
75. Halpine, S., Amino acid analysis of proteinaceous media from Cosimo Tura's "The Annunciation with Saint Francis and Saint Louis of Toulouse," *Stud. Conserv. 37*: 22–38 (1992).
76. Chiavari, G. and G. Calletti, Pyrolysis-gas chromatography/mass spectrometry of amino acids, *J. Anal. Appl. Pyrol.*, *24*(2): 123–137 (1992).
77. Chiavari G., G. Galletti, G. Lanterna, and R. Mazzeo, The potential of pyrolysis-gas chromatography/mass spectrometry in the recognition of ancient painting media, *J. Anal. Appl. Pyrol.*, *24*(3): 227–242 (1993).
78. Boon, J. J. and J. W. De Leeuv, Amino acid sequence information in proteins and complex proteinaceous material revealed by pyrolysis-capillary gas chromatography—Low and high resolution mass spectrometry, *J. Anal. Appl. Pyrol.*, *11*: 313–327 (1987).
79. Marmer, W. N. and P. Magidman, Analysis of wool proteins by pyrolysis-gas chromatography, *Preprints of 198th ACS National Meeting*, Florida, Sept., 1989.
80. Cutter, E., P. Magidman, and W. Maarmer, Pyrolysis-gas chromatography of Hercosett 125 finish on wool, *Text. Chem. Colorist*, *25*(6): 27–29.

7

Environmental Applications of Pyrolysis

T. O. Munson
ECKENFELDER INC., Nashville, Tennessee

I. INTRODUCTION

The purpose of this handbook is as a practical guide to the application of pyrolysis techniques to various samples and sample types. A subset of this overall scope, to be dealt with here, comprises the environmental applications of pyrolysis.

A narrow view of this mission would include only the use of analytical pyrolysis (e.g., pyrolysis-mass spectrometry (Py-MS) and pyrolysis-gas chromatography-mass spectrometry (Py-GC/MS)) to identify and measure "contaminants" in samples of outdoor air, soils, sediments, water, and biota. In order to include interesting and useful applications of pyrolysis techniques which otherwise might not be mentioned in this handbook, a broader view of environmental applications will be used to include such topics as: pyrolytic conversion to chemical feedstocks of solid wastes, such as used tires and waste plastics; identification of growths defacing limestone buildings; characterization of life-threatening pyrolysis products formed during building fires; analysis of particulates emitted from internal combustion engine exhausts; and measurement of chemical changes in biological materials exposed to environmental stress.

The subject areas delineated by the above discussion all have some link to environmental contamination. Another broad subject area which will only be briefly mentioned in this chapter is the use of analytical pyrolysis to gain understanding of natural environmental processes, such as the conversion of plant materials into soil, coal, and petroleum hydrocarbons. (This subject was included in a recent review paper describing the use of analytical pyrolysis for environmental research [1].)

No extensive effort has been made to uncover all published reports of pyrolysis use which might be classified as environmental. Instead, time has been devoted to identifying and including good examples of as many different types of environmental applications as possible. Another sensible criterion was to choose published reports that would be readily accessible to an interested reader.

The most useful format in which to display the gathered information did not readily present itself. An obvious choice was to arrange the applications according to the pyrolytic technique described in the report (e.g., Py-GC/MS, Py-MS, etc.). This approach was not selected for two reasons: specific pyrolysis techniques are discussed in other chapters of this handbook; and, it seems appropriate for this chapter to emphasize the applications rather than the techniques. Following this logic, it was decided to sort the applications generally by the type of environmental media to which they apply (such as air, water, soil) and/or the type of environmental problem to which they apply (such as solid waste disposal or recycling).

II. APPLICATIONS RELATED TO AIR

Since the term *pyrolysis* as used in the context of this handbook refers to the conversion of large, nonvolatile organic molecules, such as organic polymers, to smaller volatile organic molecules, it should not be surprising that the reports discussed in this section fall into two general categories: release of organic materials into air by the pyrolytic decomposition of nonvolatile material; and, the use of pyrolytic techniques for the examination of particulate material recovered from air samples.

A. Examination of Air Particulates

1. Smoke Aerosols

Tsao and Voorhees used Py-MS together with pattern recognition for the analysis of smoke aerosols from nonflaming [2] and flaming [3] combustion of such materials as Douglas fir, plywood, red oak, cotton wool, polyurethanes, polystyrene, polyvinyl chloride, nylon carpet, and polyester carpet. The driving force for these studies was to determine whether this technique could provide a means for assessing the fuels involved in building fires from the analysis of the smoke aerosols formed. Apparently this type of information can be useful in legal actions associated with catastrophic fires, such as the one at the MGM Hotel in Las Vegas, NV.

During flaming combustion, smoke aerosols form that contain, in addition to various other materials, an amorphous polymeric substance which Tsao and Voorhees speculated should contain, unchanged, a great deal of the original polymer structure. Because Py-MS had been shown useful for differentiating, classifying, and sometimes characterizing nonvolatile macromolecules, it was speculated that Py-MS combined with pattern recognition procedures might be useful in classifying aerosol materials from combustion.

The sample preparation procedures were similar for both of these studies. Smoke aerosols of the materials of interest were produced in the laboratory under either nonflaming [2] or flaming

[3] combustion conditions, and the material produced collected on glass-fiber filters. The volatile organic fraction was removed from the collected material by gentle heating (50–55°C) under vacuum for 24–48 h. A portion of the nonvolatile material was then subjected to analysis by Py-MS using a Curie-point pyrolyzer (510°C) interfaced to a quadrupole mass spectrometer via an expansion chamber.

Upon pyrolysis, under both combustion conditions, the material collected from the smoke aerosols produced a unique Py-MS spectrum for each substance. In many cases, visual inspection of the spectra from simple mixtures led to the identification of the fuel materials. For more complex mixtures, however, extensive "crunching" of the data was necessary using pattern recognition techniques before judgments could be made as to what fuel mixture might have produced the smoke aerosol analyzed. The authors did demonstrate considerable success in identifying some or all of the fuel components of complex mixtures and speculated that with further experiments and refinement in the applications of chemometric techniques, this approach could become a useful tool in fire investigations.

Those interested in these studies might also be interested in a study of the products of nonflaming combustion of poly(vinyl chloride) in which 2 g samples of the granular polymer were decomposed in a tubular flow reactor under various temperature conditions and gas mixtures [4]. The evolved products were trapped in cold hexane or styrene and identified using GC and GC/MS. An earlier study along these same lines, using phenol-formaldehyde resin foam, was reported by the same investigator [5].

There is a rich literature associated with studies of the breakdown and/or formation of specific chemical compounds and classes of chemical compounds during burning processes. A study of the thermal decomposition of pentachlorobenzene, hexachlorobenzene, and octachlorostyrene in air contained many such citations [6]. In this study, nearly pure, 10–20 mg samples of the cited chemicals were decomposed in a vertical combustion furnace and the decomposition products trapped on cooled XAD-4 resin followed by charcoal tubes. The adsorbed components were desorbed with toluene and analyzed using capillary GC and GC/MS. The decomposition products formed depended upon the applied temperature, the oxygen concentration, and the residence time in the hot zone of the combustion chamber.

2. Particulate from Vehicular Traffic

Voorhees et al. [7] reported a study of the insoluble carbonaceous material in airborne particulates from vehicular traffic using Py-MS and Py-GC/MS (in addition to thermogravimetric analysis (TGA) and elemental and radiocarbon analysis). The solvent-soluble organic compounds separated from atmospheric particulate matter (which encompasses a broad spectrum of solid and liquid particles generally ranging in size from several hundred angstroms to several hundred micrometers) had been extensively studied by numerous investigators. Because of its complexity, the insoluble carbonaceous material (ICM) in urban and rural particulate material has not been studied in depth.

Apparently this complexity had caused earlier attempts to associate ICM with possible formation sources to be highly speculative. In this study the authors proposed to characterize ICM collected under conditions that would ensure a known source—vehicular traffic.

Air particulate samples were collected on glass microfiber filters in the Eisenhower tunnel on Interstate 70 under conditions designed to minimize particulates other than those formed by the vehicular traffic passing through the tunnel. Water and volatile organics were removed from the samples by drying the filters at 45°C under vacuum for 18 h. The soluble organics were removed from the particulates by successive 18 h Soxhlet extractions with methanol, acetone, methylene chloride, and cyclohexane.

Py-GC/MS of the ICM was accomplished using a Pyroprobe (725°C set temperature, 550°C actual temperature, for 20 s) with the pyrolyzate trapped with liquid nitrogen at the head of a 15 m long by 0.32 mm i.d. bonded phase capillary GC column. The Py-MS was accomplished as described above for the study of smoke aerosols.

The capillary column pyrogram shown in Figure 1, and the peak identification list shown in Table 1—from the analysis of ICM that had not been sorted by particle size—demonstrate the immense power of this analytical technique. A 0.5 mg sample of ICM containing only a fraction of that amount of organic material (perhaps about 25 μg) was separated into 134 GC peaks, about 90 of which were presumptively identified (that is, by comparison of the unknown spectra to computer library spectra of known compounds). Py-GC/MS analysis of ICM, which was size-sorted into four size fractions (<0.6, 0.6–2.7, 2.7–10.4, and >10.4 μm, respectively), showed pyrograms of similar composition to the unsorted material.

Py-GC/MS of samples (prepared as were the ICM samples) of the four materials thought to be the major sources of the tunnel air particulates—tire rubber, road salt, diesel exhaust, and gasoline exhaust—showed that road salt gave no pyrolysis products, and the other three materials gave pyrograms of complexity similar to the ICM samples. The overlap of compound types and the absence of unique "marker" compounds made an assessment of the relative contribution of each of the three sources to the total impractical using the Py-GC/MS profiles.

Figure 2 shows the comparison of the Py-MS spectra obtained from the ICM (referred to as tunnel particles in the figure) and the three most likely contributor materials. While there are many ions in common, the patterns of ions in the spectra of the three contributor materials are distinctly different from each other. Using a statistical method for the comparison of the Py-MS spectra of the three contributor materials to the spectrum for the ICM, the authors estimated that diesel exhaust particulates, gasoline exhaust particulates, and tire wear particulates accounted for 64, 26, and 10%, respectively, of the tunnel insoluble carbonaceous material.

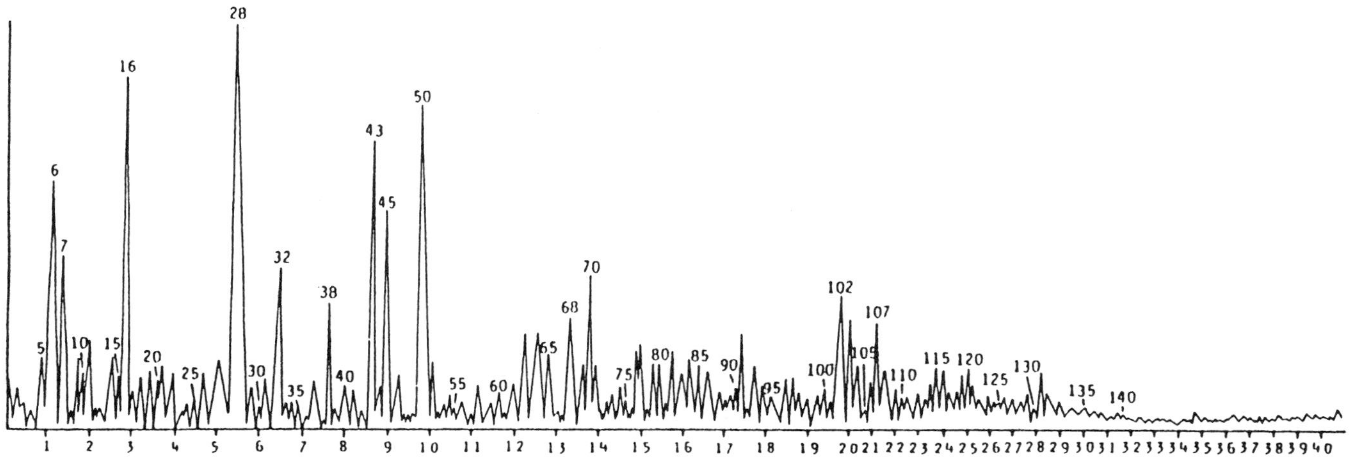

Figure 1 550°C pyrogram of insoluble carbonaceous material from vehicular traffic. (From Ref. 7.)

TABLE 1 Peak Identification: Tunnel 2

Peak no.	Compound	Peak no.	Compound
1	1,3-Butadiene	37	Unidentified
2	1,3-Butadiene	38	C_8H_{12}
3	Pentane	39	C_8H_{12}
4	2-Pentene	40	C_8H_{12}
5	Pentane	41	1-Ethylcyclohexene
6	Acetone	42	Unidentified
7	4-Methyl-2-pentene	43	Ethylbenzene
8	2,3-Dimethyl-2-butene	44	Unidentified
9	Methylpentene	45	1,2- or 1,4-Xylene
10	Hexane	46	2,4-Dimethylthiophene
11	2-Methylfuran	47	Unidentified
12	Hexene	48	Unidentified
13	2,3-Dimethylbutadiene	49	Unidentified
14	1,3,5-Hexatriene	50	Styrene
15	2-Hexene	51	Unidentified
16	Benzene	52	Unidentified
17	Hexadiene	53	Unidentified
18	Methylcyclopentene	54	1,2,3-Trimethylcyclohexane
19	1-Heptene	55	Unidentified
20	Methylhexadiene	56	Unidentified
21	2-Methylhexadiene	57	Unidentified
22	2,5-Dimethylfuran	58	1-Methylethylbenzene
23	Unidentified	59	Unidentified
24	Unidentified	60	Dimethyloctane
25	Trimethylcyclohexane	61	2,4- or 3,4-Dimethylpyridine
26	1,3,5-Heptatriene	62	2-Methylstyrene
27	2,4-Heptadiene	63	Propylbenzene
28	Toluene	64	Methylethylbenzene
29	2-Methylthiophene	65	1,3,5-Trimethylbenzene
30	Heptadiene	66	Unidentified
31	1-Octene	67	Unidentified
32	Trimethylbutene	68	1-Ethyl-3-methylbenzene
33	Unidentified	69	1-Decene
34	Unidentified	70	1,2,3-Trimethylbenzene
35	Octene	71	Decane
36	Octadiene	72	4- or 5-Decene

TABLE 1 Continued

Peak no.	Compound	Peak no.	Compound
73	Unidentified	103	1,4-Divinylbenzene
74	Unidentified	104	1,3-Divinylbenzene
75	Unidentified	105	Unidentified
76	C_4-Alkylbenzene	106	Unidentified
77	1-Methyl-4-(1-methylethyl) benzene	107	Naphthalene
78	Unidentified	108	3-Methyl-1,2-dihydronaphthalene
79	Unidentified	109	Unidentified
80	2-Propenylbenzene	110	Tridecene
81	Unidentified	111	Unidentified
82	Propynylbenzene	112	C_5-Alkenylbenzene
83	Methylpropylbenzene	113	Unidentified
84	C_4-Alkylbenzene	114	Unidentified
85	Trimethyloctane	115	C_6-Alkylbenzene
86	1-Methyl-2-propylbenzene (or 1,3)	116	C_6-Alkylbenzene
		117	C_5-Alkenylbenzene
87	1-Methyl-2-(propenyl)-benzene	118	C_6-Alkenylbenzene
88	Unidentified	119	C_{13}-Alkene
89	Unidentified	120	Alkylbenzene
90	C_4-Alkenylbenzene	121	Unidentified
91	Undecene	122	C_{14}-Alkene
92	Undecene	123	Unidentified
93	Undecene	124	Unidentified
94	Ethyldimethylbenzene	125	Unidentified
95	C_4-Alkenylbenzene	126	Unidentified
96	Tetramethylbenzene	127	Unidentified
97	C_4-Alkenylbenzene	128	Alkylbenzene
98	C_4-Alkenylbenzene	129	Alkylbenzene
99	1-Methyl-2-(1-methylethyl) benzene	130	Alkylbenzene
		131	Branched C_{14}-Alkene
100	Dimethylstyrene	132	Branched C_{14}-Alkene
101	Unidentified	133	Branched C_{15}-Alkene
102	1- or 3-Methyl-1H-indene	134	Alkylbenzene

Source: Adapted from Table 3, Ref. 7.

B. Examination of Air Pollutants from Thermal Decomposition

1. Pyrolysis of Plastic Wrapping Film

Pyrolysis has been used to examine the types of air pollutants that might be generated from various types of waste plastic materials during incineration or thermal decomposition in landfills. One such study [8] used GC and GC/MS to examine the products that were formed when two kinds of widely used plastic wrapping film [made of poly(vinylidene chloride)] were pyrolyzed.

Samples of the wrapping material were pyrolyzed at 500°C in an air flow of 300 ml/min and the products evolved trapped in ice-cold hexane. The concentrated extracts were then analyzed by capillary column GC and GC/MS. Figure 3 shows a typical capillary gas chromatogram of the pyrolysis products from one of the wrapping materials. The numbers on the peaks correspond to the numbers of the compounds identified and shown in Table 2. The identifications were carried out by comparison of the mass spectra and retention indexes of the sample peaks with mass spectra and retention indexes of authentic compounds. The original form of this table includes the retention indexes (calculated using alkanes as references) and average amounts formed calculated from eight runs.

Pyrolyzing the material of interest separately from the analysis allows a greater latitude in the selection of the pyrolysis conditions. For instance, in this case, the gas flow, which contained about 25% oxygen, would have been incompatible with the capillary GC column. On the other hand, direct coupling of the pyrolysis step to the analytical step (Py-GC and Py-GC/MS) would have allowed examination of the most volatile products formed, which could not be done here because of losses during the concentration of the hexane, and because of the huge hexane solvent peak in the early portion of the chromatogram. A modern

---

FIGURE 2 Py-MS spectra of the tunnel particles and various contributors to that sample. (From Ref. 7.)

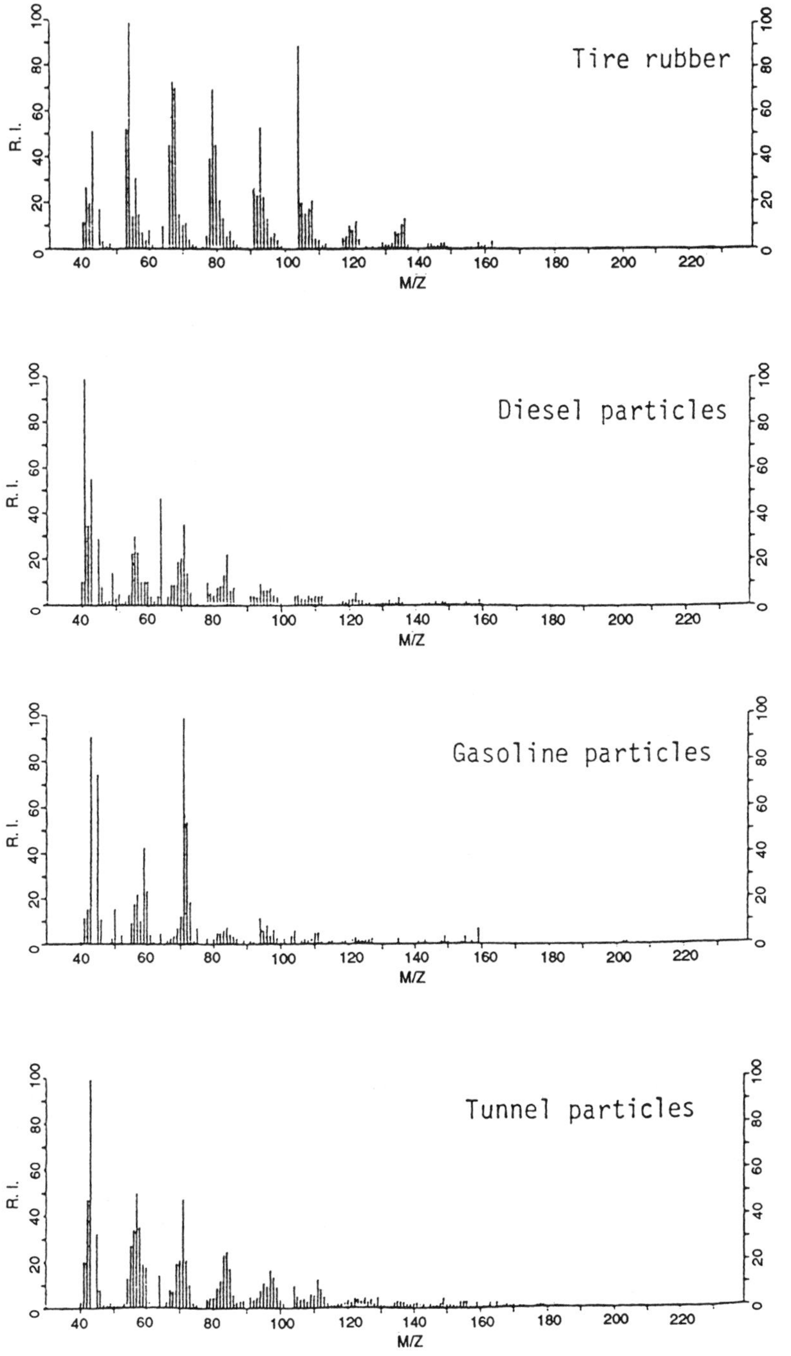
Tire rubber
Diesel particles
Gasoline particles
Tunnel particles
R. I.
M/Z

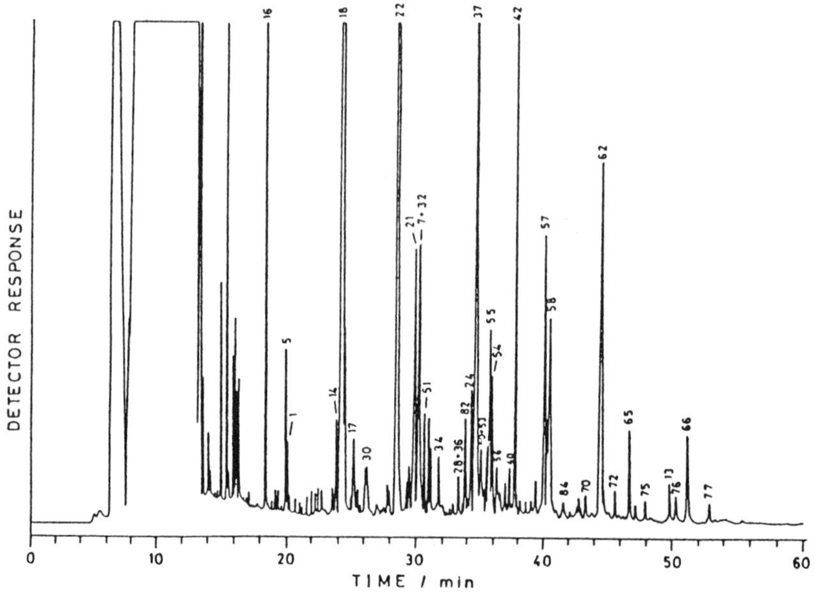

FIGURE 3 Typical gas chromatogram of pyrolysis products from a plastic wrapping film. (From Ref. 8.)

pyrolysis-concentrator unit allows the best of both arrangements by providing a gas stream other than the GC carrier gas during pyrolysis (while trapping the pyrolyzate) and then passing the pyrolyzate to the GC column under optimum conditions.

2. Waste Plastic Processing

An ideal way to recycle waste plastics is to remelt them and form new products directly, but this process is usually only possible in the case of rather homogeneous waste streams. With heterogeneous plastic containing waste streams, however, because other processing steps would be required, landfilling or incineration are the most favored disposal methods from an economic point of view. These disposal processes can lead to environmental problems, especially in the case of halogen-containing plastics such as PVC [poly(vinyl chloride)].

Table 2 Compounds Identified from Pyrolysis of Plastic Wrap

Peak no.	Compound	Peak no.	Compound
1	Trichlorobutadiene	36	Dichlorostyrene
2	Trichlorobutadiene	37	Trichlorostyrene
3	Pentachlorobutadiene	38	Trichlorostyrene
4	2,2,3-Trimethyloxetan	39	Trichlorostyrene
5	Styrene	40	Trichlorostyrene
6	Phenylacetylene	41	Trichlorostyrene
7	Naphthalene	42	Tetrachlorostyrene
8	Biphenyl	43	Tetrachlorostyrene
9	Methylphenylacetylene	44	Tetrachlorostyrene
10	Phenanthrene	45	Tetrachlorostyrene
11	Phenol	46	Tetrachlorostyrene
12	Benzaldehyde	47	2-Chlorophenol
13	Dibutylphthalate	48	2,5-Dichlorophenol
14	Benzofuran	49	2,6-Dichlorophenol
15	Dibenzofuran	50	3,5-Dichlorophenol
16	Chlorobenzene	51	Dichlorophenylacetylene
17	1,2-Dichlorobenzene	52	Dichlorophenylacetylene
18	1,3-Dichlorobenzene	53	Trichlorophenylacetylene
19	1,4-Dichlorobenzene	54	1-Chloronaphthalene
20	1,2,3-Trichlorobenzene	55	2-Chloronaphthalene
21	1,2,4-Trichlorobenzene	56	Chlorotetrahydronaphthalene
22	1,3,5-Trichlorobenzene	57	Dichloronaphthalene
23	1,2,3,4-Tetrachlorobenzene	58	Dichloronaphthalene
24	1,2,3,5- or 1,2,4,5-Tetrachlorobenzene	59	Dichloronaphthalene
		60	Dichloronaphthalene
25	4-Chlorotoluene	61	Trichloronaphthalene
26	Dichlorotoluene	62	Trichloronaphthalene
27	Dichlorotoluene	63	Trichloronaphthalene
28	Trichlorotoluene	64	Trichloronaphthalene
29	Tetrachlorotoluene	65	Trichloronaphthalene
30	Chlorostyrene	66	Tetrachloronaphthalene
31	Chlorostyrene	67	Tetrachloronaphthalene
32	Dichlorostyrene	68	2-Chlorobiphenyl
33	Dichlorostyrene	69	2,4- or 2,5-Dichlorobiphenyl
34	Dichlorostyrene		
35	Dichlorostyrene	70	2,3′-Dichlorobiphenyl

TABLE 2 Continued

Peak no.	Compound	Peak no.	Compound
71	3,5-Dichlorobiphenyl	77	2,3′,5,5′-Tetrachlorobiphenyl
72	3,4- or 3,4′-Dichlorobiphenyl	78	2,3′,4,5′-Tetrachlorobiphenyl
73	2,3,6- or 2,3′, 6-Trichlorobiphenyl	79	2,2′,3,5′,6-Pentachlorobiphenyl
74	2,3,5- or 2′,3,5-Trichlorobiphenyl	80	Chlorobenzofuran
75	2,3′,4- or 2,3′,5-Trichlorobiphenyl	81	Chlorobenzofuran
		82	Dichlorobenzofuran
76	3,4′,5-Trichlorobiphenyl	83	Dichlorobenzofuran
		84	Tetrachlorobenzfuran

Source: Adapted from Table II, Ref. 8.

Because PVC is widely used in combination with aluminum, and because ABS (a blend of acrylonitrile, butadiene, and styrene) and PET (polyethylene terephthalate) may also be used in the future, a study was undertaken to examine the use of pyrolysis to process combinations of these plastics and aluminum [9].

The thermal stability of the plastics (separately and in combination with aluminum, and under various gas atmospheres) was investigated using thermogravimetry. With this pyrolysis technique, the decomposition of a material is measured as a function of applied temperature by monitoring the weight of the sample as the temperature is increased at a programmed rate. For these experiments, the temperature was programmed from 300 to 1000 K at 0.083 K/s. If the data from such a test is plotted as the normalized mass change per unit time (based on the initial sample mass) as a function of temperature, the resulting differential thermogravimetric (DTG) curve provides a profile of the decomposition of the material as a function of temperature. The DTG curves for virgin PVC in different atmospheres (Figure 4) clearly show that the thermal decomposition of virgin PVC is a two-step process.

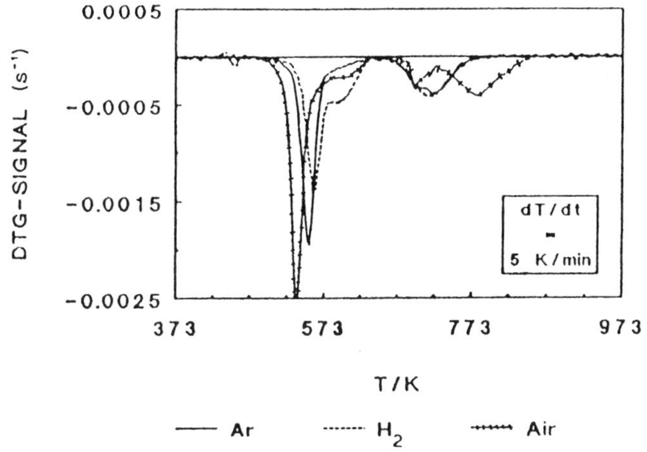

FIGURE 4 DTG curves of virgin PVC in different atmospheres. (From Ref. 9.)

In addition to the thermogravimetry experiments, batch pyrolysis experiments continuously monitoring HCl formation were performed with PVC to determine the optimum temperature for HCl formation. Under conditions of maximum HCl formation, PVC was pyrolyzed with and without oxygen in a fluidized bed reactor and the formation of polychlorinated dibenzodioxins (PCDDs) and polychlorinated dibenzofurans (PCDFs) measured. Avoiding the formation of these highly toxic compounds would be a critical element in any waste stream processing scheme.

In addition to these techniques, analytical pyrolysis experiments were performed where 0.5 mg samples were batch pyrolyzed under flowing helium gas in a tube furnace connected to a liquid nitrogen cold trap. By warming the cold trap, the pyrolysis products were transferred directly to a GC/MS system for identification.

Several important design guidelines for processing aluminum scrap, plastic waste, and combinations of the two were elucidated by these studies. For instance, thermal decomposition of PVC in the presence of oxygen (combustion) generated 10 to 1000 times as much PCDD and PCDF as did thermal decomposition in the absence of oxygen. Because anaerobic thermal decomposition of PVC does not cause PCDD and PCDF emission problems, simpler emissions scrubbing steps can be employed. Most of the organic fraction of aluminum scrap can be removed by pyrolysis below the melting point of the aluminum. If the plastic fraction of the aluminum scrap contains a considerable amount of PVC, it is more economical to incinerate the pyrolysis products on-site in combination with HCl removal and heat recovery. In the case of other plastics, it is possible to get a valuable pyrolysis oil.

3. Pyrolysis of Sewage Sludge

Disposal of sewage sludge has become a major environmental problem, particularly in areas of dense population. The increase in metals content in soil and the potential presence of toxic organic compounds has thrown the traditional utilization of sewage sludge on farmlands into question. Thermal decomposition of the sewage sludge, followed by sanitary landfill disposal of the resultant slag and ashes, represents an alternative to the agricultural use, but one must then be concerned with the fate of the inorganic and organic constituents during the thermal treatment and final disposal. A laboratory-scale pyrolysis reactor was used to determine the process conditions for a minimum flux of metals to the environment from the anaerobic thermal decomposition (pyrolysis) of sludge, and subsequent disposal of the pyrolysis products [10].

Sewage sludge containing less than 20% water was dried in 600 g batches to less than 1% water. A fraction of each batch (30–40 g) was analyzed for organic matter and metals (Cr, Ni, Cu, Zn, Cd, Hg, and Pb), and the rest was pyrolyzed at selected temperatures (350, 505, 625, 750°C) for one hour with the off-gases being passed through traps and filters.

Analysis of the solid residue (char) and the various traps and filters showed that, at all temperatures, more than 97% of the Hg was completely evaporated from the sludge. At 505°C, all of the Cd remained in the char, but at 750°C, all of the Cd was volatilized. At 625°C, the partitioning of Cd between the gas phase and the solid phase depended upon the residence time at the pyrolysis temperature. Essentially all of the other metals remained in the char at all pyrolysis temperatures.

This study demonstrated that the best temperature for the pyrolysis of sewage sludge with respect to the metals content is in the range of 500–600°C. Higher temperatures, which leave less char, are less desirable because, at $T > 600°C$, Cd and other metals with relatively high vapor pressures are transferred to the gas phase. Since Hg is very difficult to scrub from the off-gas stream and is completely volatilized at even the lowest temperatures, the most economical approach with respect to Hg seems to be to limit the input of Hg to the sewer system.

III. APPLICATIONS RELATED TO WATER

A. Analysis of Pulp Mill Effluents Entering the Rhine River

Van Loon et al. [11] used Py-MS and Py-GC/MS as analytical techniques for examining the high-molecular-weight, dissolved organic fraction from pulp mill effluents entering the Rhine River. The study had as its objectives: (a) to optimize the analytical method of ultrafiltration combined with pyrolysis-mass spectrometry (which had been reported earlier); (b) to analyze qualitatively the chlorolignins and lignosulfonates in pulp mill effluents entering the Rhine River in order to find structurally specific pyrolysis products for quantitative analysis in river water; and (c) to obtain an overview of the amounts of adsorbable organic halogens discharged by these pulp mills.

Portions of the effluent samples were treated to remove the undissolved materials and then subjected to ultrafiltration to remove the materials with molecular weights less than 1000 daltons. With repetitive treatments, water and lower-molecular-weight organic and inorganic materials were removed, effectively concentrating and desalting the high-molecular-weight, dissolved organic carbon (DOC) fractions. Volatile organic fractions were collected by vapor stripping followed by trapping on charcoal tubes, and nonpolar and moderately polar organic compounds were collected by adsorption/elution using XAD-4 resin columns.

The high-molecular-weight DOC samples were analyzed by Py-MS using the in-source platinum filament pyrolysis technique. The filament, bearing a 1–20 μg sample, was heated at 15°C/s to a final temperature of 800°C. The mass spectrometer collected one scan/s over the m/z range 20–800 amu for 1.5 min.

The Py-GC/MS analysis was performed with 20 μg samples, a Curie-point temperature of 610°C, and a pyrolysis time of 4 s. The resulting pyrolyzate was then swept onto a capillary GC column (50 m long by 0.32 mm i.d. with a 1 μm film of methyl silicones) heated from 30 to 300°C at 4°C/min for a total analysis time of about 1 h.

A discussion of the many interesting and important topics presented in this paper [11] is beyond the scope of this handbook. However, the inclusion, in detail, of some of the Py-MS, Py-GC/MS, and GC/MS data might provide some thought-provoking material for those interested in the usefulness of these techniques for the examination of such a complex organic matrix as pulp mill effluent.

Figure 5, for instance, shows the Py-MS spectra of the high-molecular-weight DOC isolated from samples from three pulp mills. A distinctive feature of the spectra is the presence of intense signals from small pyrolysis products containing sulfur or chlorine (m/z 34, hydrogen sulfide; 48, methyl sulfide; 62, dimethyl sulfide; 64, sulfur dioxide; 76, carbon disulfide; 50, methyl chloride; and 36, hydrogen chloride). Also noteworthy are the relatively low intensities of lignin pyrolysis products. These features are explained as being due to the highly sulfonated, chlorinated, and oxidized nature of the chlorolignins and lignosulfonates, which leads to macromolecules with relatively

Figure 5 Pyrolysis-mass spectra of high-molecular-weight material (MW $>$ 1000) isolated from pulp mill effluents A, B, and D. (From Ref. 11.)

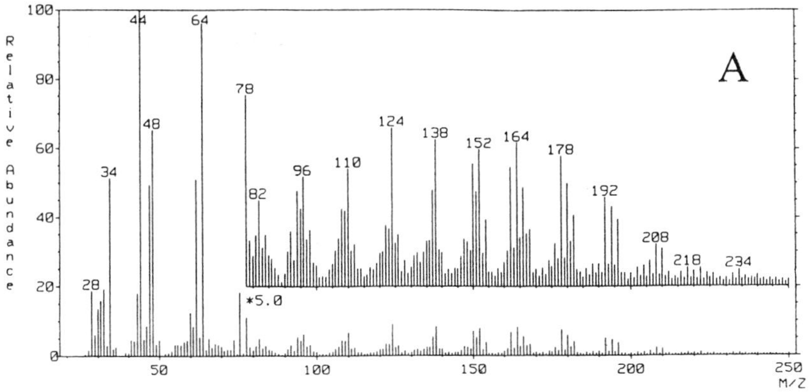
A
Relative Abundance
100
80
60
40
20
0
44
64
78
48
124
138
152
164
178
34
96
110
82
192
208
218
234
28
*5.0
50
100
150
200
250
M/Z

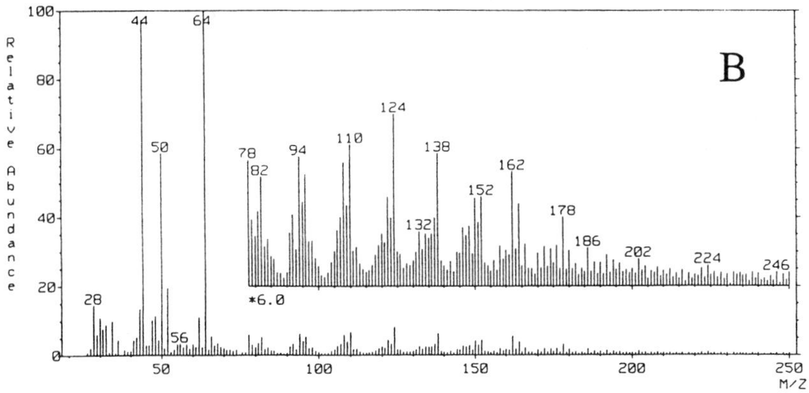
B
Relative Abundance
100
80
60
40
20
0
44
64
124
50
110
138
162
78
94
82
152
132
178
186
202
224
246
28
56
*6.0
50
100
150
200
250
M/Z

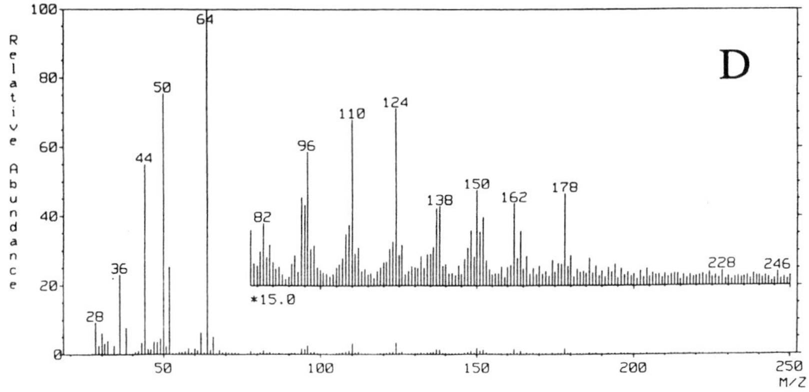
D
Relative Abundance
100
80
60
40
20
0
64
50
124
110
96
44
150
138
162
178
82
36
228
246
28
*15.0
50
100
150
200
250
M/Z

low aromatic content and high aliphatic functional group content. Comparing these spectra to the Py-MS spectra of standard materials (sodium lignosulfonate, lignosulfonic acid, sodium polystyrene sulfonate, and polystyrene sulfonic acid) provided some insights into mechanisms for the formation of the small pyrolysis products containing sulfur or chlorine.

Figure 6 presents the Py-GC/MS profile of the high-molecular-weight DOC from pulp mill effluent B. The peak numbers refer to the compound identifications shown in Table 3, a listing of all of the compounds found in the pulp mill effluents. Most likely the peak identifications in Table 3 were presumptive identifications obtained by matching spectra against known spectra in a computer database, and clearly, some of the identifications were incorrect; for instance, peaks 78 and 86 were both identified as 4-methylguaiacol. Nevertheless, the types of compounds identified, and the relative amounts of the various types of compounds, allowed the authors to reach some useful conclusions about the chemical composition of the high-molecular-weight DOC. The original table presented considerably more information than that shown here in Table 3 (such as relative retention

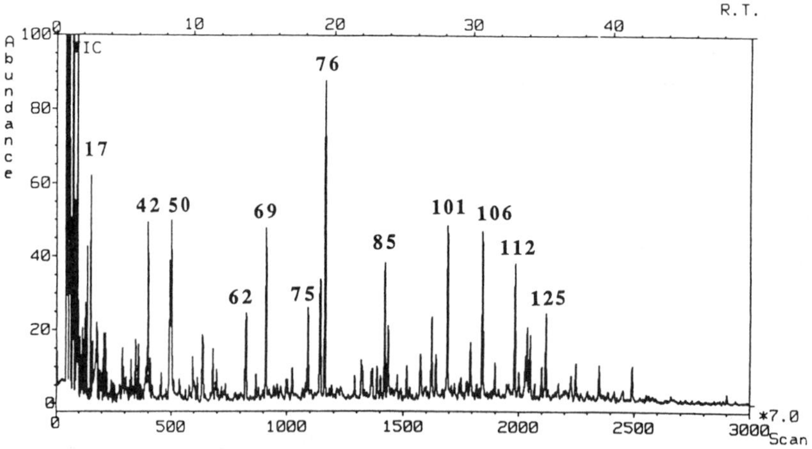

FIGURE 6 Curie-point pyrolysis-GC/MS total ion current profile of high-molecular-weight material (MW > 1000) from pulp mill effluent B. (From Ref. 11.)

TABLE 3 Curie-Point Py-GC/MS Pyrolysis Products of High-Molecular-Weight Material in Pulp Mill Effluent of Mills A, B, C, D, and E[a]

Peak no.	Compound	Peak no.	Compound
1	Carbon dioxide	35	(Methylthio)methanol
2	Sulfur dioxide	36	2-Vinylfuran
3	Methyl chloride	37	1-Methyl-1H-pyrrole
4	Acetaldehyde	38	2,3-Dihydro-3-methylfuran
5	Butene		
6	Methanethiol	39	2-Methylpyrrole
7	Acetonitrile	40	Pyridine
8	2-Propenal	41	Dimethyldisulfide
9	2-Propanone	42	2-Butenoic acid, methyl ester
10	Furan		
11	Dimethylsulfide	43	Toluene
12	Acetic acid, methyl ester	44	3,4-Dihydro-2H-pyran
13	Carbon disulfide	45	Methylthiophene
14	Cyclopentadiene	46	2,3-Dihydrofurfural
15	1-Butanal	47	3-Furaldehyde
16	2-Methyl-2-propenal	48	3-Acetylfuran
17	2,3-Butanedione	49	2,5-Dimethyl-1,4-dioxane
18	2-Butanone	50	Dimethylsulfoxide
19	2-Methylfuran	51	2-Furaldehyde
20	3-Methylfuran	52	Methyl-1H-pyrrole
21	Propionic acid, methyl ester	53	Methyl-1H-pyrrole
		54	4-Methylpyridine
22	2-Butenal	55	Ethylbenzene isomer
23	3-Methylbutanal	56	Ethylbenzene isomer
24	Acetic acid, methyl ester	57	Dimethylbenzene isomer
25	2-Methylbutanal	58	2,4-Dimethylthiophene
26	Benzene	59	CH_3-S-CO-SH
27	Thiophene	60	3-Methyl-2-cyclopenten-1-one
28	Hydrochloric acid		
29	Methylisothiocyanate	61	Dimethylbenzene isomer
30	2,3-Pentadione	62	2-Acetylfuran
31	4-Thiol-3-butyn-2-one	63	5-Methyl-2-furaldehyde
32	1-Thiol-2-propanone	64	2,3-Dimethylpyrazine
33	2,5-Dimethylfuran	65	Methoxybenzene
34	Acetic acid	66	CH_3-SO-CH_2-CO-CH_3

TABLE 3 Continued

Peak no.	Compound	Peak no.	Compound
67	3-Methyl-2-cyclopenten-1-one	95	Methyldihydroxybenzene isomer
68	5-Methyl-2-furaldehyde	96	Methyldihydroxybenzene isomer
69	2-Furancarboxylic acid, methyl ester	97	Monochloroguaiacol isomer
70	Phenol	98	4-Ethylguaiacol
71	Benzofuran	99	Monochloroguaiacol isomer
72	Trimethylbenzene	100	Methyldihydroxybenzene
73	2-Hydroxy-3-Methyl-2-cyclopenten-1-one	101	Monochloroguaiacol isomer
74	1H-indene	102	4-Vinylguaiacol
75	2-Methylphenol	103	Chloro-4-methylguaiacol
76	4-Methylphenol	104	2-Methylnaphthalene
77	Guaiacol (2-methoxyphenol)	105	Syringol (2,6-dimethoxyphenol)
78	4-Methylguaiacol	106	4-(1-propenyl)guaiacol
79	Dimethylphenol isomer	107	4-Formylguaiacol
80	Dimethylphenol isomer	108	Chloro-1,2-dimethoxybenzene
81	1-Methyl-1-formyl-1H-imidazole	109	Chloro-4-methylguaiacol
82	1-Methyl-1H-indene	110	4-(2-propenyl)guaiacol
83	Ethylphenol isomer	111	Chloro-1,2-dihydroxybenzene
84	Ethylphenol isomer	112	4-Methylsyringol
85	Dimethylphenol isomer	113	4-(2-Propenyl)guaiacol
86	4-Methylguaiacol	114	1,2-Dimethoxy-4-benzaldehyde
87	Monochloroguaiacol isomer	115	Chloro-4-ethylguaiacol
88	Naphthalene	116	4-Propylguaiacol
89	Methylguaiacol	117	Dichloroguaiacol isomer
90	1,2-Dihydroxybenzene	118	4-Ethanalguaiacol
91	Monochloroguaiacol isomer	119	4-C_3H_3-guaiacol
92	4-Vinylphenol	120	4-C_3H_3-guaiacol
93	Dimethylguaiacol		
94	Methoxydihydroxybenzene		

TABLE 3 Continued

Peak no.	Compound	Peak no.	Compound
121	4-(1-Propenone)guaiacol	136	4-Ethyoxyguaiacol
122	Chloro-4-vinylguaiacol isomer	137	4-Formylsyringol
123	4-(3-Hydroxypropenyl) guaiacol	138	3-Methoxy-2-naphthalenol
124	4(2-Propanone)guaiacol	139	Chloro-4-(2-propanone)-guaiacol
125	4-Ethylsyringol	140	4-C_3H_3-syringol
126	4(1-Propanone)guaiacol	141	Biphenol isomer
127	Chloro-4-vinylguaiacol isomer	142	C_3H_3-syringol
128	Tetrachloro compound	143	Trichloroguaiacol isomer
129	Trichloroguaiacol isomer	144	4-(1-Propenyl)syringol
130	4-(Propane-1,2-dione)-guaiacol	145	4-(Prop-1-en-3-al)-guaiacol
131	Trichloroguaiacol isomer	146	4-Ethanalsyringol
132	4-(2-Chloro-2-ethanone) guaiacol	147	4-(2-Propanone)syringol
133	4-Vinylsyringol	148	C_{16}-fatty acid methyl ester
134	4-(1-propenyl)syringol	149	1,2-Diguaiacylethene
135	Chloro-4-propenylguaiacol		

[a] Pulp mill identities are provided in the text.
Source: Adapted from Table 3 of Ref. 11.

times, the m/z of the base peak and molecular ions in each spectrum, and the pulp mill effluents found to contain each compound).

It is interesting to compare the compounds found in the pyrolyzate of the high-molecular-weight DOC (Table 3) and the types of compounds identified by GC/MS analysis of the low-molecular-weight DOC (Table 4). Many of the same compounds are present, leading to the observation that pyrolytic and chemi-

TABLE 4 Low-Molecular-Weight Compounds Identified with GC/MS in Effluent from Pulp Mills A, B, C, and D[a]

Peak no.	Compound	Peak no.	Compound
1	Benzene	25	4-Ethylguaiacol
2	Pentanol	26	4-Formylguaiacol
3	Toluene	27	Syringol (2,6-dimethoxyphenol)
4	Dimethylfuran		
5	2-Furancarboxaldehyde	28	1,7,7-Trimethyl-bicyclo-[2.2.1]heptan-2-ol
6	Dimethylsulfide		
7	2-Methanolfuran	29	C_{11}-Alkane
8	Hexanal	30	1,1,1-Trichloro-2-propanone
9	2-Acetylpyrrole	31	4-Propenylguaiacol
10	5-Methyl-2-furancarboxaldehyde	32	4-Ethanalguaiacol
		33	4-Acetylguaiacol
11	Trichloromethane	34	C_{12}-Alkane
12	C_3-Benzene	35	4-(2-Propanone)guaiacol
13	Guaiacol(2-methoxyphenol)	36	4-(Ethanoic acid)guaiacol
14	2-Acetylfuran	37	C_{13}-Alkane
15	C_4-Benzene	38	Dichloroguaiacol (3 isomers)
16	4-Propylphenol		
17	C_{10}-Alkane	39	1,1,3,3-Tetrachloro-2-propanone
18	Dichlorobenzene		
19	4-Vinylguaiacol	40	Trichlorophenol
20	4,6,6-Trimethyl-bicyclo-[3.3.1]hept-3-en-2-one	41	C_{14}-Alkane
		42	4-(2-Propanone)syringol
21	1,7,7-Trimethyl-bicyclo-[2.2.1]heptan-2-one	43	C_{15}-Alkane
		44	4-(Propanedione)syringol
22	1,7,7-Trimethyl-bicyclo-[2.2.1]heptan-2-ol	45	C_{16}-Alkane
		46	C_{14}-Fatty acid
23	1,3,3-Trimethyl-bicyclo-[2.2.1]heptan-2-one	47	C_{17}-Alkane
		48	C_{18}-Alkane
24	2,6,6-Trimethyl-bicyclo-[3.1.1]heptan-3-one	49	Tetrachloroguaiacol

[a] Pulp mill identities given in the text.
Source: Adapted from Table 4, Ref. 11.

cal degradation can lead to monomeric products with a high structural similarity.

This paper presents a good example of how the two techniques, Py-MS and Py-GC/MS, can be used in a complementary fashion to elucidate various aspects of a very complex analytical problem. Each method has its particular strengths and weaknesses. Py-MS, for instance, does not provide the wealth of detail concerning the various individual compounds that constitute the pyrolyzate of the material being studied, but has the great advantage of speed and simplicity. If a particular feature of the Py-MS spectrum represents a marker for a component of interest, many samples can be examined for this marker in a relatively short time. The authors suggest that sulfur dioxide may be such a marker that could be used to determine lignosulfonates in river water. (For further reading, consult Refs. 12–15.)

B. Analysis of Groundwater Contamination

Voorhees et al. [16] described an environmental application that is not a pyrolysis application, per se, but rather an environmental application of pyrolysis equipment which seemed too clever not to mention here.

Groundwater contamination by organic chemicals is frequently measured by drilling wells into aquifers to obtain samples of the groundwater for subsequent purge-and-trap GC/MS analysis. When used in a reconnaissance mode, this process can be very expensive. These organic chemicals can migrate vertically to the surface as vapors but their concentrations are often too low for direct measurement. The trapping procedure described by these authors, however, provides a viable approach.

Static traps were prepared by coating about 1 cm of the end of 358°C Curie-point pyrolysis wires with finely powdered activated charcoal. After being precleaned by heating to the Curie point under vacuum, the traps were transported to the field test site in sealed culture tubes. At the field test site (a site known for groundwater contamination by tetrachloroethylene, PCE), the static traps were placed in 25–35 cm deep holes, covered with inverted aluminum cans, and the soil replaced in the holes. After three days the traps were removed for analysis.

Analysis was performed by desorbing the organics from the traps with a Curie-point pyrolyzer unit in series with a quadrupole mass spectrometer. The data produced was similar to Py-MS data, although, quite likely, thermal desorption was taking place rather than pyrolysis. The typical mass spectrum obtained from the contaminated areas was dominated by the major ions of PCE. Table 5 shows the various compounds that were identified in spectra obtained from the 25 samples spaced around the contaminated area. Table 6 shows the compounds identified by static trapping from a particular location and by purge-and-trap GC/MS analysis of water from an adjacent well.

This static trapping technique could be an excellent reconnaissance technique for many volatile organic compounds. In this study, the ion counts for the PCE were proportional to the surface fluxes from the plume contamination, falling off sharply at the plume edges. An attractive feature of the technique is the short analysis time—that is, a few minutes per sample. However, as pointed out by the authors, the technique is limited in some instances by the lack of separation of the adsorbed components prior to the measurement by mass spectrometry. Apparently, the overlapping ions from the mixture of components can make some components unidentifiable in the mixed spectrum. This can be a serious limitation in some cases; for instance, in the comparison

Table 5 Compounds Identified by Static Trapping/MS Analysis for a Denver Industrial Site

Benzene	Dichlorobenzene
Toluene	Chloroform
Xylene	Trimethylbenzene
Phenol	Naphthalene
Trichloroethylene	Carbon
Tetrachloroethylene	tetrachloride

Source: Adapted from Table I, Ref. 16.

Table 6 Compounds Identified by Static Trapping/MS and by Purge-and-Trap GC/MS

Compound	Static trapping/ MS above well 23179 (ion counts)	Purge-and-trap GC/ MS Well 23179 (μg/liter)
Dichloroethylene	ND[a]	12
Benzene	ND	22.1
Toluene	1052	<2
Xylene	853	4.3
Trichloroethylene	352	9.5
Tetrachloroethylene	616	75.2
Chloroform	—[b]	3440
Carbon tetrachloride	110	<3

[a] ND = not detected.
[b] Other compounds interfered with m/z 83 and 85; identification could not be made.
Source: Adapted from Table II, Ref. 16.

shown in Table 6, a very high chloroform concentration in the purge-and-trap GC/MS analysis was undetected in the static trapping/MS analysis because of a high background of ions from other substances masking the chloroform ions. The authors suggest that combining this Py-MS technique with Py-GC or Py-GC/MS could be an improvement.

C. Measuring the Organic Carbon in Wastewater

In 1970, Nelson et al. [17] described the use of a newly designed instrument for measuring the total organic load in sewage treatment plant wastewater samples. This instrument employed flash pyrolysis followed by quantitation of the total pyrolyzate with a flame ionization detector (FID). Two tests of the organic load in wastewater, chemical oxygen demand (COD) and biological oxygen demand (BOD), are used to provide data for making adjustments to processes within sewage treatment plants. Because COD tests take two hours and BOD tests can take up to five days, this pyrolysis-FID (Py-FID) technique was proposed as a rapid alternative procedure to provide the needed information.

The design of the pyrolysis reactor was such to convert all of the organic material, either suspended or dissolved, in a 50 μl wastewater sample to volatile organic compounds which would then be measured as a single peak by the FID. A standard mixture of glucose and glutamic acid was pyrolyzed to calibrate the unit. By entering the BOD value of the standard mixture into the final calculation, the results obtained by Py-FID could be directly compared to the BOD values for the same samples.

The relative percent differences between the organic loads measured by Py-FID and BOD for raw sewages, and for primary treated effluents, were 4.1 and 5.6%, respectively. Apparently, the organic material in both these wastewaters was mostly biodegradable, judging from the close agreement between the Py-FID and BOD values. A comparison of Py-FID and BOD values for secondary treated effluents, however, yielded values nearly twice as high by Py-FID. Apparently, the bacterial degradation of organic material during the secondary treatment shifts the total organic material to a higher ratio of nonbiodegradable organic matter. Since Py-FID measures the total organic material and BOD only the biodegradable, this large difference is understandable.

IV. ANALYSIS OF SOIL AND SEDIMENT

A. Thermal Distillation-Pyrolysis-GC

The use of thermal distillation-pyrolysis-GC (TD-Py-GC) and GC/MS to examine marine sediments and suspended particulates for anthropogenic input was reported by Whelan et al. [18,19]. The TD-Py-GC technique differed from typical Py-GC in that the temperature of the sample was raised slowly (e.g., 60°C/min) to 800°C, evolving two distinct peaks of organic material, well separated in time. The first, a low-temperature peak (P_1) in the range 100–200°C, contained unaltered, absorbed volatile organic compounds; and the second pyrolyzate peak (P_2), which emerged in the 350–600°C range, consisted of compounds thermally "cracked" from high-molecular-weight organic materials. The TD-Py-GC apparatus is shown diagrammatically in Figure 7.

As described in the first report [18], a 0.5–50 mg sample of wet (or frozen) sediment was placed in a quartz tube which was then placed in the platinum coil of a pyrolysis probe. The probe was inserted into the cooled interface chamber of the reaction system which was then purged with helium carrier gas. The interface chamber was heated to 250°C and the pyrolysis probe temperature ramp started. About 10% of the helium carrier stream was split off to a thermal conductivity-flame ionization detector series which provided profiles of products being evolved from the sample. The main portion of the helium carrier flow was directed via switchable multiport valves through a pair of Tenax traps for the appropriate time periods to trap the P_1 and P_2 groups of evolved organic materials.

The GC-FID analysis of the Tenax-trapped portions of P_1 and P_2 was accomplished sequentially. By flash desorbing the adsorbed organics from each of the Tenax traps in turn, P_1 and P_2 were sent sequentially to the GC for analysis. One column (a micropacked column—3 m long with a 3% OV-17 stationary phase coated on a 160–180 mesh support material) of a dual-column, dual-FID GC connected to the system was used for this analysis. To improve chromatographic peak shape, the desorbed organic materials were cryofocused with liquid nitrogen prior to being transferred to the GC column. Qualitative analysis of the low-molecular-weight portions of P_1 and P_2, which were not trapped by the Tenax traps (e.g., C_1–C_5 hydrocarbons), was accomplished by sweeping the helium effluent from the interface chamber through the Tenax traps (via switchable multiport valves) to a liquid nitrogen trap on the first coil of the other column of the GC (*n*-octane/Porasil C, 6 ft long by 0.085-in i.d.).

Rather than analysis by GC-FID, the P_1 and P_2 fractions could be analyzed by GC/MS by being trapped on small Tenax traps. These traps would then be desorbed by being placed in the heated inlet of the GC of the GC/MS system, with the desorbed

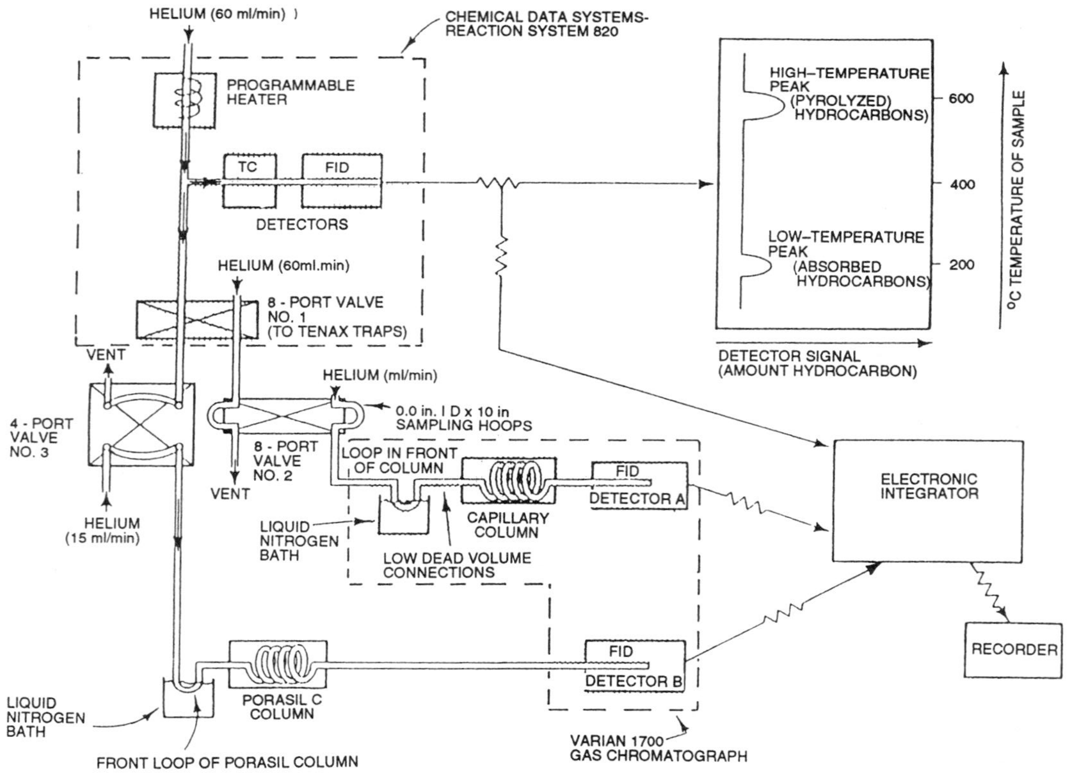

FIGURE 7 Diagram of the thermal distillation-pyrolysis-GC apparatus. (From Ref. 18.)

organics being cryofocused with liquid nitrogen at the head of the GC column.

In the later work [19], the micropacked GC column was replaced with a 50 m long, narrow-bore capillary column with probably more than ten times as many theoretical plates. Comparing the P_2 profile (micropacked GC column) in Figure 8 with the P_2 profile (narrow-bore capillary GC column) in Figure 9 illustrates the tremendous additional amount of detail that can be obtained through the use of high-resolution capillary columns for the separation of pyrolyzates.

Whether they were generated using a packed column or a narrow-bore capillary column, the GC profiles from the TD-Py-GC technique proved quite useful for examining the distribution

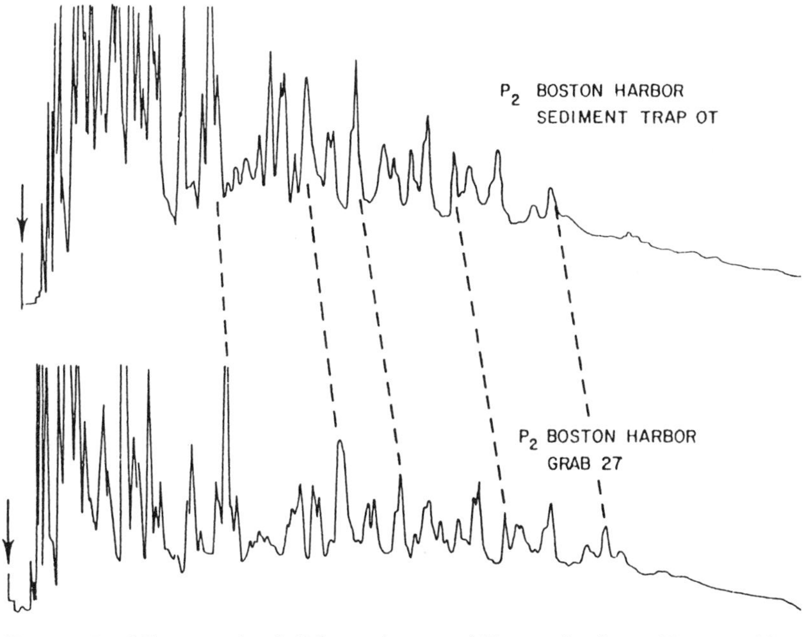

FIGURE 8 Micropacked GC analyses of P_2 peaks from Boston Harbor particles. OT is a midwater sediment trap sample; GRAB 27 is a surface sediment sample. (From Ref. 18.)

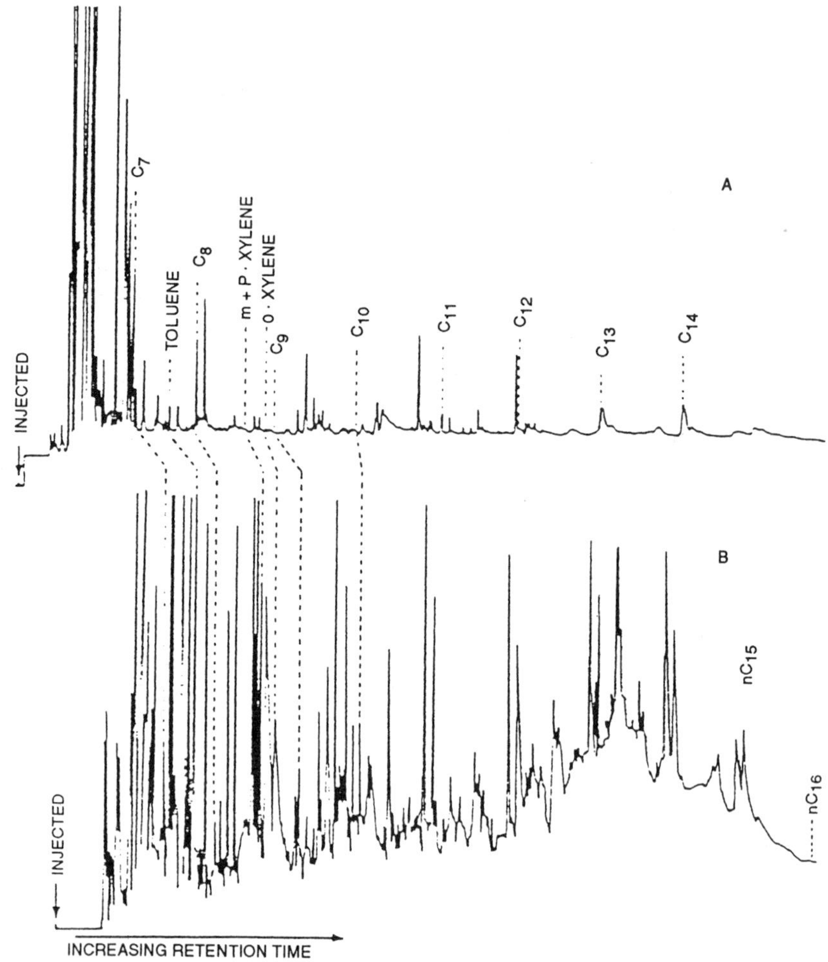

FIGURE 9 P_2 capillary GC patterns: seston (A) and sediment trap particles (B). (From Ref. 19.)

of organic materials in the various "compartments" of the marine ecosytem. Because of the relatively small sample size required, the method was especially useful for examining organic matter in several types of samples, such as marine particles and parts of small marine animals. Analysis of sediments and various particles, including sewage sludge from the Boston Harbor area, provided interesting insights into sediment deposition processes in this area and the fate of the anthropogenic organic matter.

B. Flash Evaporation-Pyrolysis-GC and GC/MS

Several groups of researchers noticed that nonpolymeric organic compounds that are reasonably volatile at elevated temperatures do not fragment upon pyrolysis but simply volatilize [20,21]. They realized that pyrolysis could be a rapid way of extracting these compounds from a complex sample matrix and transferring them directly into the analytical system, thus providing a rapid screening method and avoiding the often lengthy extraction and cleanup procedures. One group [21] referred to this procedure as flash evaporation pyrolysis (EV-Py). This technique differs from TD-Py, already described, in that the sample is elevated to the pyrolysis temperature in milliseconds rather than the slow temperature ramp used in TD-Py.

McMurtrey et al. [20] evaluated EV-Py-GC/MS as a rapid method for screening soils for polychlorinated biphenyls (PCBs). Five milligram samples of dried lake sediment spiked with PCB (Aroclor 1254) were pyrolyzed at 1000°C for 10 s with a probe pyrolyzer and the evolved organics swept onto an 80°C packed GC column (3% OV-1, 6-ft long by 2-mm i.d.), which was then temperature programmed to 275°C. This method proved easily capable of demonstrating the presence of the spiked PCB at the 10 ppm level using the MS in the full scan mode. Although they did not do the experiments, the authors theorized that the detection limit could be reduced considerably by operating the MS in the selected ion monitoring mode. They realized that Py-GC using the much more sensitive electron-capture detector would be a more attractive screening method for PCBs than Py-GC/MS, but their preliminary attempts in that direction did not seem very promising because pyrolysis of sediments released so much electron-capturing material that the peak patterns of the PCBs were obscured and, hence, unrecognizable.

De Leeuw et al. [21] demonstrated the usefulness of EV-Py-GC/MS as a rapid screening technique for polycyclic aromatic hydrocarbons (PAHs), halogenated organics, aliphatic hydrocarbons, heteroaromatics, elemental sulfur, cyanides, and pyrolysis products of synthetic polymers. A 200 μg sample (either polluted soil or sediment) was heated to 510°C with a Curie-point pyrolyzer and the evolved organics swept onto a high-resolution capillary column (CP-SIL 5, 25 m long by 0.22 mm i.d.) at 0°C. The GC temperature was programmed to 275°C at 3°C/min. The MS scanned the range 50–550 amu at 1.5 s/scan.

Figure 10 displays a total ion current profile (TICP) of EV-Py-GC/MS analysis of a polluted soil sample. A central portion of the TICP has been expanded 8× to emphasize the wealth

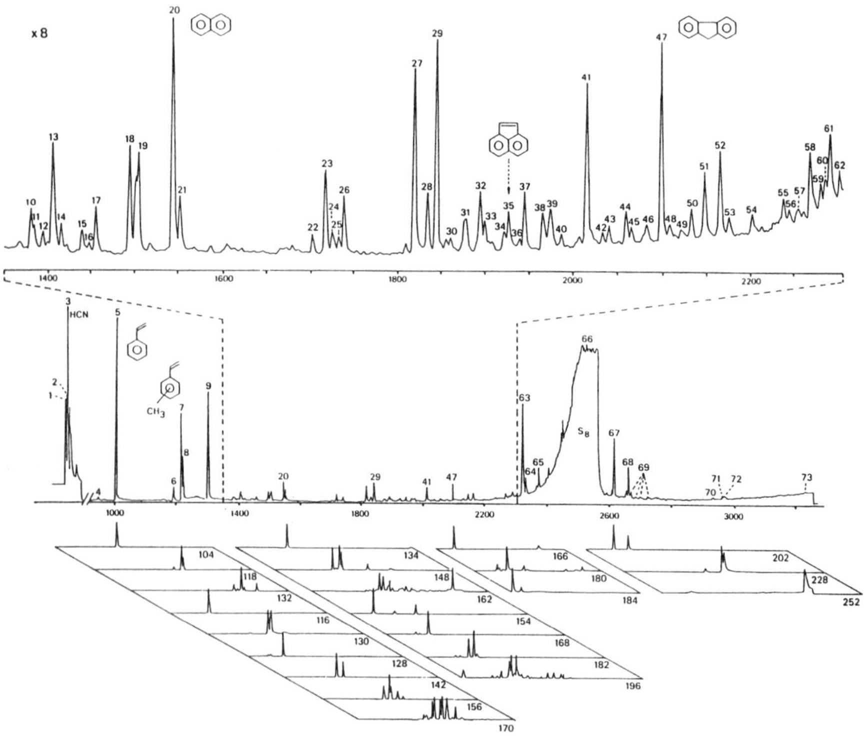

FIGURE 10 TIC of EV-Py-GC/MS analysis of polluted soil sample. The upper trace represents a part of the TIC magnified eight times. The figure details are explained in the text. (From Ref. 21.)

of compounds present (upper trace). The numbers in the mass chromatograms (lower third of the figure) represent the m/z values indicative of the following classes of compounds: 104, 118, 132 (C_0–C_2 styrenes); 116, 130 (C_0–C_1 indenes); 128, 142, 156, 170 (C_0–C_3 naphthalenes); 134, 148, 162 (C_0–C_2 benzo[*b*]thiophenes; 154 (biphenyl and acenaphthene); 168, 182, 196 (C_1–C_2 biphenyls and C_0–C_2 dibenzofurans); 166, 180 (C_0–C_1 fluorenes and 9-fluorenone); 184 (dibenzothiophene); 202 (fluoranthene and pyrene); 228 (benzo[*c*]phenanthrene, benz[*a*]anthracene, chrysene, and triphenylene); and 252 (benzo[*e*]pyrene and benzo[*a*]pyrene). The *x* axes of the mass chromatograms correspond exactly with the appropriate parts of the TIC *x* axis directly above them (e.g., the major peak in the mass chromatogram of m/z 168 corresponds with peak 41 in the total ion current profile). The peak numbers in Figure 10 correspond to the peak numbers of the presumptively identified compounds listed in Table 7. Relating the response for phenanthrene (peak 63) and benzo[*a*]pyrene (peak 73) to the 200 μg applied to the pyrolysis wire, it was estimated that the detection limit for PAHs using this method was about 5 ppm.

The authors pointed out that, because no pretreatment of the samples was carried out, the peaks present in the TICP reflected both components generated by pyrolysis of primary sample compounds ("real pyrolysis products") and components that were present as such in the sample and simply evaporated ("free products"). In this soil sample, for instance, they saw four different groups of anthropogenic compounds: HCN and dicyanogen (pyrolysis products), elemental sulfur (present as such), PAH (mainly unsubstituted, present as such), and styrenes and phenyl ethers (pyrolysis products). They then went on to speculate about the previous industrial activities that took place in the area where the soil sample was collected which might have led to the suite of compounds identified by the EV-Py-GC/MS analysis.

The authors presented an analytical scheme (Figure 11) to use flash evaporation pyrolysis with a variety of detectors in addition to MS to provide greater selectivity and/or sensitivity for common classes of pollutants, but only data for MS were presented in this report.

TABLE 7 Identified Evaporation and Pyrolysis Products of the Soil Sample

Peak no.	Compound	Peak no.	Compound
1	H_2S, CO_2, CO	32	1,3-Dimethylnaphthalene
2	Dicyanogen	33	1,7- and/or 1,6-Dimethylnaphthalene
3	Hydrogen cyanide	34	2,3- and/or 1,4-Dimethylnaphthalene
4	Ethylbenzene	35	Acenaphthalene
5	Styrene	36	1,2-Dimethylnaphthalene
6	α-Methylstyrene	37	(C_1-phenyl)ethyl *tert*-butyl ether (tentative)
7	3-Methylstyrene	38	(C_1-phenyl)ethyl *tert*-butyl ether (tentative)
8	4-Methylstyrene	39	Acenaphthene + 4-methylbiphenyl
9	Indene	40	3-Methylbiphenyl
10	α,3-Dimethylstyrene	41	Dibenzofuran
11	3-Ethylstyrene	42	C_3-Naphthalene
12	α,4-Dimethylstyrene	43	C_3-Naphthalene
13	3,5-Dimethylstyrene	44	C_3-Naphthalene
14	α,2-, or 2,5- or 2,4-Dimethylstyrene	45	C_3-Naphthalene
15	Phenyl ethyl ether	46	C_3-Naphthalene
16	2,3-Dimethylstyrene	47	Fluorene
17	3,4-Dimethylstyrene	48	C_3-Naphthalene + dimethylbiphenyl
18	Methylindene	49	Dimethylbiphenyl
19	Isomeric methylidenes	50	Unknown organic sulfur compound
20	Naphthalene	51	Methylbenzofuran
21	Benzo[*b*]thiophene	52	Methylbenzofuran
22	Methylbenzo[*b*]thiophene	53	Methylbenzofuran
23	2-Methylnaphthalene	54	α-1-Phenyl ethyl ether (tentative)
24	Methylbenzo[*b*]thiophene	55	2-Methylfluorene
25	Methylbenzo[*b*]thiophene	56	1-Methylfluorene
26	1-Methylnaphthalene	57	C_2-Benzofuran
27	1-Phenylethyl *tert*-butyl ether (tentative)	58	9-Fluorenone
28	Biphenyl	59	C_2-Benzofuran
29	Unknown organic sulfur compound		
30	1-Ethylnaphthalene + dimethylbenzo[*b*]thiophene		
31	2,6- and/or 2,7-dimethylnaphthalene		

TABLE 7 Continued

Peak no.	Compound	Peak no.	Compound
60	C_2-Benzofuran	67	Fluoranthene
61	Dibenzothiophene	68	Pyrene
62	C_2-Benzofuran	69	Isomeric naphthobenzofurans
63	Phenanthrene	70	Benzo[*c*]phenanthrene
64	Anthracene	71	Benzo[*a*]anthracene
65	Bis(1-phenylethyl) thioether (tentative)	72	Chrysene + triphenylene
		73	Benzo[*a*]pyrene + benzo[*e*]pyrene
66	Elemental sulfur		

Source: Adapted from Table I, Ref. 21.

C. TD-Py-FID Applied to Marine Sediments

Kennicutt, et al. [22] described the use of Py-FID for assessing the areal distribution of drilling fluids in surficial marine sediments around drilling platforms. Unlike the Py-FID, which utilized flash pyrolysis (described in Sec. III.C), their technique subjected the samples to a 30°C/min temperature ramp from ambient to 700°C, with the evolved organic compounds being swept into an FID for measurement. This technique would better be described as thermal distillation pyrolysis-FID (TD-Py-FID) rather than Py-FID.

Freeze-dried, ground, and sieved surficial sediment samples (upper few centimeters) were heated from ambient temperature to 700°C at 30°C/min in a helium atmosphere, and the evolved organics swept into an FID with the area under the resulting peak (or peaks) integrated and digitized. The resulting FID area units were "normalized" by being divided by the organic carbon content (measured by combustion in an oxygen atmosphere—after carbonate removal—and measurement of the evolved CO_2) of the sample and expressed as nanograms to produce values referred to as pyrolysis ratios (PRs).

The data presented clearly demonstrated that different materials, such as planktonic debris, ancient shales, leaves, and

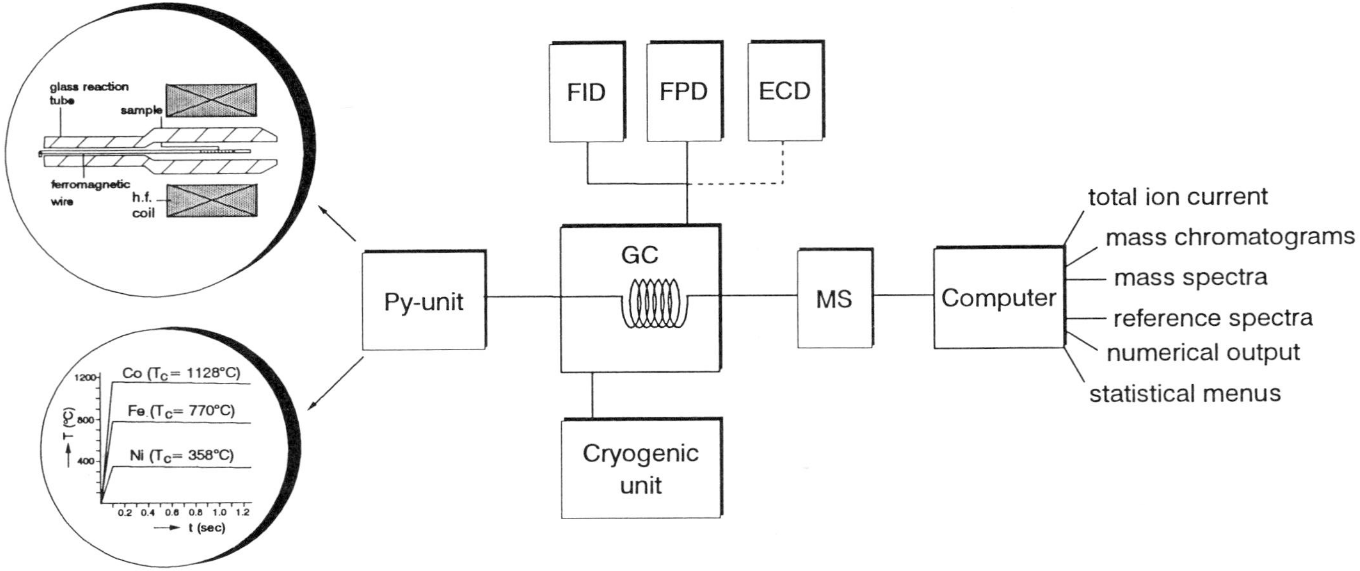

FIGURE 11 Instrumental setup for screening analysis by evaporation/pyrolysis gas chromatography. (From Ref. 21.)

wood, show widely different PRs. Just what the ratio between the amount of organic carbon as measured by TD-Py-FID and the amount of organic carbon as measured by an oxidative technique actually represented was not at all clear. Nevertheless, the isopleths showing PRs as a function of the distance from drilling platforms indicated that PRs could be used to delineate the areal extent of materials added to surficial sediments by drilling operations.

V. OTHER ENVIRONMENTAL APPLICATIONS

A. Ancient Limestone: Examination by Py-GC and Py-GC/MS

Py-GC and Py-GC/MS analysis played an interesting role in the characterization of a green layer of material 1 mm below the surface of the limestone on the south-facing exterior wall of a fourteenth century building in Tongeren, Belgium [23]. This green layer consisted of a discrete band (about 0.2–0.5 mm thickness) beneath the surface of the limestone, following the surface contours of the stone. Various bits of evidence suggested that this green layer comprised a cryptoendolithic ecological niche being filled by a moss, and two species of cyanobacteria. Presumably, these photosynthetic organisms gained some protection from the covering limestone surface, but were able to use the moisture and light that passed through it.

A small portion of the green layer was ground to a powder, applied to the wire of a Curie-point pyrolysis unit, and heated within 0.1 s to 610°C for 10 s. The resulting pyrolysis products were separated on a 25 m long by 0.32-mm i.d. capillary GC column temperature programmed from 0°C to 310°C at 3°C/min. The mixture of compounds volatilized from the sample and/or generated during the pyrolysis was resolved into more than 150 separate peaks by capillary GC, yielding a pyrogram (not shown here) not greatly dissimilar in appearance from some of those shown earlier in this chapter. Table 8, however, lists the compounds presumptively identified from this mixture by Py-GC/MS. The list of compounds, deriving from the pyrolysis of living

TABLE 8 Compounds Identified from Pyrolysis of a Green Layer in Limestone from a Building in Tongeren, Belgium

Peak no.	Compound	Peak no.	Compound
1	Sulfur dioxide	36	C_3-Alkylfuran
2	1,3-Butadiene + but-1-ene	37	C_3-Alkylfuran
3	Methanethiol	38	Benzaldehyde
4	*trans*-But-2-ene	39	C_3-Alkylbenzene
5	Acetone	40	5-Methyl-2-furaldehyde
6	Furan	41	3-Methyl-2-cyclopenten-1-one
7	Cyclopentadiene	42	C_3-Alkylbenzene
8	2-Methylpropanal	43	C_3-Alkylbenzene
9	2,3-Butanedione	44	α-Methylstryene
10	2-Butanone	45	Benzofuran
11	2-Methylfuran	46	C_3-Alkylbenzene
12	3-Methylfuran	47	n-Dec-1-ene
13	2-Butenal	48	Phenol
14	3-Methylbutanal	49	n-Decane
15	Benzene	50	C_3-Alkylbenzene
16	2-Methylbutanal	51	$C_{3:1}$-Alkylbenzene
17	Pentane-3,4-dione	52	C_3-Alkylbenzene
18	Cyclopentanone	53	Indene + 2-hydroxy-3-methyl-2-cyclopenten-1-one
19	Pentane-2,3-dione		
20	2,5-Dimethylfuran	54	Cyanopyridine
21	2,4-Dimethylfuran	55	C_4-Alkylbenzene
22	Vinylfuran	56	C_4-Alkylbenzene + $C_{4:1}$-alkylbenzene
23	*N*-Methylpyrrole + pyridine		
24	Toluene	57	Tolualdehyde
25	3-Furaldehyde	58	*m*/*p*-Cresol
26	2-Furaldehyde	59	C_4-Alkylbenzene
27	2-Methyl-2,3-dihydrofuran-3-one	60	*p*-Cresol
28	Ethylbenzene	61	n-Undec-1-ene
29	*m*/*p*-Xylene	62	n-Undecane
30	1-Acetoxypropan-2-one	63	Benzylcyanide + C_4-alkylbenzene
31	Styrene		
32	*o*-Xylene	64	C_4-Alkylfuran
33	2-Methyl-2-cyclopenten-1-one	65	Methylindene
34	C_3-Alkylfuran	66	Methylindene
35	C_3-Alkylfuran	67	Ethylphenol

TABLE 8 Continued

Peak no.	Compound	Peak no.	Compound
68	C_2-Alkylphenol	100	n-Tetradec-1-ene
69	C_4-Alkylbenzene	101	Ileu-ileu sequence
70	Naphthalene	102	n-Tetracecane
71	C_2-Alkylphenol	103	Ileu-leu sequence
72	n-Dodec-1-ene	104	Ileu-ileu sequence
73	n-Dodecane	105	Ileu-leu sequence + leu-ileu sequence
74	Methylbenzylcyanide		
75	Val-val sequence + C_3-alkylphenol	106	Leu-leu sequence
		107	Leu-ileu sequence
76	C_3-Alkylphenol + C_2-Alkylindene	108	Leu-leu sequence
		109	C_8-Alkylbenzene
77	Val-val sequence	110	Levoglucosane
78	C_2-Alkylindene	111	C_x-Methylketone
79	Quinoline	112	1,2-Diphenylethane
80	Methylnaphthalene	113	n-Pentadec-1-ene
81	$C_{5:1}$-Alkylbenzene	114	n-Pentadecane
82	C_3-Alkylphenol	115	Isoprenoid hydrocarbon
83	$C_{5:1}$-Alkylbenzene + methylnaphthalene	116	Methyl pentadecane
		117	Dodecanoic acid
84	$C_{5:1}$-Alkylbenzene	118	n-Hexadec-1-ene
85	n-Tridec-1-ene	119	n-Hexadecane
86	n-Tridecane	120	1,3-Diphenylpropane
87	Val-ileu sequence	121	Methylhexadecane
88	Ileu-val sequence	122	Biphenyl
89	Val-leu sequence + $C_{5:1}$-alkylbenzene	123	9,10-Dihydroanthracene
		124	1,3-Diphenyl-3-methyl-1-propene
90	Val-ileu sequence		
91	Val-leu sequence	125	n-Heptadec-1-ene
92	Ileu-val sequence	126	n-Heptadecane
93	Leu-val sequence	127	Prist-1-ene
94	$C_{6:1}$-Alkylbenzene	128	7-Methylheptadecane
95	Leu-val sequence	129	Tetradecanoic acid
96	$C_{6:1}$-Alkylbenzene	130	Pristane
97	Methylindole	131	n-Octadec-1-ene
98	$C_{6:1}$-Alkylbenzene	132	n-Octadecane
99	C_{14}-Aldehyde	133	Dialkyl phthalate

TABLE 8 Continued

Peak no.	Compound	Peak no.	Compound
134	Phytadiene	139	n-Nonadecane
135	Phyt-1-ene	140	Dialkyl phthalate
136	Pentadecanoic acid	141	Hexadecanoic acid
137	Phytadiene	142	n-Eicosane
138	Phytadiene		

Source: Adapted from Table 3 of Ref. 23.

material, is quite unlike the lists shown in the earlier tables in this chapter.

The substances identified support the assertion that the green layer contained biogenic material, because the vast majority of the evaporation/pyrolysis products obtained were related to polysaccharides, proteins, and lipids, substances that are the main cellular components of organisms such as cyanobacteria and mosses.

B. Digested and Undigested Pollens: Discrimination by Py-MS

The so-called yellow rain in southeast Asia in the early 1980s led to this application of Py-MS. A high pollen count was reported for some of the yellow samples and it was suggested that pollens might be used as a support or carrier for the distribution of chemical agents. As part of the investigations, it became important to distinguish between pollen and bee feces (undigested and digested pollen, respectively). Because Py-MS had been used extensively for the characterization of biological and other polymeric materials, and combined with pattern recognition techniques had been successful for providing compositional information for a number of different materials, a study was undertaken to evaluate whether Py-MS analysis coupled with pattern recognition data analysis procedures could distinguish between bee feces and pollen samples [24].

The U.S. Army Chemical Research and Development Center provided 11 southeast Asian samples for this study as unknowns, that is, with no sample history or identification known to the analysts at the time of analysis. The analytical task was to determine whether these samples were pollen, bee feces, or some other unrelated material. For purposes of comparison, a set of known materials was constructed consisting of bee feces, pollens, beeswax and honey—20 samples of different origins obtained from various sources.

The samples were applied to Curie-point wires as methanol suspensions of about 10 mg/ml, and pyrolyzed at 610°C with a rise time of 100 ms. The pyrolyzer was interfaced to a quadrupole mass spectrometer set to accumulate low-energy (15 eV) electron-impact ionization spectra across the scan range of 45–245 amu with a scan speed of 1200 amu/s. For the subsequent data analysis, 30 spectra were summed together to produce a Py-MS spectrum for each sample. The 31 samples were run three times, as three sets of 31, with each set arranged in random order before the beginning of analysis.

The four Py-MS spectra—for pollen, bee feces, and two of the unknowns—presented as a figure in this report (but not shown in this chapter), clearly showed distinct differences, but the authors found that the complexity present in the 93 spectra made visual classification of the samples impossible. A discussion of the data analysis techniques used for the classification of these samples is beyond the scope of this chapter, but a brief summary of the results can be made. (Those readers with interest in statistical methods for analysis of data generated by analytical pyrolysis-mass spectrometry should find this paper interesting and may also want to read more recent work [25,26] in this area.)

Unsupervised learning (that is, factor analysis and nonlinear mapping) showed that the Py-MS data contained enough distinguishing chemical information that digested and undigested pollens could be differentiated. Also, samples that were quite different from pollen, such as waxes and honeys, were identified as nonpollen samples through unsupervised learning techniques. Discriminant analysis was the most successful supervised statistical procedure used. Employing blind digested and undigested pollen standards, a classification success rate of 95% was achieved. Of the 11 southeast Asian samples submitted blind, the four of known composition were correctly classed as either digested or undigested pollen. Of the five yellow rain samples in the set, three were classified as digested pollens (bee feces) and two as undigested pollen.

C. Spruce Needles: Examination by Py-FIMS

Field ionization-mass spectrometry was initially developed for molecular-weight determinations and mixture analyses. This soft ionization technique is particularly well suited for the mass spectrometric examination of the extremely complex mixture of compounds resulting from the pyrolysis of macromolecules because predominantly characteristic, high-mass molecular ions are produced [27].

As part of a research project on forest damage, spruce needles were examined by pyrolysis-field ionization mass spectrometry (Py-FIMS) using high-resolution mass spectrometry, time-resolved high-resolution mass spectrometry, and Curie-point Py-GC/MS [28]. Subsequent chemometric analysis of the data using pattern recognition techniques led to the conclusion that, in the geographical region studied, the impact of acids and water stress were the major causes of the observed tree damage [1].

Presentation and discussion of the data obtained from the elegant analytical mass spectrometric techniques utilized for these studies are beyond the scope of this chapter. However, for those with access to the necessary instrumentation, Py-FIMS appears to offer unique insights into difficult environmental problems.

D. Pyrolysis for Thermal Recycling of Polymers

Thermal recycling of polymers includes a range of activities from thermally recycling used plastic items into new plastic items by a variety of processes (such as extrusion, injection molding, compression molding, and foam forming) to burning the used plastic items for fuel [29]. Residing somewhere between these two extremes is recycling using pyrolysis to decompose the plastic polymers into usable petrochemicals.

Table 9, taken from a recent review of plastic recycling using pyrolysis [30], compiles the various pyrolysis processes for decomposition of plastic wastes, rubber, and scrap tires,

TABLE 9 Processes for Pyrolysis of Plastic Wastes

Name of process	Process	Products	Size, site
DBA (BKM)	Rotary kiln, indirect heated, 450–500°C	Energy	6 t/h, Günzburg, Germany
Ebara	Two fluidized beds	Energy	4 t/h, Yokohama, Japan
Kobe Steel	Rotary kiln, indirect heated, 500–700°C	Oil, gas, energy	1 t/h, Kobe, Japan
Tosco II	Rotary kiln, indirect heated, 500–550°C	Energy, carbon black	Pilot plant, Golden, CO, U.S.
KWU	Rotary kiln, indirect heated, 450–500°C	Energy	3 t/h, Ulm, Germany
Dr. Otto Noell	Rotary kiln, indirect heated, 650–700°C	Oil, gas	6 t/h, Salzgitter, Germany
Mitsubishi Heavy Industries	Melting vessel, indirected heated, 550°C	Oil	100 kg/h, Japan
Energas	Rotary kiln, indirect heated, 650–750°C	Energy	150 kg/h, Gladbeck, Germany
Tsukishama Kikai	Two fluidized beds, one oxidizing	Energy	3 × 6.25 t/h, Funabashi, Japan
Hamburg	Fluidized bed, indirect heated, 600–900°C	Oil, gas, carbon	20–60 kg/h, University of Hamburg; 0.5 t/h Ebenhausen; 1 t/h Grimma; all in Germany

Source: Adapted from Table 1, Ref. 30.

which have reached the stage of industrial testing. The fluidized bed reactor has excellent heat and mass transfer, as well as constant temperature throughout, special advantages that yield better control during the pyrolysis process. Typical yields are high. Using waste tires, for instance, up to 50% of the input material may be retrieved in the form of a liquid which is 95% aromatic petroleum hydrocarbon material. Table 10 lists products from the pyrolysis of various plastic wastes in a fluidized bed reactor.

At the present time, operating a pyrolysis recycling facility profitably requires an appropriate balance between supply of waste plastic materials, design of the pyrolysis facility, and price of the end products. Profitability largely depends upon the price of petroleum. As petroleum prices continue to rise, as well as the costs of landfill disposal of waste plastics, one would expect pyrolytic recycling to become more attractive.

E. Analysis of Intractable Environmental Contamination

In the United States, most of the field investigation, remediation, and monitoring of environmental chemical contamination utilizes specific analytical methodologies, primarily those set forth by the U.S. Environmental Protection Agency. These methods are directed almost entirely to measuring the presence and amounts of the several hundred chemical elements and compounds on the various US EPA lists of chemicals of concern. Most of the methods for organic compounds involve a primary extraction followed by a concentration step, or dissolution followed by dilution, with subsequent separation of the organic mixtures by gas chromatography prior to measurement with any one of a variety of detectors.

These methods not only fail to identify or measure most synthetic polymer materials, but are sometimes greatly hampered by the presence of these materials within the sample matrix. Probably every environmental laboratory has suffered from this problem although it may have gone unrecognized. The experience of having a single sample suddenly "trash" an analytical system, rendering it incapable of meeting analytical quality as-

TABLE 10 Pyrolysis of Plastic Waste in a Fluidized Bed: Different Feeds and Their Products

Feed[a]	Pyrolysis (°C)	Gas (wt %)	Oil (wt %)	Residue (wt %)	Others (wt %)
Polyethylene	760	55.8	42.4	1.8	
Polypropylene	740	49.6	48.8	1.6	
Polystyrene	580	9.9	24.6	0.6	64.9 Styrene
Mixture PE/PP/PS	750	52.0	46.6	1.4	
Polyester	768	50.8	40.0	7.1	2.1 H_2O
Polyurethane	760	37.9	56.3	0.5	5.0 H_2O; 0.3 HCN
Polyamide	760	39.2	56.8	0.6	3.4 HCN
Polycarbonate	710	26.5	46.4	24.6	2.5 H_2O
Poly(MMA)	450	1.25	1.4	0.15	97.2 MMA
PVC	740	6.8	28.1	8.8	56.3 HCl
Poly(tetrafluoroethylene)	760	89.3	10.4	0.3	
Medical syringes	720	56.3	36.4	5.8	1.5 Steel
Plastic from household waste separation	787	43.6	26.4	25.4	4.6 H_2O
Plastic from car shredding	733	29.9	26.7	27.6	14.0 Metals, 1.8 H_2O
EPDM rubber	700	32.3	19.2	47.5	1.0 H_2O
SB rubber	740	25.1	31.9	42.8	0.2 H_2S
Scrap tires	700	22.4	27.1	39.0	11.5 Steel
Lignin	500	3.4	29.9	49.3	17.4 H_2O
Cellulose (wood)	700	47.1	23.0	18.6	11.3 H_2O
Sewage sludge	600	34.3	27.7	33.2	4.8 H_2O

[a] Abbreviations: PE: polyethylene; PP: polypropylene; PS: polystyrene; MMA: methyl methacrylate; PVC: poly(vinyl chloride); EPDM: ethene-propene-diene monomers; SB: styrene-butadiene
Source: Adapted from Table 2, Ref. 30.

surance criteria without a major system overhaul, unfortunately, is not uncommon. Many of these occurrences are probably due to synthetic polymers that do not pass through the analytical system as the volatile and semivolatile analytes do, but rather "coat" the system and change its analytical performance characteristics.

Py-GC and Py-GC/MS could find useful places in the routine environmental analytical laboratory to help with the nonroutine samples. Abandoned drums of "goo," or the collected containers of household waste chemicals that no longer bear labels possibly could be handled with some sort of thermal vaporization/pyrolysis analytical scheme.

Using analytical pyrolysis for such samples would be just one component of a recently proposed "systems approach to multimedia analysis" [31], a multifaceted approach to the examination of environmental samples centered upon the use of a purge and trap/headspace/pyrolysis/gas chromatography system utilizing multiple detection systems. Although still in the formative stage, a critical facet of this approach appears to be innovations in analysis of the analytical data.

REFERENCES

1. N. Simmleit and H.-R. Schulten, *J. Anal. Appl. Pyrol.*, *15*: 3 (1989).
2. R. Tsao and K. J. Voorhees, *Anal. Chem.*, *56*: 368 (1984).
3. K. J. Voorhees and R. Tsao, *Anal. Chem.*, *57*: 1630 (1985).
4. A. Alajbeg, *J. Anal. Appl. Pyrol.*, *12*: 275 (1987).
5. A. Alajbeg, *J. Anal. Appl. Pyrol.*, *9*: 255 (1986).
6. W. Klusmeier, P. Vogler, K.-H. Ohrback, H. Weber, and A. Kettrup, *J. Anal. Appl. Pyrol.*, *14*: 25 (1988).
7. K. J. Voorhees, W. D. Schulz, L. A. Currie, and G. A. Klouda, *J. Anal. Appl. Pyrol.*, *14*: 83 (1988).
8. A. Yasuhara and M. Morita, *Environ. Sci. Technol.*, *22*: 646 (1988).
9. A. B. J. Oudhuis, P. De Wit, P. J. J. Tromp, and J. A. Moulijn, *J. Anal. Appl. Pyrol.*, *20*: 321 (1991).
10. R. C. Kistler, F. Widmer, and P. H. Brunner, *Environ. Sci. Technol.*, *21*: 704 (1987).
11. W. M. G. M. van Loon, J. J. Boon, and B. de Groot, *J. Anal. Appl. Pyrol.*, *20*: 275 (1991).
12. D. van de Meent, J. W. de Leeuw, and P. A. Schenck, *J. Anal. Appl. Pyrol.*, *2*: 249 (1980).

13. W. M. G. M. van Loon and J. J. Boon, *Anal. Chem.*, *65*: 1728 (1993).
14. E. R. E. van der Hage, M. M. Mulder, and J. J. Boon, *J. Anal. Appl. Pyrol.*, *25*: 149 (1993).
15. M. Kleen and G. Lindblad, *J. Anal. Appl. Pyrol.*, *25*: 209 (1993).
16. K. J. Voorhees, J. C. Hickey, and R. W. Klusman, *Anal. Chem.*, *56*: 2604 (1984).
17. K. H. Nelson, I. Lysyj, and J. Nagano, *Water Sewage Works*, *117*: 14 (1970).
18. J. K. Whelan, J. M. Hunt, and A. Y. Huc, *J. Anal. Appl. Pyrol.*, *2*: 79 (1980).
19. J. K. Whelan, M. G. Fitzgerald, and M. Tarafa, *Environ. Sci. Technol.*, *17*: 292 (1983).
20. K. D. McMurtrey, N. J. Wildman, and H. Tai, *Bull. Environ. Contam. Toxicol.*, *31*: 734 (1983).
21. J. W. de Leeuw, E. W. B. de Leer, J. S. Sinninghe Damsté, and P. J. W. Schuyl, *Anal. Chem.*, *58*: 1852 (1986).
22. M. C. Kennicutt II, W. L. Keeney-Kennicutt, B. J. Bresley, and F. Fenner, *Environ. Geol.*, *4*: 239 (1983).
23. C. Saiz-Jimenez, J. Garcia-Rowe, M. A. Garcia Del Cura, J. J. Ortega-Calvo, E. Roekens, and R. Van Grieken, *Sci. Total Environ.*, *94*: 209 (1990).
24. S. J. DeLuca, K. J. Voorhees, and E. W. Sarver, *Anal. Chem.*, *58*: 2439 (1986).
25. R. S. Sahota and S. L. Morgan, *Anal. Chem.*, *65*: 70 (1993).
26. R. Goodacre, A. N. Edmonds, and D. B. Kell, *J. Anal. Appl. Pyrol.*, *26*: 93 (1993).
27. H.-R. Schulten, N. Simmleit, and R. Müller, *Anal. Chem.*, *59*: 2903 (1987).
28. H.-R. Schulten, N. Simmleit, and R. Müller, *Anal. Chem.*, *61*: 221 (1989).
29. W. Kaminsky, *J. Anal. Appl. Pyrol.*, *8*: 439 (1985).
30. W. Kaminsky, *Emerging Technologies in Plastics Recycling* (G. E. Andrews and P. M. Subramanian, eds.), ACS Symposium Series 53, American Chemical Society, Washington, D.C. (1992), p. 60.
31. J. E. Bumgarner, pers. commun. (1993).

8

Examination of Forensic Evidence

JOHN M. CHALLINOR
Chemistry Centre (W.A.), East Perth, Western Australia

I. INTRODUCTION

A. The Analysis Problem

The chemical examination of forensic evidence from a crime scene or criminal has different requirements to many other chemical analyses. The quantity of material for examination is often limited to minute traces found at the scene, hence high sensitivity is important. The material under scrutiny must be characterized as comprehensively as possible to ensure maximum discrimination from other material in the same class. Forensic laboratories are multiinstrument facilities required to deal with many types of physical evidence found at a crime scene and therefore the routine methods used should preferably employ relatively inexpensive instrumentation. In order to protect the integrity, samples should preferably be analyzed "as received" and any work-up minimized. The method should preferably not be labor intensive. Pyrolysis gas chromatography (Py-GC) has proved to be an effective means of satisfying these requirements in many forensic science laboratories [1–3].

B. Types of Evidence

Py-GC is an appropriate characterization technique for such crime scene material as paint from breaking-and-entering crimes and hit-and-run traffic accidents, and adhesives from insulation tapes and improvised explosive devices. Rubbers in automobile tires, foams in carpets and clothing, fibers from garments, plastics from household goods, and automobile parts are examples of commodities that may be identified. Other applications are as diverse as the identification of blood stains [4], propellants in ammunition [5], human hair [6], or chewing gum [7]. The numerous uses of analytical pyrolysis, including forensic applications, have been reviewed in a selected bibliography [8]. The technique is essential for the effective operation of any crime lab required to examine these and other organic materials found as evidence.

C. Development of Py-GC

Py-GC was adopted by forensic scientists in the 1970s [9]. Packed Carbowax phase GC columns were accepted as the standard in many forensic science laboratories [10]. Reproducible pyrograms enabled a reliable data base to be established. However, packed columns had a limited life span and were unsatisfactory for chromatographing very polar and higher-molecular-weight compounds. These compounds are often diagnostic for many polymers of forensic interest. As a comparative technique, Py-GC was excellent [2,11–13]. The limitations of packed column GC prompted the use of capillary GC columns, particularly high-resolution vitreous (fused) silica types. A simple system for interfacing a Curie-point pyrolyzer to a gas chromatograph equipped with a medium polarity phase capillary column and forensic applications were reported and comparison was made with the results obtained from a packed column [14]. Pyrolysis *capillary* column GC has become the standard adopted in most forensic science laboratories and discussion of the applications in this chapter will be restricted to this technique.

D. Pyrolysis Derivatization: Modifications to the Pyrolysis Process

In order to obtain greater information about the composition of polymers and macromolecular material, modifications to pyrolysis have been developed. These have included on-line pyrolysis hydrogenation of polyolefins using hydrogen carrier gas and a palladium catalyst [15]. Polyacetals have been pyrolyzed in the presence of a cobalt catalyst to produce cyclic derivatives, which afford greater information about the structure of the polymer [16]. Another recent development in pyrolysis techniques has involved high-temperature derivation reactions that gave more information about the structure of polymers [17]. In the pyrolysis derivatization process, tetraalkylammonium hydroxide, when reacted with polymers having hydrolyzable groups, gives alkyl derivatives that reflect the composition of the polymer. Resins used in drying oil-modified alkyd enamels and saturated and unsaturated polyesters are particularly suitable for this procedure. Pyrolysis profiles are generally simplified and easier to interpret. The precursors of the polymer or other resin components may be identified providing data about the polymer not obtainable by conventional Py-GC.

II. PAINT

Paint, because of its variability and complexity, can be important evidence in a forensic investigation [18]. The resin or binder varies widely within a class. Py-GC is an effective method for identifying and differentiating the organic binder of paint. In some cases, additives may be detected and identified. The Py-GC identification of organic pigments in a resin matrix is a challenging prospect for the future.

The Py-GC examination of paint, in the context of an overall forensic analysis of this type of evidence, has been described [19].

A. Automotive Paint

Motor vehicle traffic-related crime can include hit-and-run, willful damage, and homicide-related incidents. Automotive paint binder types can be identified on microgram-sized samples of topcoat [20–22]. Some examples of pyrograms of acrylic lacquer, acrylic enamel, and alkyd enamel types, typically found in original paint systems, are shown in Figure 1.

The variability in pyrolysis profiles of the different classes is self-evident. The interpretation of the composition revealed is as follows: The acrylic lacquer (General Motors) is a methyl methacrylate, butyl methacrylate copolymer plasticized with butylbenzyl phthalate. The acrylic enamel (Ford) is a styrene-ethylhexyl acrylate-methyl methacrylate terpolymer. The alkyd enamel (Honda) pyrolysis profile indicates that the paint resin is an orthophthalic alkyd containing a butylated-amino resin crosslinking component.

1. Repainted Vehicles

Acrylic lacquer formulations are commonly used for refinishing post-accident damage. These are often based on a methyl methacrylate monomer but are plasticized by the incorporation of monomers that produce "softer" polymers, such as butyl methacrylate and/or external plasticizers, such as the phthalates. Pyrograms of three different refinishing acrylic lacquers are shown in Figure 2.

2. Alkyd-Based Enamels

Alkyd enamels occurring as original baked enamels or spraying enamels may be identified by a modification of the Py-GC technique termed simultaneous pyrolysis methylation (SPM) [23]. The term was subsequently modified in later reports to thermally assisted hydrolysis and methylation (THM) to avoid possible

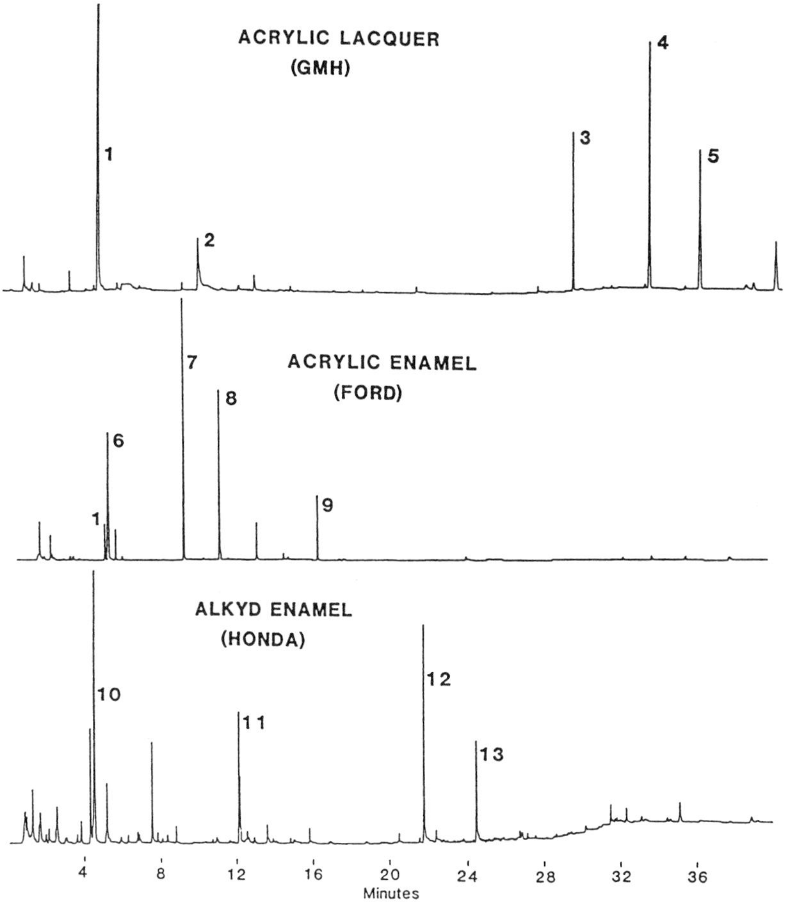

FIGURE 1 Pyrograms of an acrylic lacquer, acrylic enamel, and an alkyd enamel. The identities of the major peaks are 1 = methyl methacrylate, 2 = methacrylic acid, 3 = dibutyl phthalate, 4 = butyl cyclohexylphthalate, 5 = butylbenzyl phthalate, 6 = butanol, 7 = styrene, 8 = butyl methacrylate, 9 = 2-ethyl acrylate, 10 = isobutanol, 11 = vinyltoluene, 12 = phthalic anhydride, 13 = phthalimide.

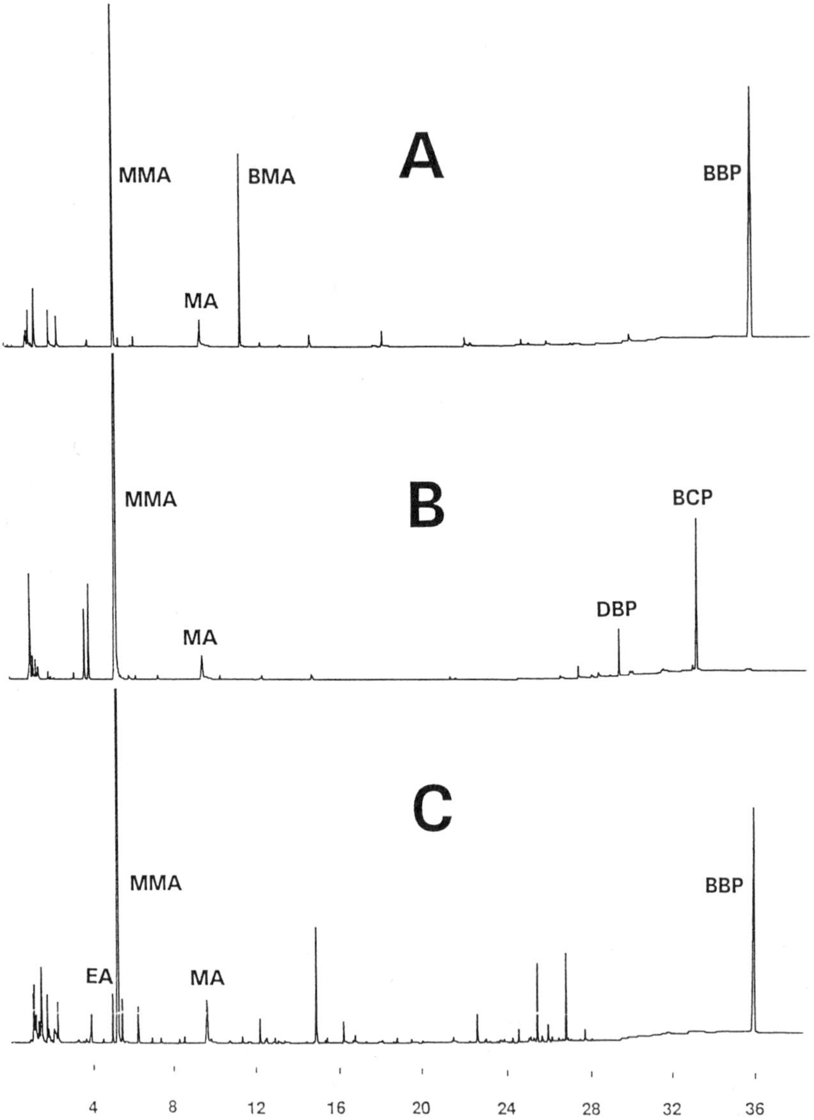

FIGURE 2 Pyrograms of three acrylic lacquers of different composition. (A) Methyl methacrylate (MMA), butyl methacrylate (BMA), methacrylic acid (MA) terpolymer plasticized with butylbenzyl phthalate (BBP). (B) MMA-MA copolymer plasticized with a mixture of dibu-

misunderstanding of the reaction mechanism. This simple but useful procedure involves a high-temperature hydrolysis and methylation reaction using tetramethylammonium hydroxide (TMAH). These polyesters are converted to methyl derivatives of the polyol, polybasic acid, and drying oil. More information than that obtained by Py-GC can be gained from the SPM-GC analysis of alkyd resins. For example, Figure 3 shows the pyro-

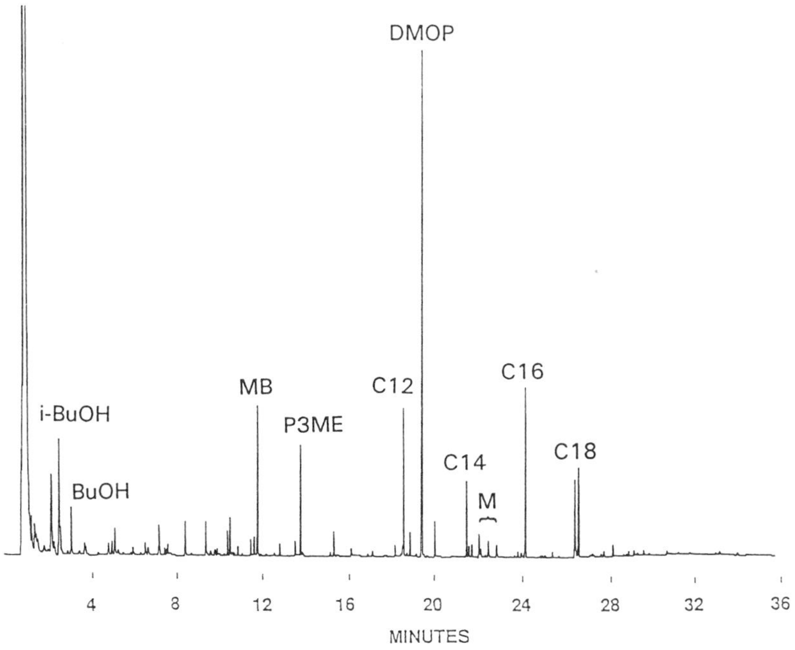

Figure 3 Simultaneous pyrolysis methylation profile of baked alkyd enamel paint smear. i-BuOH = isobutanol, BuOH = butanol, MB = methyl benzoate, P3ME = pentaerythritol trimethyl ether, DMOP = dimethyl orthophthalate, M = triazines from melamine, and C12, C14, C16 and C18 = respective fatty acid methyl esters.

---

tyl phthalate (DBP) and benzene cyclohexylphthalate (BCP). (C) MMA-EA (ethyl acrylate) long-chain methacrylate-type polymer plasticized with BBP.

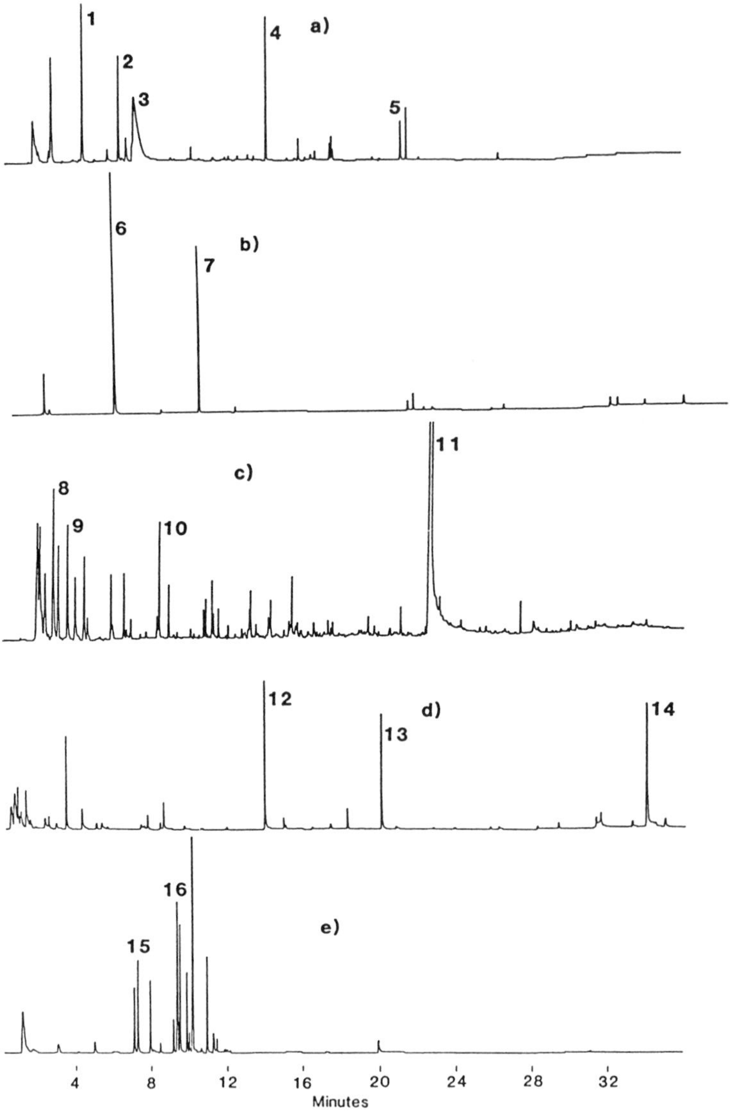

FIGURE 4 Pyrograms of (a) vinyl acetate, (b) acrylic, (c) alkyd enamel, (d) epoxy, and (e) chlorinated rubber-type architectural paints. The identities of the major peaks are as follows: 1 = benzene, 2 = isooctene, 3 = acetic acid, 4 = 2-ethylhexyl acrylate, 5 = 2,2,4-trimethyl-

gram of a blue paint smear found on a crowbar allegedly used to damage a blue automobile. The paint is a baked alkyd enamel.

The interpretation from the pyrogram is that the paint is a pentaerythritol-orthophthalic acid-baked alkyd enamel having a coconut nondrying oil crosslinked with a butylated melamine formaldehyde resin.

B. Architectural Paint

Break-in tools used to gain access to premises may well carry traces of paint which have abraded from painted surfaces at the point of entry. Commonly encountered household paint types include polyvinyl acetate (PVA), acrylics, alkyd enamels, epoxies, and chlorinated rubbers. Py-GC distinguishes these different classes. (See Fig. 4.)

Further, their polymer class and composition may be determined. For example, the PVA type is a vinyl acetate-2-ethylhexyl acrylate copolymer binder. The acrylic is a methyl methacrylate-butyl acrylate copolymer. The alkyd enamel is a linoleic-rich pentaerythritol orthophthalic alkyd and the epoxy is a bisphenol-A type. In some cases, paint additives may be identified. The presence of a latex-coalescing agent, trimethyl pentanediol monoisobutyrate (Texanol), may also be detected (peak 5). Studies of the differentiation of alkyd enamels by conventional Py-GC have been carried out [24,25]. As discussed previously, more structural information about alkyd enamels may be obtained by the SPM technique [23]. Polybasic acid, polyhydric alcohol, drying oil composition, oil length, degree of cure, and rosin modification can be determined. Discrimination of alkyd enamels is improved. For example, the alkyd enamel whose pyrogram is shown in Figure 4 gives the following SPM profile, as shown in Figure 5.

1,3-pentanediol monobutyrate, 6 = methyl methacrylate, 7 = butyl methacrylate, 8 = acrolein, 9 = methacrolein, 10 = hexanal, 11 = phthalic anhydride, 12 = phenol, 13 = isopropenylphenol, 14 = bisphenol A, 15 = xylenes, 16 = trimethylbenzenes.

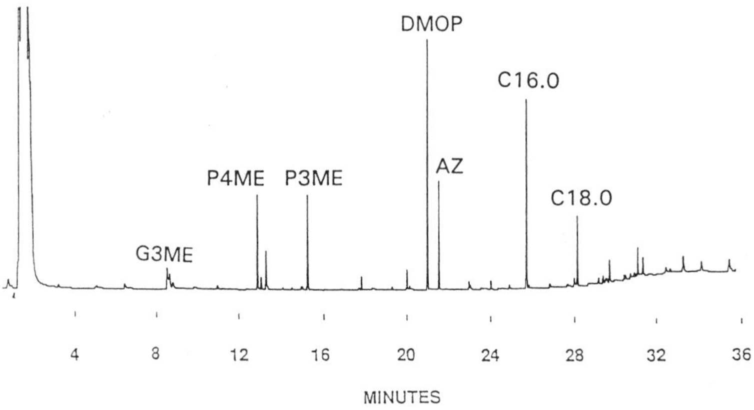

Figure 5 Simultaneous pyrolysis methylation pyrogram of an architectural alkyd enamel. G3ME = glycerol trimethyl ether, P4ME = pentaerythritol tetramethyl ether, P3ME = pentaerythritol trimethyl ether, DMOP = dimethyl orthophthalate, AZ = methyl azelate, C16.0 = methyl palmitate, C18.0 = methyl stearate.

In this case, the paint can be identified as a pentaerythritol *o*-phthalic alkyd consistent with having a high proportion of a soya bean drying oil (from the C16.0:C18.0 ratio). The total absence of unsaturated C18 acid methyl esters and significant azelaic acid indicates that the paint has a high degree of cure. No rosin or other modification is detected.

C. Industrial Paints

This category includes coatings used on goods manufactured on a factory production line, such as domestic appliances, furniture, tools, and products derived from sheet metal. Organic binders are either heat cured, air dried, catalyzed air dried, or radiation cured. Heat-cured coatings or baking or stoving enamels are the most commonly found industrial coatings. The binders are alkyd enamels, thermosetting acrylics, polyesters, epoxies, silicone acrylics, vinyls, and the heat-resisting polyimides and fluorocarbons. The air-dried coatings include acrylic, vinyl, and nitrocellulose lacquers, alkyd resins, urethanes, and latex types. Polyurethanes and epoxy resins are examples of catalyzed air-dried binders. Circumstances where these coatings occur as evidence involve items such as crowbars, tire levers, baseball bats, domestic appliances and furniture. Typical examples of Py-GC characterization of two of the wide variety of these coatings are shown in Figure 6.

From these data we interpret that the beverage can coating is a methyl methacrylate-ethylhexyl acrylate copolymer-modified polyester. The polyester is an adipic acid-modified neopentyl glycol iso-/orthophthalic acid type. The electrical appliance coating is a butylated amino resin crosslinked pentaerythritol-orthophthalic thermosetting alkyd enamel.

Conventional Py-GC gives either a misleading diagnosis or data from which it is impossible to obtain structural information. Pyrolysis derivatization provides more useful data about composition [26].

Other applications of Py-GC to synthetic resins include the identification of photocopy toners and surface coatings on currency notes.

III. ADHESIVES

There is a wide diversity of commercial adhesives available to the retail trade for industrial applications [27]. Animal- and vegetable-based glues, including casein and starch types, are less commonly used as a consequence of the development of more versatile synthetic products. Case situations involving adhesives are many and various. Their identification by Py-GC in improvised explosive devices has been reported [28]. Fraudulently resealed packages and mail are examined for the presence of foreign adhesive used for resealing purposes. Adhesives and sealers are becoming more frequently used in automotive construction and, therefore, occur in traffic-related crimes. Hot melt adhesives are becoming more common in the sealing of various types of packaging. Pyrolysis mass spectrometry has also been used for the identification of adhesives [29].

Pyrograms of some typical adhesives obtained by Curie-point pyrolysis using a medium polarity phase vitreous silica column are shown in Figure 7.

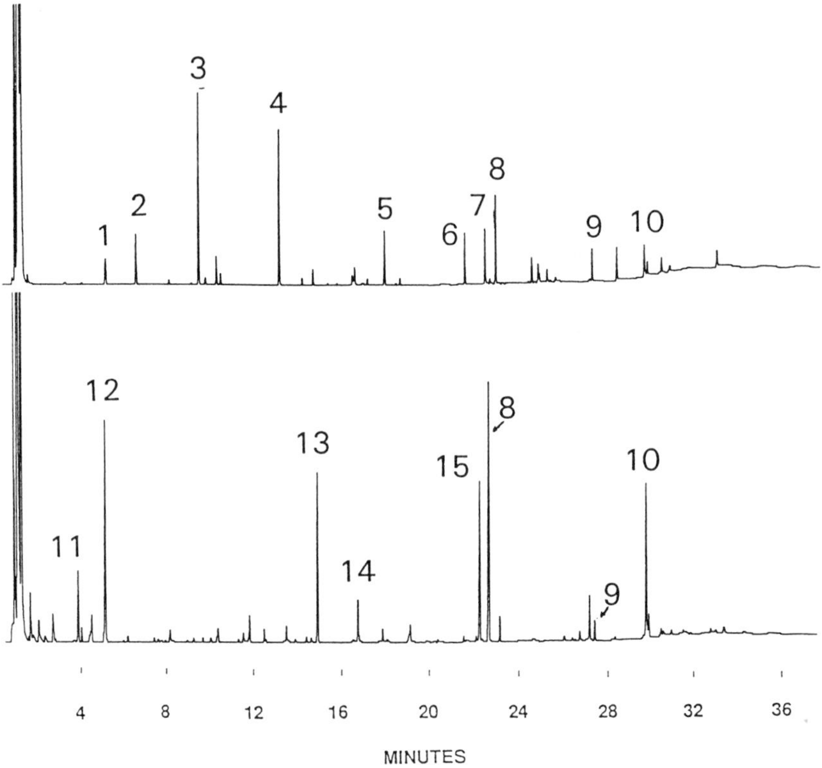

FIGURE 6 Simultaneous pyrolysis methylation pyrograms of two industrial finishes on a beverage can (top) and an electrical appliance (bottom) 1 = methyl methacrylate, 2 = neopentylglycol dimethyl ether, 3 = neopentylglycol monomethyl ether, 4 = ethylhexyl alcohol, 5 = dimethyl adipate, 6 = methyl laurate, 7 = dimethyl orthophthalate, 8 = dimethyl isophthalate, 9 = methyl palmitate, 10 = methyl stearate, 11 = isobutanol, 12 = *n*-butanol, 13 = methyl benzoate, 14 = pentaerythritol trimethyl ether, 15 = *N*-methyl phthalimide.

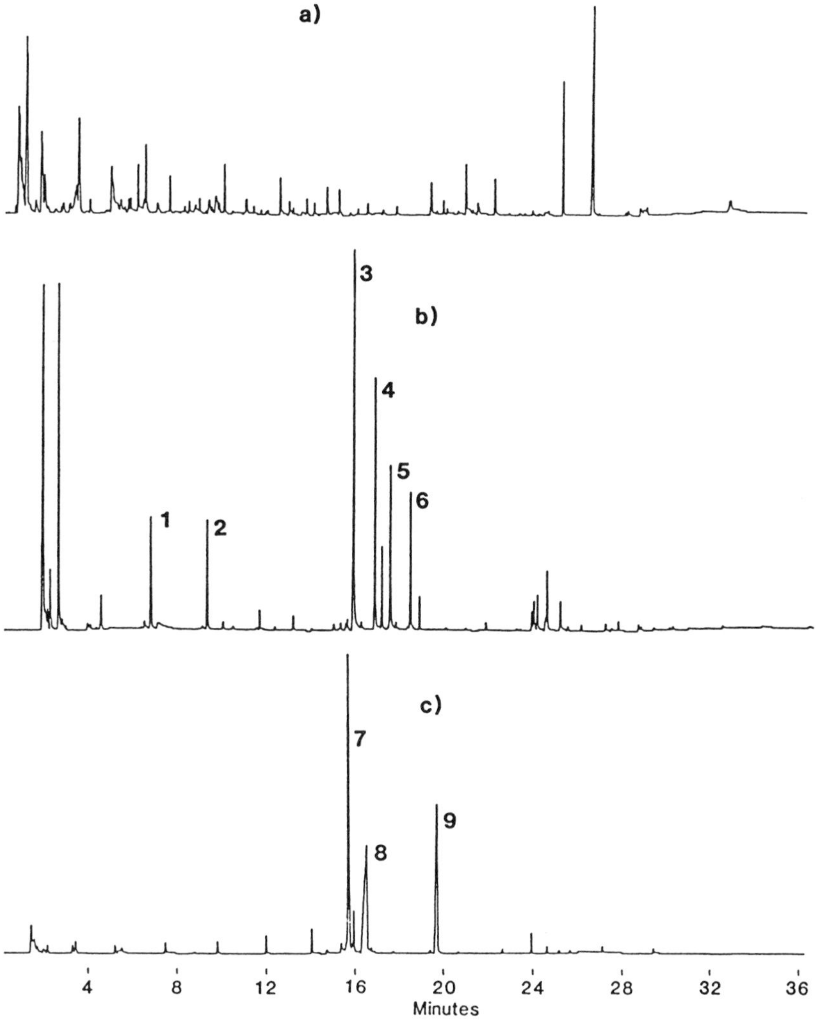

FIGURE 7 Pyrograms of (a) starch-based, (b) phenol-resorcinol formaldehyde, and (c) vinylpyrrolidone-type adhesives. The identities of the major peaks are 1 = toluene, 2 = xylene, 3 = phenol, 4 = ortho-cresol, 5 = para-cresol, 6 = xylenol, 7 = vinylpyrrolidone, 8 = pyrrolidone, 9 = caprolactam.

The major diagnostic pyrolysis products found in some frequently found adhesives are tabulated as follows:

Adhesive resin	Diagnostic pyrolysis products
Vinyl Acetate	Acetic acid
Acrylic	Ethyl acrylate, (methyl methacrylate) butyl acrylate, 2-ethylhexyl acrylate, ethylhexyl alcohol, isooctene
Epoxy	Phenol, isopropenyl phenol, bisphenol A
Polyindene	Indene, vinyl toluene
Polyisoprene (natural rubber)	Dipentene, isoprene, dimethylvinylcyclohexene
Neoprene-novolac	Chloroprene oligomers, substituted phenols
Phenol formaldehyde	Phenol, cresols, dimethyl phenols
Silicone	Methyl-substituted cyclic siloxanes
Starch	Furans, levoglucosan
Vinylpyrrolidone	Vinylpyrrolidone, caprolactam

A. Intraclass Differentiation

Within-class discrimination is an important factor in the forensic examination of materials, and Py-GC may be used to distinguish closely related polymers. For example, vinyl acetate polymers may be plasticized internally or externally. Copolymers include ethylhexyl acrylate, while phthalate plasticizers, dibutyl or diisobutyl phthalate, may be used as external plasticizers in commercial products.

IV. RUBBERS

Rubbers have physical characteristics and chemical composition that precludes their successful identification by infrared spectroscopy because of their inherent elasticity and highly filled composition. In contrast, no such difficulties are encountered with Py-GC. Crime scene rubber evidence from automotive tires and rubber vehicle components is found in hit-and-run cases and in soles of shoes worn by offenders in crimes against property. Discrimination of vehicle bumper rubbers by Py-GC has been reported [30]. Volatile and polymeric components of rubbers and other polymers have been analyzed by Py-GC and the inorganic residue recovered for subsequent analysis [31]. The technique may also be used to quantitate rubber blends by measuring ratios of characteristic pyrolysis products.

Figure 8 shows examples of the pyrograms of three common types of rubber.

Peak ratios of 2 (from butadiene), 3, (styrene), and 4 (from polyisoprene) reflect the relative proportions of butadiene, styrene, and isoprene in the rubber blend.

V. PLASTICS

Motor vehicles, fire scene debris, and household utensils from crime scenes are materials that may contain plastics. Their identification and comparison are often necessary to further the criminal investigation. A wide range of thermoplastic and thermosetting plastics may be encountered as crime scene evidence. These include polyolefins, polyacrylates, polyamides, polyesters, epoxies, and vinyl polymers. Pyrograms of some of these are shown in Figure 9.

The pyrogram of polyethylene was obtained on a nonpolar methyl silicone phase capillary column.

Additional structural information can also be derived by Py-GC. Stereoisomerism, crystallinity, and sequence distribution data can be obtained. Isotactic and syndiotactic polypropylene may be identified by the ratio of oligomer components [32]. High- and low-density polyethylene can also be determined by the proportion of branch-chain alkane pyrolysis products [15] using a novel pyrolysis hydrogenation technique.

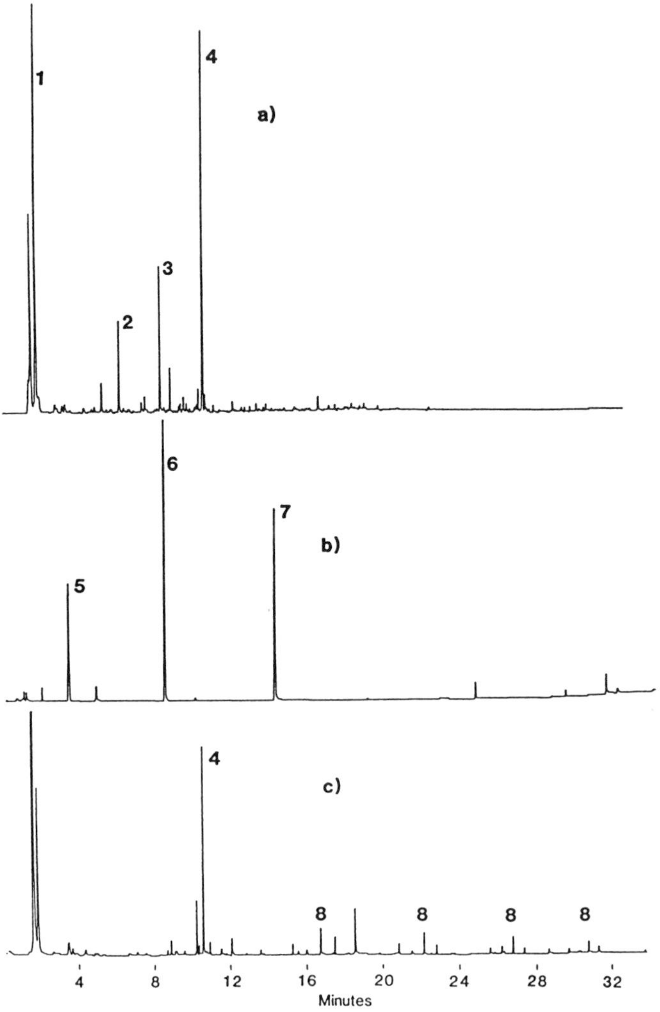

FIGURE 8 Pyrograms of (a) natural rubber/styrene butadiene blend, (b) polyurethane, and (c) butyl rubbers. The identities of the major peaks are 1 = isoprene, 2 = vinylcyclohexene, 3 = styrene, 4 = dipentene, 5 = tetrahydrofuran, 6 = cyclopentanone, 7 = butanediol, 8 = isobutene oligomers.

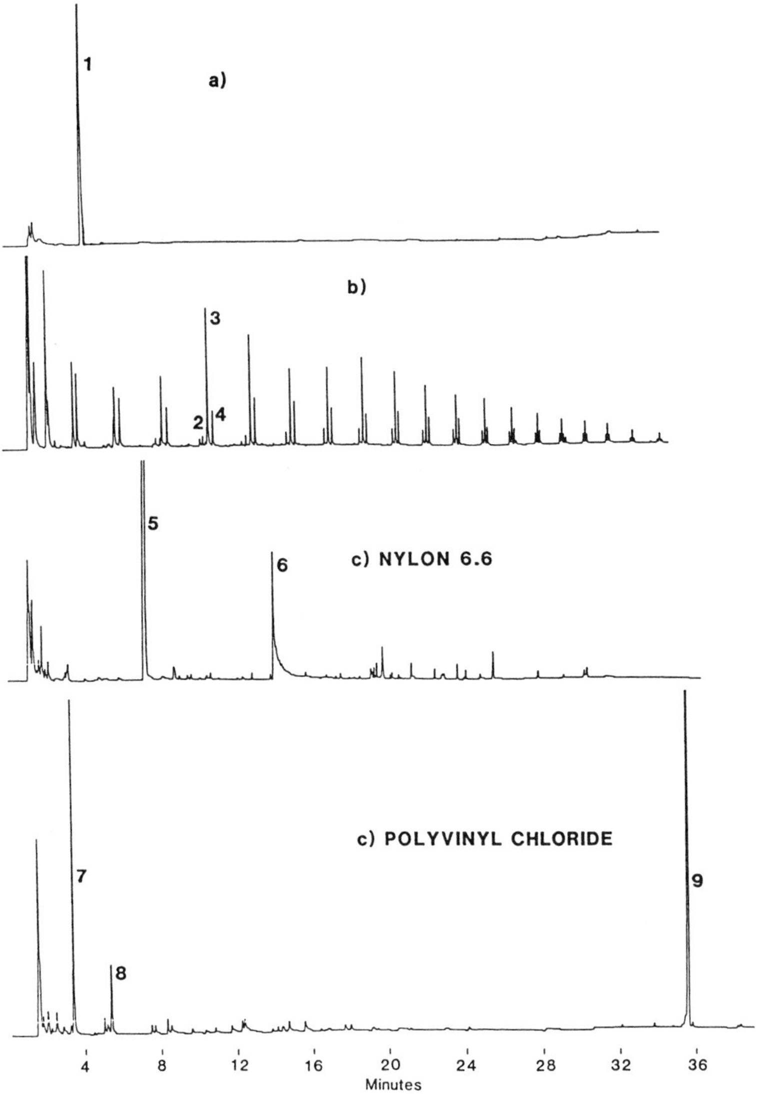

FIGURE 9 Pyrograms of (a) polymethyl methacrylate, (b) polyethylene, (c) nylon 6.6, and (d) polyvinyl chloride. The identities of the major peaks are as follows: 1 = methyl methacrylate, 2 = α-, ω-decadiene, 3 = 1-decene, 4 = decane, 5 = cyclopentanone, 6 = hexamethylene diamine, 7 = benzene, 8 = toluene, 9 = butylbenzyl phthalate.

VI. MOTOR VEHICLE BODY FILLERS

Body fillers from repainted vehicles in traffic accidents may be left at the scene. Identification and subsequent comparison to material from the vehicle of the offender may be required to establish that the vehicle was at the scene. Auto body fillers are usually a mixture of easily sanded inorganic fillers and styrenated unsaturated polyester resins. Composition of the resins may vary according to the application, resin supplier, and the relative cost of raw materials at the time of manufacture. The resins may vary in polyol, polybasic acid, or additive composition. The SPM-GC procedure provides the most diagnostic information about the structure of the resin. A pyrogram of a typical body filler is shown in Figure 10.

From these data it is apparent that the resin in the body filler is a styrenated diethylene glycol-isophthalic acid polyester crosslinked with maleic anhydride/fumaric acid and modified with adipic acid to give the resin flexibility. It might be expected that it would be possible to determine the degree of cure by monitoring the concentration of residual maleic anhydride/fumaric acid, detected as dimethyl maleate in the SPM procedure.

In a survey of body filler fillers [33], it was found that resin composition varies between products in the proportion of styrene, the phthalic acid isomer, and the presence of adipic acid. Most resins contained diethylene glycol and also propylene glycol in some cases.

VII. FIBERS

Textile fibers can provide strong evidence in criminal activity because of the wide variety in color, dye, and organic composition found in their populations. Fibers may be transferred between garments in person-to-person contact during assaults, transferred to vehicles in hit-and-run traffic incidents, and encountered as ropes or twine in cases of kidnapping or deprivation of liberty.

The forensic examination of natural and synthetic fibers employs optical and scanning electron microscopy for characteriz-

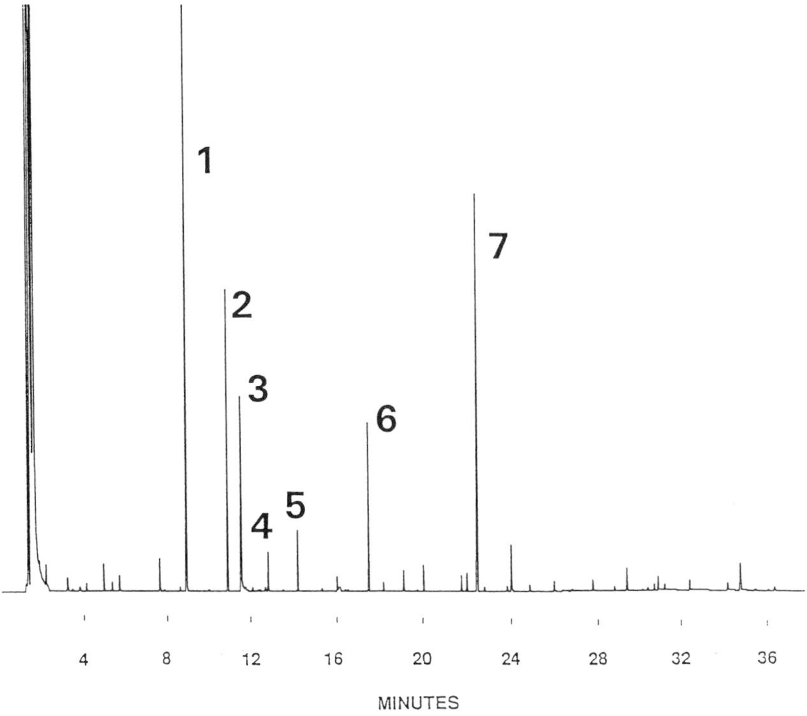

FIGURE 10 Simultaneous pyrolysis methylation pyrogram of an automotive body filler. 1 = styrene, 2 = diethylene glycol dimethyl ether, 3 = diethylene monomethyl ether, 4 = dimethyl adipate, 5 = methyl benzoate, 6 = dimethyl adipate, 7 = dimethyl isophthalate.

ing color, morphology, surface features, and elemental composition. The organic composition of fibers can be determined by Fourier-Transform infrared spectroscopy (FT-IR) and Py-GC. A single fiber is often sufficient to obtain an identification of the polymer class by both techniques. The Py-GC technique has been criticized for lack of sensitivity. However, with contemporary pyrolyzers and efficient gas chromatographic systems it is often possible to identify submicrogram quantities of fibers by this method. Using pyrolysis derivatization and selected ion

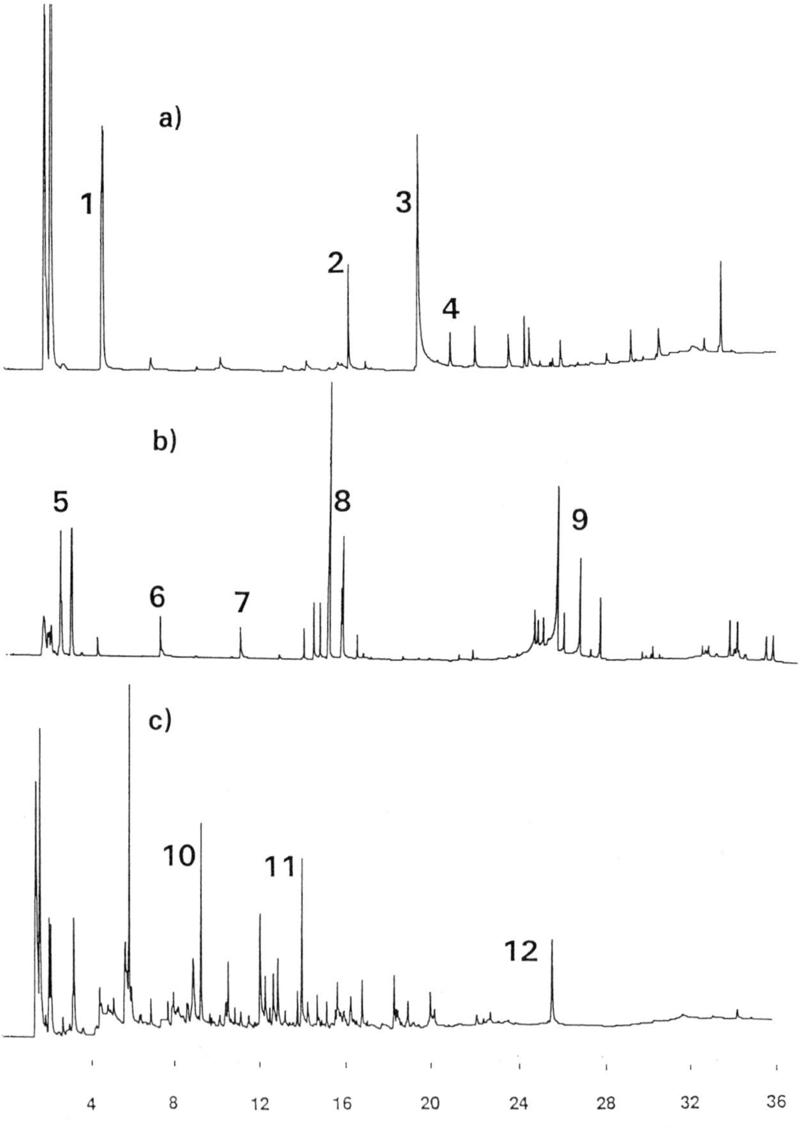

FIGURE 11 Pyrograms of (a) polyester, (b) acrylic, and (c) cotton fibers. The identities of the major peaks are 1 = benzene, 2 = vinyl benzoate, 3 = benzoic acid, 4 = biphenyl, 5 = acrylonitrile, 6 =

monitoring-mass spectrometry techniques, it is possible to lower the limit of detection by an order of magnitude. The application of Py-GC to the forensic examination of textile fibers has been reviewed [34]. Examples of the selectivity of Py-GC to several fiber types were described. The limitations in sensitivity of the technique and direction to be made to alleviate this restriction and possible solutions were outlined.

Py-GC can therefore be used as an effective means of determining the chemical composition of man-made homopolymer and copolymer fibers, natural fibers, fiber blends, or partly degraded fibers. Pyrograms of polyester, acrylic, and cotton fibers show how these fibers can readily be identified (Fig. 11).

Polyester fibers are composed of linear chains of polyethylene terephthalate (PET), which produce benzene, benzoic acid, biphenyl and vinyl terephthalate on pyrolysis. Acrylic fibers comprise chains made up of acrylonitrile units, usually copolymerized with less than 15% by weight of other monomers, e.g., methyl acrylate, methyl methacrylate, or vinylpyrrolidone. Thermolysis results in the formation of acrylonitrile monomer, dimers, and trimers with a small amount of the copolymer or its pyrolysis product. In this case the acrylic is Orlon 28, which contains methyl vinyl pyridine as comonomer. Residual dimethyl formamide solvent from the manufacturing process is also found in the pyrolysis products.

Cotton, which is almost pure cellulose, comprises chains of glucose units. The pyrolysis products of cellulose, identified by GC-MS, include carbonyl compounds, acids, methyl esters, furans, pyrans, anhydrosugars, and hydrocarbons. The major pyrolysis products are levoglucosan (1,6-anhydro-*B*-D-glucopyranose) and substituted furans. Further, Py-GC examination of synthetic polymer fibers can often provide more data than other techniques in cases where there are minor differences in compo-

---

dimethyl formamide, 7 = methyl vinyl pyridine, 8 = acrylonitrile dimers, 9 = acrylonitrile trimers, 10 = 2-furaldehyde, 11 = dihydromethylfuranone, 12 = levoglucopyranose.

sition within a class. In contrast, fibers that are chemically very similar are difficult to differentiate by IR and Py-GC. Cotton and viscose rayon, polyesters based on PET, and wool and regenerated protein are examples of these.

The pyrolysis derivatization approach may also be used. SPM results in the formation of dimethyl terephthalate when PET fibers are subjected to the procedure. Tetrabutylammonium hydroxide may be used to replace TMAH in simultaneous pyrolysis butylation to confirm the presence of vinyl acetate in acrylonitrile-vinyl acetate copolymers. The derivatized product is butyl acetate [17].

VIII. OILS AND FATS

Vegetable oils and animal fats are found alone or combined with other ingredients in proprietary products. They are usually identified and compared as their fatty acid derivatives or their triglycerides. The well-established methods usually depend on GC as a means of identification. Pyrolysis derivatization procedures developed more recently [26] provide a method for characterizing these materials. Microgram quantities of the triglycerides are reacted with tetramethylammonium hydroxide (TMAH) at high temperature to yield fatty acid methyl esters without employing multistep procedures. The method is proving to be reliable, reproducible, and particularly suitable for forensic samples. Vegetable oils can be considered to belong to classes that include oleic; linoleic; linolenic, and ricinoleic-rich types. Examples of SPM-GC of these types indicate the degree of discrimination that can be achieved on submicrogram quantities without prior sample preparation (Fig. 12). A high polarity capillary column and 100°C oven starting temperature was used for the GC separations.

FIGURE 12 SPM-GC profiles of olive oil (oleic rich), soya bean oil (linoleic rich), linseed oil (linolenic rich), and castor oil (ricinoleic rich) using a cyanopropyl (50%)-methyl silicone phase capillary column.

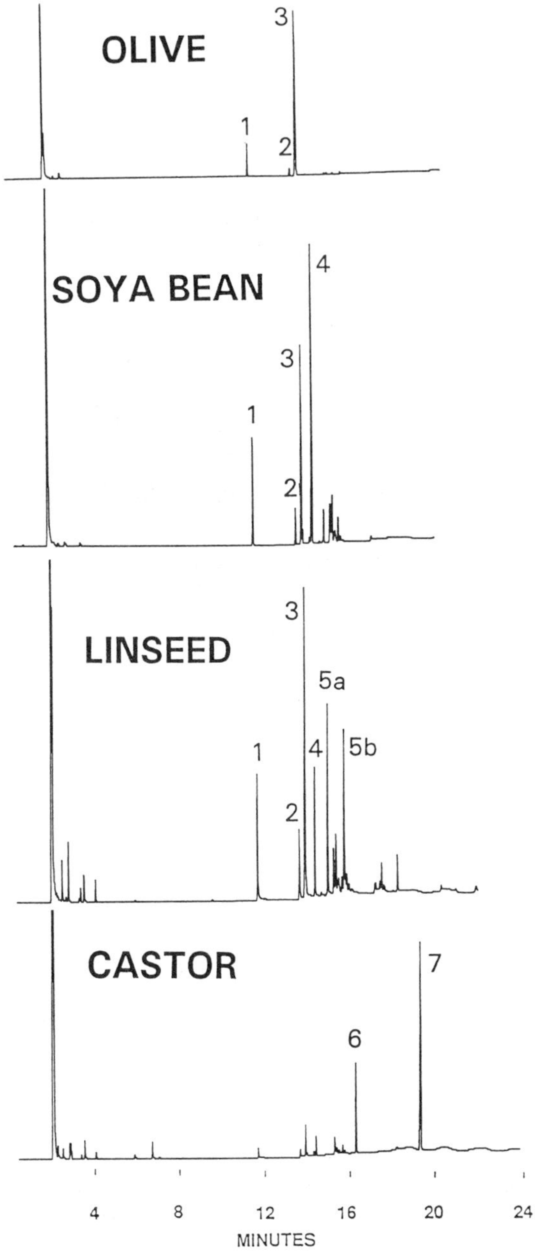
OLIVE
1
2
3
SOYA BEAN
1
2
3
4
LINSEED
1
2
3
4
5a
5b
CASTOR
6
7
4
8
12
16
20
24
MINUTES

Olive oil and other oleic-rich oils are recognized by their relatively high content of oleic acid detected as its fatty acid methyl ester (FAME) (peak 3). Palmitic and stearic acid FAME (peaks 1 and 2) are also detected. Soya bean oil and safflower, sunflower, and dehydrated castor oils have approximately 60% linoleic acid, detected as FAME (peak 4). Linseed oil is characterized by its high linolenic FAME (peak 5) concentration. Some thermal isomerization of the polyunsaturated fatty acids takes place when TMAH is employed for the SPM procedure. These artifacts are not produced when TMAH is replaced by the commercially available reagents MethPrep 1 and MethPrep 2, which are trifluoromethylphenyl trimethylammonium hydroxide solution in water and methanol. Castor oil, which contains a high proportion of ricinoleic acid (9-hydroxy octadecenoic acid) in triglyceride form, may be identified by the presence of its free hydroxy FAME (peak 7) or as the methoxy derivative (peak 6), in which both the hydroxy and the carboxylic acid groups are both methylated. While these examples are not intended to give a comprehensive description of the determination of fatty acids in vegetable oils, they do indicate the degree of differentiation obtainable with the technique.

IX. COSMETICS

Traces of cosmetics, such as lipstick smears, face creams, and body lotions, periodically provide valuable evidence to link an offender to a victim at the scene of a crime.

Lipstick bases comprise a blend of waxes, oils, and emollients. While the dyes and pigments present may be identified by techniques such as thin-layer and high-performance liquid chromatography and electron microscopy, Py-GC techniques may be used to characterize the lipstick bases [36]. SPM-GC techniques are useful for the identification of triglycerides and polyols formulated into the product. Fatty acids in lipids are converted to methyl esters. Aliphatic alcohols are converted to their methyl ethers. The triglyceride of ricinoleic acid is the major component of castor oil, the most common medium found in lipstick bases. Saponification and derivatization of the triglyceride by the SPM process is necessary to facilitate its identification. Using the pyrolysis apparatus to volatilize the organic components of the cosmetic product into the GC, Py-GC facilitates the identification of mineral oils, aliphatic alcohols, and simple fatty acid esters used as emollients. Figure 13 shows an SPM-GC profile of a typical lipstick base.

Volatilization of the lipstick using the Py-GC process indicates the presence of cetyl acetate (CA) and isopropyl myristate (IPM). Heptanal (C-7AL), a pyrolysis product of castor oil, is a major product. The SPM profile is more complex and gives more information about the composition of the product. Cetyl acetate is converted to the methyl ether of cetyl alcohol (C-16-OME), while IPM is partly converted to the methyl ester. The major components, RIC-OME and RIC, are the methyl ether of ricinoleic acid methyl ester and ricinoleic acid methyl ester having a free OH group, respectively. These compounds result from hydrolysis and complete or partial methylation of the major component of castor oil. C8.0 and C10.0 are fatty acid methyl esters, suggesting the presence of coconut oil in the formulation. This product was differentiated from more than 60 lipsticks examined in the study on the basis of composition of the organic components.

Body lotions usually comprise oil/water emulsions, which may contain long-chain fatty alcohols, glycerol, long-chain fatty acids, vegetable oils, mineral oils, and emollients. Approximately 50 body lotions have been characterized by these techniques [35], and it has been possible to differentiate all of these products. Figure 14 shows an SPM-GC profile of body lotion traces found at a crime scene.

The fatty acid methyl esters C16.0, C18.0, and C18.2 indicate the presence of a vegetable oil. However, the detection of isopropyl palmitate (IPP) and isopropyl stearate (IPS) in the Py-GC experiment (not shown) means that there would also be a contribution of their fatty acids to this pyrogram. The methyl ethers of cetyl alcohol (C16-OME) and stearyl alcohol (C18-OME) suggest the presence of the respective free fatty alcohols and this is confirmed by an inspection of the pyrogram from the Py-GC experiment (not shown). Myristyl alcohol methyl ether

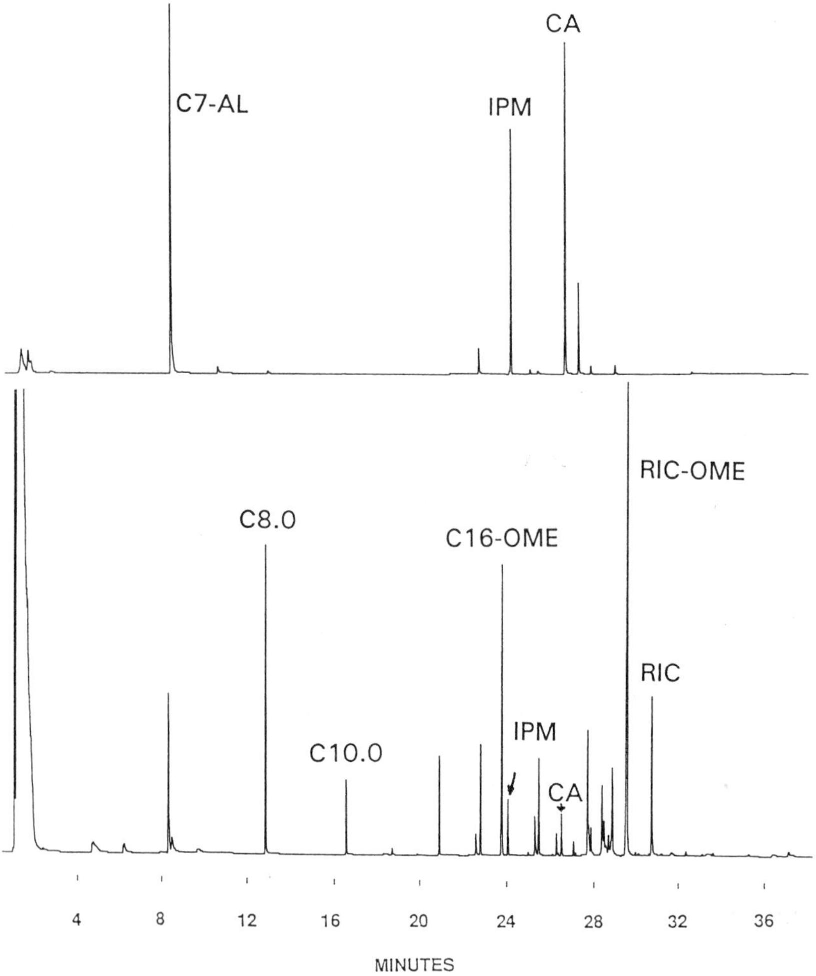

FIGURE 13 Py-GC (upper) and SPM-GC (lower) profiles of a lipstick base.

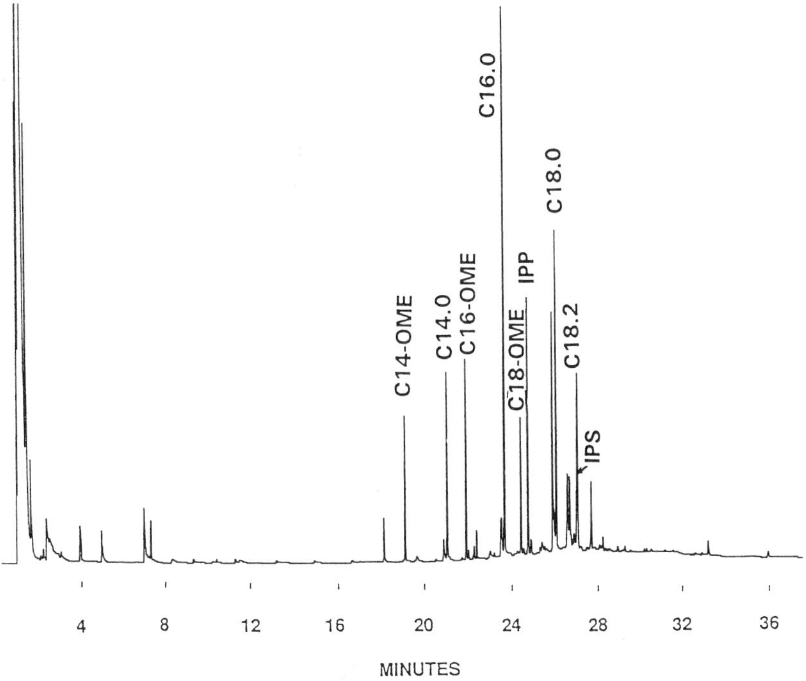

FIGURE 14 SPM-GC profile of a body lotion.

(C14-OME) and myristic acid methyl ester (C14.0) originate from myristyl myristate in the formulation.

X. NATURAL RESINS: ANCIENT AND MODERN APPLICATIONS

In order to establish authenticity of oil paintings and works of art, an examination of the surface-coating resin and pigment may be required. In older objects, resins may be oleoresins, wood rosin, amber, dammar, shellac, oriental lacquers, or other such natural resins.

Oleoresinous varnishes contain vegetable drying oils such as linseed and tung oils. Wood rosin comprises a mixture of rosin acids. Shellac is the flaked form of purified lac, the natural secretion of the insect *Laccifer lacca kerr*. It contains 46% aleuritic acid, a trihydroxy-substituted hexadecanoic acid, 27% shelloic acid, and a dihydroxy polycyclic aliphatic dicarboxylic acid together with other components, many of which are undetermined. The main components of oriental lacquer are urushiol, laccol, and thitsiol, which are C-15 and C-17 alkyl- and alkenyl-substituted benzene-1,2-diol compounds. Printing inks on questioned document letterheads may also contain natural resins in combination with contemporary synthetic modifiers, including phenol-formaldehyde resins.

Considering the polar nature of these constituents, it is not surprising that SPM-GC would be an appropriate technique to characterize these resins. SPM-GC pyrolysis profiles of shellac, a Burmese lacquer, and a printing ink resin are shown in Figure 15.

Shellac displays an SPM profile distinctly different from the Py-GC pyrogram (not shown), indicating a largely hydrolyzable component in the resin. There is some evidence of C14, C16, and C18 FAME (peaks 1 and 2), 6-hydroxy C18 FAME (peak 3), an oxirane group containing FAME (peak 4), and a polymethoxy substituted C18 FAME (peak 5) in the products. SPM-GC analysis of the cured Burmese lacquer results in a series of straight-chain and branched-chain alkyl benzenes ranging from toluene to dodecyl benzene, and C6- to C17-alkanes and alkenes. These products are similar in composition to the products obtained by conventional pyrolysis. Compounds corresponding to peaks marked with an asterisk are fatty acid methyl esters, which are products of the SPM reaction and become apparent as the lacquer cures. These two natural resins are therefore clearly distinguished by this method.

In the SPM-GC profile of the printing ink resin [36], dimethyl fumarate (peak 6) is detected, suggesting the presence of maleic or fumaric acid. Pentaerythritol is indicated by the presence of the tetra- and trimethyl ethers (peaks 7 and 8, respectively). Tertiary butyl phenol and methyl and dimethyl tertiary butyl phenols are detected as their methyl ethers (peaks 9, 10, and 11, respectively). Rosin acid methyl esters are detected with

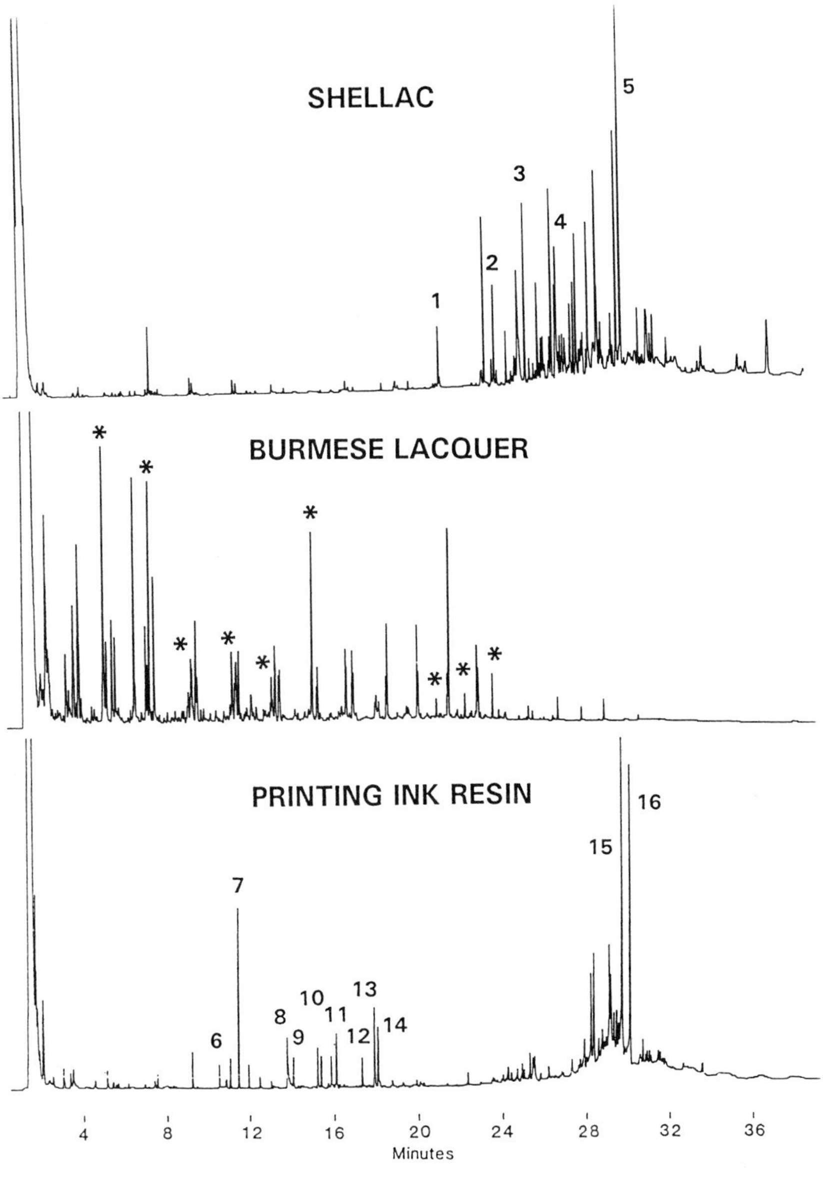

Figure 15 SPM-GC profiles of shellac, a Burmese lacquer, and a printing ink.

dehydroabietic acid and abietic acid methyl esters predominating (peaks 15 and 16, respectively). This resin is therefore diagnosed as a tertiary butyl phenol-formaldehyde condensate-modified pentaerythritol rosin maleic ester type.

XI. INKS

Forensic document examiners are asked to identify altered ballpoint ink and in some cases to determine when such an entry has been made. Ballpoint inks are composed of an intimate blend of resin, pigment or dye, and solvent.

In principle, the age of a ballpoint ink entry may be determined by its dye component composition. With a sufficiently wide historical database of such information, it has been possible to "date" a ballpoint entry. An alternative approach to determine the composition and relative age is by Py-GC methodology.

Py-GC using the SPM technique has shown some potential both for identifying solvent composition and resin type [35]. An SPM-GC profile of a typical ballpoint ink is shown in Figure 16.

From these data the solvents are identified as phenoxy ethanol (PE), benzyl alcohol (BA), and propylene glycol (PG). The methyl ethers, which are also detected, are suffixed by ME. It would appear that the resin is a polyvinylpyrrolidone type as indicated by the detection of the dimer (PVP2) and trimer (PVP3). Mass spectral data suggests that the peak marked DIAZ is derived from a diazo dye. The relative age of the ink on the document can be assessed by monitoring the reduction in concentration of individual solvent components. In alkyd resin binder inks, it may be possible to determine the age of the ink entry by monitoring changes in the fatty acid composition as discussed previously.

XII. MISCELLANEOUS

There are a group of evidence types that either have not been widely examined by Py-GC, or their development is in its infancy.

A discussion of the applications of Py-GC to forensic chem-

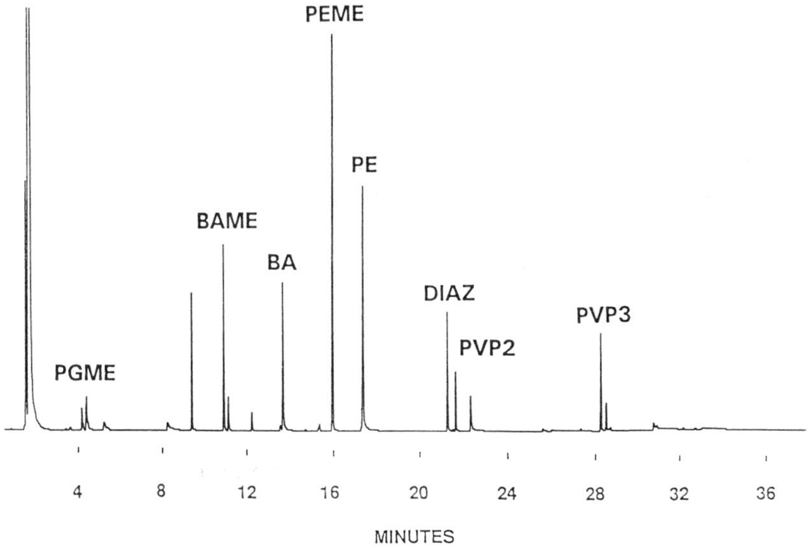

FIGURE 16 SPM-GC profile of a typical ballpoint ink.

istry would not be complete without including the role of the technique to the characterization of proprietary drugs and other organic substances used illegally. This topic has been adequately described by Irwin [3]. In spite of the advantages, Py-GC has not been widely adopted for this purpose in forensic laboratories. The pyrolysis derivatization techniques previously described has potential for identifying drugs in the form of esters and carboxylic acids (such as Ibuprofen) and their excipients, including vegetable oils and sugars and preservatives [26]. Soils containing high levels of organic matter may be characterized by Py-GC techniques. The presence of humic acids in soils infers that there should be potential for soil differentiation by pyrolysis alkylation techniques.

The identification of wood is an area of forensic interest. Microscopy methods on the basis of morphology are generally used for this purpose. However, a recent development indicates that there is potential for the chemical taxonomy of wood species based on the pyrolysis of TMAH extractives [37].

XIII. CASE STUDY

During a police investigation of a multiple murder at the home of the victims, a greasy substance was found on a bedroom door. A fingerprint was found on the grease.

The forensic laboratory was asked to identify the substance so that appropriate fingerprint development techniques could be used to visualize the print. Some of the greasy substance was carefully removed from the door and analyzed by Py-GC using the simultaneous pyrolysis methylation technique described previously. The substance was a cosmetic body lotion (see Fig. 15). The print matched that of a person known to have been in the house several days previously and was therefore not necessarily incriminatory.

However, this was considered to be suspicious and the laboratory asked for swabs from three of the victims to be forwarded for comparison as some sexual activity had been known to have taken place. Methylene chloride extracts of the swabs were taken and evaporated on to the pyrolysis wires and the experiments repeated. The results were compared to those of the body lotion taken from the door. There was a high degree of correspondence between the composition of the body lotion on the door and the composition of the components in the swabs, taking into account that some adsorption of some of the components could have taken place on the skin surfaces. The link between the suspect's fingerprint and the victims was established. As a result the suspect was questioned again and confessed to all aspects of the crime.

XIV. CONCLUSIONS

It is evident that Py-GC may be used to characterize many different types of material likely to be associated with a crime scene. It is a versatile and discriminatory technique which may be used for the forensic examination of materials by:

High-temperature pyrolysis to determine the structure of macromolecular materials. This may lead to products that do not directly relate to the parent polymer, particularly in cases where the polymer is derived from nonpolar monomers.

Elevated temperature reaction pyrolysis in which the analyte is reacted with a chemical reagent. The method is appropriate for those materials that contain hydrolyzable bonding or free reactive groups, e.g., acids, alcohols, phenols. Usually, the "pyrograms" more clearly reflect the composition of the parent material.

A system for thermally desorbing chemical compounds from the sample. This is useful for introduction by volatilization of the total sample to the GC system and compounds adsorbed on an adsorbent, such as activated charcoal or porous polymer. The latter method can be used for the analysis of hydrocarbon accelerants from fire scenes in cases of arson or environmental or toxicological monitoring.

An effective interpretation of data cannot be made without a knowledge of the identity of the pyrolysis products and hence with good fortune a diagnosis of the composition of the original polymer. Therefore, every effort should be made to identify and interpret the significance of, at least, the major products by mass spectrometry, relative retention times, or infrared spectroscopy (in the case of GC-FTIR). While this discussion has been limited to Py-GC, other pyrolysis techniques should not be forgotten. Py-MS, in its simplest form, compared to other more sophisticated techniques such as field ionization-MS, laser ablation-MS, and fast atom bombardment, provide data which, with sufficiently wide databases, can give useful information about structure. Future directions include the application of other separation techniques, such as supercritical fluid chromatography [38] and development of novel high-temperature chemolytic methods which may yield even more information about the structure of macromolecules.

REFERENCES

1. B. B. Wheals, *J. Anal. Appl. Pyrol.*, *2*: 277 (1980).
2. B. B. Wheals, *J. Anal. Appl. Pyrol.*, *8*: 503 (1985).

3. W. J. Irwin, *Analytical Pyrolysis, A Comprehensive Guide* (Chromatographic Series Vol. 22.) Marcel Dekker, New York (1982).
4. P. K. Clausen and W. F. Rowe, *J. For. Sci.*, *25*(4): 765 (1980).
5. R. O. Keto, *J. For. Sci.*, *34*(1): 74 (1989).
6. T. O. Munson and J. Vick, *J. Anal. Appl. Pyrol.*, *8*: 493 (1985).
7. J. B. F. Lloyd, K. Hadley, and B. R. G. Roberts, *J. Chrom.*, *101*: 417 (1974).
8. T. P. Wampler, *J. Anal. Appl. Pyrol.*, *16*: 291 (1989).
9. P. R. De Forest, *J. For. Sci.*, *19*: 113–120 (1974).
10. R. W. May, E. F. Pearson, and D. Scothern, *Pyrolysis Gas Chromatography*, Analytical Science Monographs No. 3, The Chemical Society (1977).
11. T. P. Wampler and E. J. Levy, *Crime Lab. Digest*, *12*: 25 (1985).
12. A. Parabyk, The characterization of plastic automobile bumper bars using FTIR, PGC and SEM-EDX, M.S. thesis, George Washington University (1988).
13. R. Saferstein, *Polymer and GC Analysis* (S. A. Liebman and E. J. Levy eds.) Marcel Dekker, New York. p. 339 (1985)
14. J. M. Challinor, *For. Sci. Int.*, *21*: 269 (1983).
15. S. Tsuge, Y. Sugimura, and T. Nagaya, *J. Anal. Appl. Pyrol.*, *1*: 221 (1980).
16. Y. Ishida, K. Abe, H. Ohtani, and S. Tsuge, *Macromolecules*, to be submitted.
17. J. M. Challinor, *J. Anal. Appl. Pyrol.*, *16*: 323 (1989).
18. J. I. Thornton, *Forensic Science Handbook* (R. Saferstein, ed.) Prentice-Hall, N.J. (1982).
19. J. M. Challinor, *Expert Evidence, Advocacy and Practice*, (I. Freckleton and H. Selby, eds.) Law Book Company, Melbourne, Australia.
20. P. J. Cardosi, *J. For. Sci.*, *27*: 695 (1982).
21. K. Fukuda, *For. Sci. Int.*, *29*: 227 (1985).
22. D. McMinn, T. L. Carlson, and T. O. Munson, *J. For. Sci.*, *30*(4): 1064.
23. J. M. Challinor, *J. Anal. Appl. Pyrol.*, *18*: 233 (1991).
24. J. M. Challinor, Proceedings of the 13th Meeting of the International Association of Forensic Sciences, Oxford, U.K., 1984.
25. J. W. Bates, T. Allinson, and T. S. Bal, *For. Sci. Int.*, *40*: 25 (1989).

26. J. M. Challinor, *J. Anal. Appl. Pyrol.*, *20*: 15 (1991).
27. I. Skeist, *Handbook of Adhesives*, 2nd ed., Van Nostrand Reinhold, New York (1977).
28. N. L. Bakowski, E. C. Bender, and T. O. Munson, *J. Anal. Appl. Pyrol.*, *8*: 483 (1985).
29. J. C. Hughes, B. B. Wheals, and M. J. Whitehouse, *For. Sci. 10*; 217 (1977).
30. R. D. Blackledge, *J. For. Sci.*, *26*: 557 (1981).
31. J. Chi-an Hu, *J. Chrom. Sci.*, *19*: 634 (1991).
32. S. Tsuge and H. Ohtani, *Analytical Pyrolysis, Techniques and Applications* (K. J. Vorhees, ed.), Butterworths (1984), pp. 407–427.
33. J. M. Challinor, unpublished.
34. J. M. Challinor, *The Forensic Examination of Fibers* (J. Robertson, ed.), Ellis Horwood (1992), p. 219.
35. J. M. Challinor, Proceedings of the 12th International Association of Forensic Sciences, Adelaide, Australia, 1990.
36. J. M. Challinor, *J. Anal. Appl. Pyrol.*, *25*: 349 (1993).
37. J. M. Challinor, Proceedings of the 12th Analytical Chemistry Symposium of the Royal Australian Chemical Institute, Perth, Western Australia, Oct. 1993.
38. P. R. DeForest et al., *Gas Chromatography in Forensic Science* (I. Tebbett, ed.), Ellis Horwood (1992), pp. 165–187.

9

Characterization of Microorganisms by Pyrolysis-GC, Pyrolysis-GC/MS, and Pyrolysis-MS

Stephen L. Morgan, Bruce E. Watt, and
Randolph C. Galipo
The University of South Carolina, Columbia, South Carolina

I. INTRODUCTION

Detecting, identifying, and characterizing microorganisms is vital to solving important environmental, biological, and medical problems. In monitoring a food production process for pathogenic microorganisms or surveying a battlefield for biological warfare agents, detection may require recognizing that microbial species other than those expected in the normal environment are present. A diagnosis of pneumonia for a patient may require the ability to classify a microbial sample as one of several closely related species. Microorganisms play a central role in soil biodegradation processes and analysis and characterization of environmental microbial communities can contribute to the design and control of waste bioremediation processes. Characterization of different organisms might also provide supporting evidence for or against relatedness of species for taxonomic purposes.

Bergey's *Manual of Determinative Bacteriology* [1] catalogs over 250 bacterial genera. Traditional microbiological identification of bacterial species is based on appearance under a microscope (shape, size, presence of particular structures), response to staining (e.g., the classic gram stain), or indirect characteristics (growth under aerobic or anaerobic conditions, generation of specific enzymes or biochemical products, etc.). Morphology is closely related to taxonomy but its use for microbial characterization is not definitive [2,3]. Structural characteristics of "gram type" (e.g., thick peptidoglycan in gram-positive organisms versus lipopolysaccharide in gram-negative organisms) do not always correlate with the results of the gram stain. Other cell types (such as acid-fast species) also exist [4,5]. The presence of specific enzymes, the response of an organism to growth substrates, or susceptibility of an organism to specific antibiotics is often due to metabolism rather than structure. Although valuable for identification, the use of these largely phenotypic characteristics requires culturing and isolation of viable cells, which can be time-consuming and may even fail for fastidious organisms [6].

The complex mixture of microbial cellular and extracellular components is an appealing target for chemotaxonomic analysis by gas chromatography (GC), mass spectrometry (MS), or the combination of the two techniques together (GC/MS). Such instrumental methods focus on chemical structures present in a sample and are not dependent, as are biological tests, on the microorganism being viable. Additionally, since only microgram amounts of sample are required, instrumental methods offer enhanced sensitivity.

Chemical components characteristic of specific microorganisms or microbial groups are usually enclosed in or part of a polymeric cellular matrix. Such nonvolatile and intractable biological samples present some difficulties for their direct characterization by GC, MS, or GC/MS. For these techniques to be useful, chemical components characteristic of the microbial sample must be released intact or a related compound must be generated before GC or MS analysis. Depolymerization is usually performed off-line by acid hydrolysis, methanolysis, saponification, or other reactions. One or more chemical derivatization steps might then be employed to produce volatile and thermally stable derivatives suitable for GC or MS.

When analytical pyrolysis is applied to a microbial sample, depolymerization and volatilization of bacterial components is accomplished simultaneously. The volatile thermal products may be separated on-line by capillary GC with flame ionization detection (Py-GC-FID), separated by GC and detected by MS (Py-GC/MS), or detected directly by MS (Py-MS). In contrast to derivatization-based methods and like other direct approaches, microbial characterization by analytical pyrolysis requires minimal sample pretreatment and short total analysis times.

The use of analytical pyrolysis for biomedical taxonomy was proposed by Zemany [7] in 1952. A single strain of an unidentified microorganism was used in fundamental studies of pyrolysis temperatures and residence times by Oyama [8] in 1963. The first comprehensive work was that of Reiner [9] who used pyrolysis-GC in 1965 to identify different strains of *E. coli*, *Shigella* sp., *Streptococcus pyogenes*, and *Mycobacteria*. Although most early work was performed with packed columns and nonselective flame ionization detection, the potential of analytical pyrolysis for making sensitive discriminations between various microorganisms samples was well established by the late 1970s.

In 1979, Gutteridge and Norris [10] reviewed the application of pyrolysis methods to the identification of microorganisms. Significantly, this review appeared in a bacteriology journal and listed 148 references. Meuzelaar, Haverkamp, and Hileman [11] discussed pyrolysis-MS methods for biomaterials in their 1982 text, describing nearly 30 applications in clinical microbiology, quality control, and other biological areas. Irwin [12] included a chapter on microbial taxonomy in his comprehensive 1982 guide to analytical pyrolysis and listed 124 references. The use of analytical pyrolysis in clinical and pharmaceutical microbiology was described by Wieten et al. [13] in 1984 and over 100 references were cited. In 1985, Bayer and Morgan [14] reviewed the analysis of biopolymers by pyrolysis-GC and listed 80 literature citations on microorganisms. Also in 1985, Fox and Morgan [6] discussed a chemotaxonomic approach to characterizing microorganisms based on analysis of chemical signatures detected by analytical pyrolysis and derivatization GC/MS. Wampler's 1989 bibliography of analytical pyrolysis references [15] listed 29 microbial applications along with other examples of biopolymer analysis. Analytical methods for microorganisms, including pyrolysis, were also surveyed by Eiceman, Windig, and Snyder [16]. More recently, Morgan et al. [17,18] reviewed their work on identification of chemical markers for microbial differentiation produced by pyrolysis.

Improvements in analytical capability for the analysis of complex pyrolyzate mixtures have appeared during the last decade: high-resolution capillary GC with more polar and selective stationary phases coated on inert fused silica columns; coupling of capillary GC with sensitive, selective, and lower-cost mass spectrometric detectors; enhanced pyrolysis-MS techniques, hyphenated analysis methods, including GC-Fourier transform infrared spectroscopy (GC/FTIR) and tandem MS; and better strategies for handling complex multidimensional pyrolysis data. The present chapter reviews the known chemotaxonomy of microorganisms, summarizes practical considerations for the use of pyrolysis in microbial characterization, and critically discusses selected applications of analytical pyrolysis to microbial characterization.

II. MICROBIAL CHEMOTAXONOMY

Chemotaxonomy is the study of chemical differences in organisms and the use of those differences to identify and classify them. As suggested in the last section, direct chemical analysis provides a powerful supplement to the more traditional methods of classifying microorganisms by morphological and biochemical characteristics. Although variations in growth, treatment, and sampling conditions can influence their structure and composition, microorganisms also exhibit many invariant structures that can be employed for differentiation.

Bacteria and other prokaryotes lack the complex intracellular structures (endoplasmic reticulum, Golgi apparatus, lysosomes, mitochondria, etc.) found in plant and animal cells (eukaryotes). Bacterial morphology (shape and size) is limited to a few simple forms, such as rods, cocci, chains, and spirals [5,19]. The chemical composition of the cell walls, outer membranes, and capsules of bacteria is sufficiently diverse to offer potential for taxonomic discrimination [3,4]. For example, the chemical makeup of a single colon bacillus (*Escherichia coli*) has been estimated to contain 2000–4500 different types of small molecules distributed as 120 different amino acids and their precursors and derivatives, 250 different carbohydrates and their precursors, 50 different fatty acids and their precursors, 100 different nucleotides and their precursors and derivatives, and 300 quinones, polyisoprenoids, porphyrins, vitamins, and other small organic molecules [20]. Many of these molecules are common to all organisms and not well suited for discrimination among microorganisms. Some of these molecules are specific to various major groups of organisms and can serve as chemical markers for their detection or identification.

The cell envelope usually consists of a cell membrane, a cell wall, and an outer membrane [21,22]. Figure 1 shows schematic representations of the structure of gram-positive and gram-negative cell envelopes. The cell wall consists of the peptidoglycan (PG) layer and associated structures. PG is the only substance common to almost all bacteria (except *Mycoplasma* and *Chlamydia*) and not present in nonbacterial matter [23]. PG and its associated chemical components may account for up to 10–40% of the dry weight of the cell [5]. As seen in Figure 2, PG consists of a polysaccharide backbone that is a repeating polymer of *N*-acetylglucosamine and *N*-acetylmuramic acid. Attached covalently to the lactyl group of muramic acid are tetra- and pentapeptides (composed of repeating L- and D- amino acids) cross-

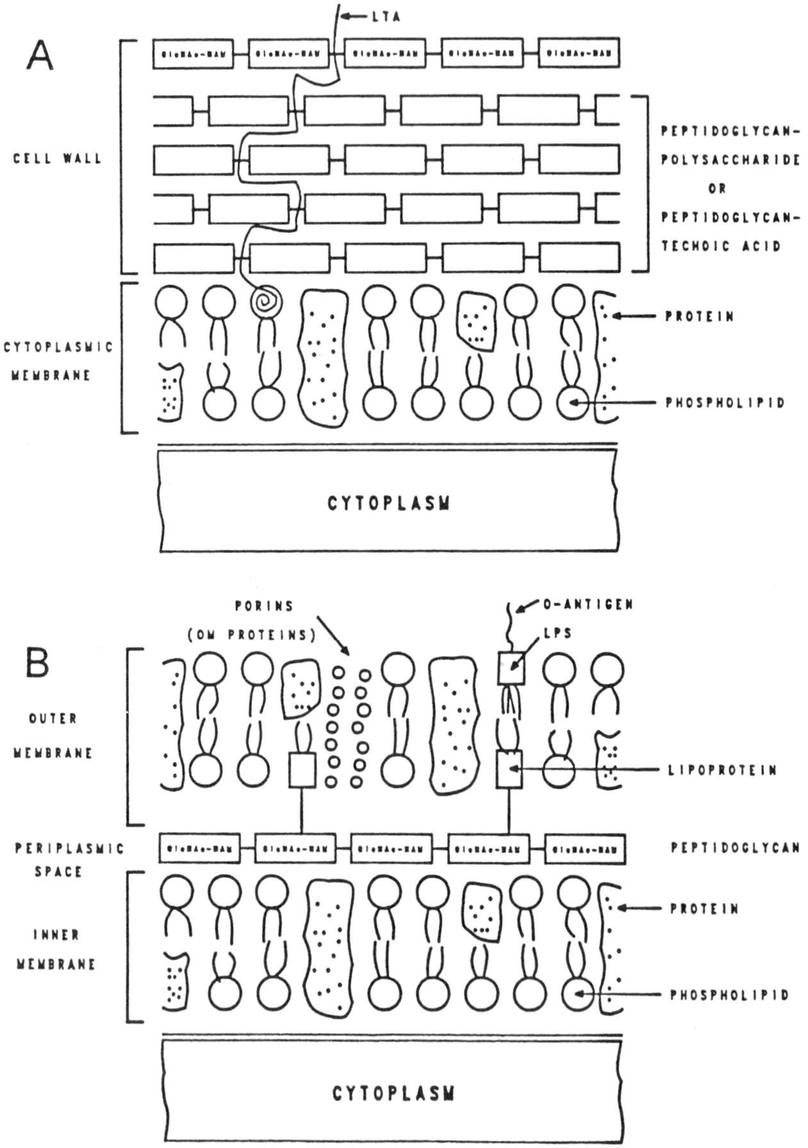

Figure 1 Generalized structure of bacterial cell envelope. (A) Gram-positive organism. (B) Gram-negative organism.

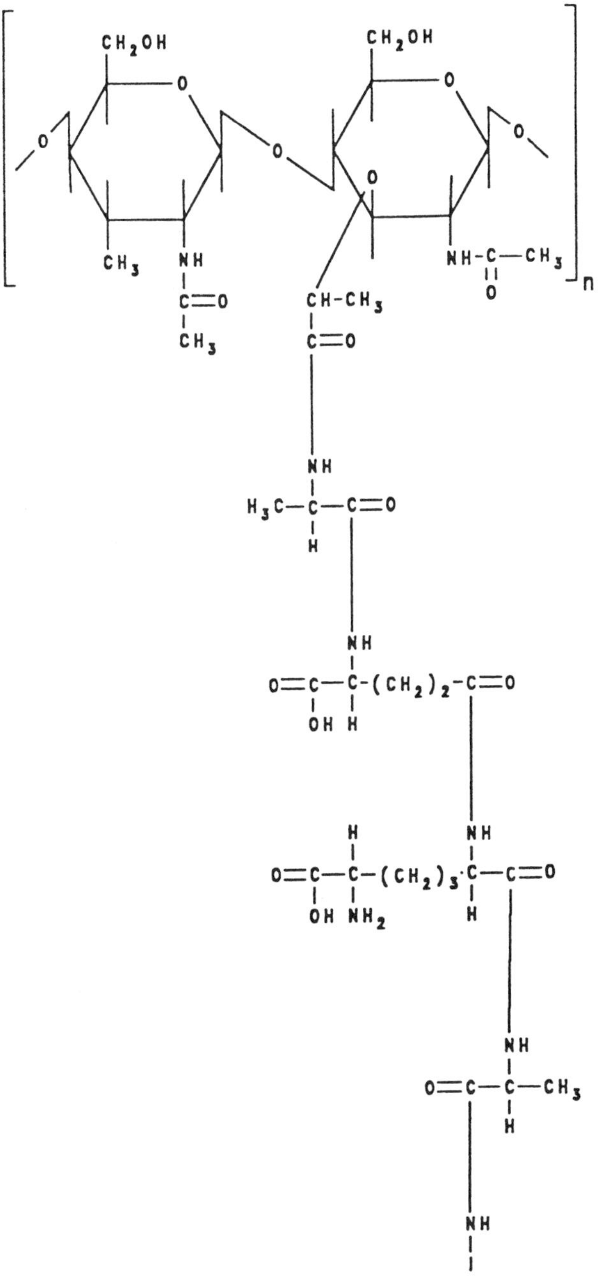

FIGURE 2 Structure of bacterial peptidoglycan.

linked by peptide bridges. The amino sugar muramic acid (3-*O*-α-carboxethyl-D-glucosamine) is a fairly definitive marker for bacteria. Other chemical markers in PG include D-alanine, D-glutamic acid, and diaminopimelic acid [24,25]. The D-amino acids are sometimes found in other bacterial components, but are not synthesized by mammals. Different bacteria may vary in the sequence of the amino acids in the peptide side chains and crossbridges.

The PG layer is thicker in gram-positive bacterial cell envelopes than in gram-negative bacteria (Fig. 1A). As a result, chemical markers for PG are found in higher amount in gram-positive cells. Teichoic acids (with ribitol or glycerol phosphate backbones and side chains of variable amino acid composition) are covalently bound to the thick peptidoglycan layer in some gram-positive species [26]. Lipoteichoic acids (LTA) composed of glycerophosphate polymers terminated in glycolipid are not bound to the cell wall but are sometimes linked through the cell wall to muramic acid [5,26]. Teichuronic acids (glucuronic acid polymers with variable side chains) and neutral polysaccharides are also bound to PG.

Gram-negative cell envelopes usually have a thin PG layer (Fig. 1B). PG is attached to the outer membrane by lipoprotein containing phospholipids and other hydrophobic substances. A variety of phospholipids and proteins are found on the inner side of the outer membrane; some of these (porins) spanning the outer membrane. The external surface of the outer membrane of gram-negative bacteria contains its primary endotoxin, a unique lipopolysaccharide (LPS), consisting of an outer O-antigen, a middle core, and an inner lipid A region [27]. The lipid A region contains a glucosamine disaccharide with covalently bound 2- and 3-hydroxy fatty acids. Glucosamine is common, but 3-hydroxy and 2-hydroxy fatty acids are unusual. Although the fatty acid composition of LPS varies among gram-negative bacteria, β-hydroxymyristic acid is a chemical marker for LPS.

Fatty acid profiling by GC is routine in some clinical reference laboratories, particularly for identification of anaerobic bacteria [28,29]. Fatty acids and lipids are bonded to proteins, carbo-

hydrates, or other chemical entities in microbial cell walls and membranes. Fatty acids of chain length from C_9 to C_{20} are useful for identifying gram-negative organisms at the species and genus level. Perhaps the only automated GC-based microbial characterization system that is commercially available is a microbial analysis system based on derivatization GC of fatty acid methyl esters (Microbial ID, Inc., Newark, NJ) [30].

The core region in the LPS of gram-negative bacteria contains two unusual sugars, ketodeoxyoctonic acid (KDO) and L-D-glycero-mannoheptose. KDO and heptoses are not commonly found in structures other than LPS and are potential markers for gram-negative organisms. Levels and types of sugars (e.g., rhamnose and fucose) within the core and O-antigen regions of LPS differ among bacterial species. Carbohydrate profiling using derivatization followed by GC/MS is an excellent tool for bacterial differentiation [31].

Various groups of gram-positive bacteria also contain cell wall-associated carbohydrates. For example, rhamnose-containing polysaccharides are major components of streptococcal cell walls and are used for their serological classification (Lancefield grouping). The group A-specific polysaccharide consists of a polyrhamnose backbone with single *N*-acetylglucosamine side chains, while the group B-specific polysaccharide consists of a backbone of rhamnose and glucitol phosphate residues with trisaccharide side chains composed of rhamnose, galactose, and *N*-acetylglucosamine [32].

Mycobacteria, nocardiae, corynebacteria, and some related organisms have unusual cell envelopes. Not all *Mycoplasma* contain peptidoglycan [5] and PG in *Chlamydia* does not contain muramic acid [33]. Although mycobacteria and nocardiae stain as gram-positive organisms, some resist decolorization by acid alcohol after staining (acidfastness). Acidfastness is believed to be related to the presence of long-chain mycolic acids [5]. Mycolic acids are α-alkyl, β-hydroxy fatty acids ranging in size from C_{30} to C_{86} and may constitute up to 60% of the dry weight of these bacterial envelopes. Mycolic acids in corynebacteria are relatively short chained. Mycolic acids have a variety of struc-

tural forms within the cell envelope and some mycolic acids are covalently bound to PG by a polysaccharide. Other compounds that contain mycolic acid form a thick waxy layer around the outside of the cell wall.

Under certain circumstances many bacteria produce capsules outside their cell envelopes. Capsules are usually made of polysaccharide, however some *Bacillus* capsules are composed of D-glutamic acid polypeptide. Certain gram-positive bacteria, most notably strains of *Bacillus* and *Clostridium*; produce modified cells (endospores) capable of surviving in adverse environments [5]. Spore PG is found between an inner and outer membrane and differs from that in the normal vegetative cell: muramic acid is mostly in the lactam form; the spore PG has fewer peptide cross links; and the cell exterior is coated with keratin. The spore also contains large quantities of calcium dipicolinate, a substance involved in microbial heat resistance.

Poly-β-hydroxybutyrate (PHB), a polymer of 3-hydroxybutyric acid, serves as a major energy and carbon storage compound. Diverse bacteria usually accumulate PHB in intracellular granules having an approximate composition of 98% PHB, 2% proteins, and small amounts of lipids. In genetically competent bacteria, PHB is in the plasma membrane or covalently linked to extracellular polysaccharides. Some *Bacillus* and *Pseudomonas* species accumulate up to 30% of their weight at PHB [5].

Many of the characteristic components mentioned in this section are common to diverse bacterial species. For example, chemical markers for peptidoglycan (muramic acid, etc.) are ubiquitous in bacteria. Other chemical components have the potential to be chemical markers whose presence or absence differentiates dissimilar (or even closely related) microbial species. The bacterial cell envelope (membrane, surrounding wall, and outer membrane) in particular offers a rich assortment of unusual polymers that have distinctive monomeric constituents. Whether any of these chemical components from specific microorganisms generates identifiable and unique pyrolysis signature has been the focus of much research over the past decade.

II. PRACTICAL ASPECTS

When analytical pyrolysis is subjected to the same rigorous controls and quality checks as any other analytical technique, reproducible results can be achieved both within and between laboratories. Biological variability and heterogeneity of sampling tend to have deleterious effects on reproducibility—these problems will plague any analytical technique applied to microorganisms. Because of the great intraspecies variability of bacteria, it is important to employ representative samples of more than one strain of each organism if possible. Ruggedness, simplicity, and ease of interpretation of results are other performance characteristics relevant in developing pyrolysis methods for microbial characterization.

A. Microbial Sample Handling

The first step in the handling of any microbial sample, whether taken from a clinical specimen or from the environment, is usually to culture the organism in growth median under controlled conditions. This step serves several purposes: the number of cells of the organism(s) is increased thereby increasing the detectability, and the organism(s) present may be isolated as nearly pure strains thus increasing the chances of identification. Such cleanup or isolation of strains can be important. Cultures containing mixed populations of microorganisms can present difficulties for identification schemes, even those based on chemotaxonomic markers.

The formulation of growth media, inoculation, incubation times and temperatures, and sampling of a microorganism should be carefully standardized and controlled so as to not introduce undesired variability into later attempts at characterization. Media should be formulated in a standardized manner and be of known composition. Inoculation of the microbial cells into the culture media should be done with the same quantity of cells in the same growth stage. All strains should then be cultured for the same period of time and at the same temperature to provide uniformity. Ideally, organisms should be grown under exactly the same conditions (media, temperature, stage of growth). Unfortunately, this is not always practical. More often the isolation of an unknown microbiological sample might require a battery of different media appropriate for different growth types. Related strains can be grown under identical conditions to the same growth stage. Organisms having very different physiological requirements for growth clearly cannot be grown under the same conditions (e.g., obligate anaerobes and aerobes). In any case, standardization and quality control are required.

Changes in the type or composition of growth media may influence the production of extracellular components (enzymes, metabolic products, etc.) as well as some cellular components. For example, Gutteridge and Norris [34] studied the effects of different growth conditions on the discrimination of three different bacteria by Py-GC. Pyrograms of the same organism grown on different media were more different from one another than pyrograms of different organisms grown on the same media. The effect of culture time and incubation temperature was less obvious, producing variations for two organisms but not for the other organism. Engman et al. [36] also examined the effect of growth media, incubation time, and sample storage time on the pyrograms of five bacterial species of the genus *Flavobacterium*. Pyrograms of organisms grown on two different but similar media such as *Flavobacterium* medium and trypticase soy agar (both consisting of a protein hydrolyzate agar with no added carbohydrate) showed few differences. When glucose was added to the growth medium, however, several new peaks appeared in the pyrograms. Samples incubated for different times also showed few differences except for two peaks that increased in height upon longer incubation. It is desirable, of course, for differences in the amounts of discriminating pyrolysis products to be representative of relatively stable genetic or structural differences between organisms that do not vary appreciably with culture conditions.

Depending on its origin, incubation time may affect the levels of a particular pyrolysis product. Watt et al. [36] profiled the production of (*trans*) 2-butenoic acid, a pyrolysis product of poly-β-hydroxybutyrate (PHB) as a function of incubation time for *Legionella pneumophila*. The amount of 2-butenoic in the pyrograms followed the expected growth curve for PHB. PHB is found in the plasma membrane of *E. coli* and is only present in high amounts when growth is limited and cells are genetically competent. Pyrolysis products of PHB were not found in *E. coli* pyrograms, perhaps because the growth conditions were not optimal to force production of PHB.

After harvesting bacteria, it is usually desirable to wash the samples several times with sterile distilled water to remove possible contamination from the media. Samples are often autoclaved or heat-killed (at 60 to 120°C in air, depending on the susceptibility of the organism) for safety reasons, they lyophilized (freeze-dried) to a dry powder. Heating may be performed under an inert gas to avoid the possibility of oxidation. Since the sample is exposed to much more energetic degradation conditions during pyrolysis, most researchers have assumed that heat-killing and lyophilization do not appreciably change the chemical structure of the microbial sample. Organisms should be prepared and killed in a similar fashion whenever possible. Preparing samples in widely different fashions may not permit hypotheses concerning observed pyrolysis patterns to be tested. The sample may then be refrigerated for storage before analysis.

In addition to being cultured in a well-defined manner, strains should also be systematically characterized by traditional microbiological tests. Only in this manner can quality control be assured. Traditional microbiological characterization should be performed before killing as most of these tests require viable cells. In a pyrolysis study of streptococci, a sample labeled as a Group F strain was found to contain a chemical marker (due to glucitol phosphate) previously detected only in Group B organisms [37]. Routine tests (hippurate hydrolysis and serological grouping) identified the organism as a group B streptococcus. It was later confirmed that this sample had been incorrectly identified by the originating laboratory. Bacteria are not simple pure chemical substances; incorrect assignments and sample contamination are not uncommon when routine microbiological control is not carried out. This is not a criticism of interlaboratory studies as much as a comment on the complexity of adequate sample and information transmittal between laboratories.

Knowledge of chemical composition with respect to relevant carbohydrate, amino acid, or fatty acid chemical markers is often critical for proper interpretation of the pyrolysis of a microbial sample. Sometimes this information is already available in the literature regarding specific bacteria. More often, however, complementary analyses for that particular class of components by an independent GC, GC/MS, or spectroscopic method may be needed.

Finally, researchers working with potentially pathogenic microorganisms should be aware of local and government regulations pertaining to operating a safe and healthy workplace [38]. In the United States, this involves compliance with the Occupational Safety and Health Act of 1970 and regulations of the Department of Labor, including OSHA document 29 CFR 1910 [39]. Employers are required to inform and train employees regarding the specific hazards associated with their work. Safety techniques and devices should be used to isolate personnel from biohazards. Specific examples include use of protective laboratory clothing, aseptic technique to avoid environmental spread of the pathogenic agent, and safety cabinets to minimize spread of aerosols. Containment considerations also include: use of nonporous surfaces for laboratory bench tops; type of door used to separate the laboratory from public areas; the type of ventilation system in use; and accessibility of autoclaves for sterilization. Biohazard labels should be affixed to waste containers, refrigerators, freezers, and transport containers holding potentially infectious materials. Specific procedures for storage, decontamination, and disposal of biohazardous agents should also be documented.

B. Analytical Pyrolysis Techniques for Microorganisms

Ease of sample handling is often cited as an advantage in pyrolysis-based methods. This advantage generally refers to the fact that lengthy sample pretreatment, derivatization, or clean-up steps are not required. As pointed out by Windig [40], biological samples are often heterogeneous and the preparation of a representative suspension involving microgram amounts of bacteria can be tedious. Additionally, Montaudo [41] points out that the assumption that *any* sample pretreatment should be avoided is naive. Indeed, if the characteristic discriminating information for a microbial application is known a priori to reside in a certain cellular component, a carefully designed pretreatment step may be able to provide a suitable isolated cell fraction. This approach will generally give better reproducibility because the background of extraneous chemical components not relevant to decision-making is eliminated.

To prepare a microbial sample for analysis, an amount of the lyophilized powder is weighed and added to a measured volume of water to give a solution of known concentration (e.g., 10 $\mu g/\mu l$). The solution should be vortexed or sonicated to produce a homogeneous suspension before sampling onto the pyrolysis sampling device.

Alternatively, a microbial sample may be transferred by a plastic inoculation loop directly from the culture media into a quartz sampling tube or onto a metal wire. Care should be taken to avoid carrying media with the sample; blank pyrolysis runs on media alone should also be done to check for background contamination. Sample homogeneity (composition, shape, and amount) is desirable for reproducible results and this may be difficult to achieve. In some instances, however, analytical pyrolysis may be conducted directly on original samples with little or no sample pretreatment. Gilbart et al. [42] discriminated group B *Streptococci* from groups A, C, F, and G following swabbing of the samples direct from sheep blood agar plates and suspension of the sample in distilled water to a consistent opacity.

When a resistively heated filament pyrolyzer (e.g., the Pyroprobe from CDS Analytical, Inc., Oxford, PA) is used, the sample may be placed directly on a platinum filament or may be placed in a quartz tube or boat inside a platinum coil. In either case, the placement of sample with respect to the sampling tube or the ribbon should be the same for all samples. For liquid sample suspensions placed on a ribbon or in a coil, the solvent is evaporated prior to pyrolysis. Solid microbial samples can be sandwiched between quartz wool plugs inside the quartz sampling tube so as to reduce extraneous nonvolatile material from leaving the sampling tube during pyrolysis. With quartz the sample never comes into direct contact with the pyrolyzer filament, as it does when sample is coated directly on a thin ribbon filament. Ribbon filaments sometimes exhibit a memory effect (particularly with polar components), are harder to clean, and typically have a shorter lifetime. Quartz tubes may be reused after cleaning.

The temperature actually experienced by the sample in a resistively heated probe may be different from the pyrolysis temperature setting. Factors influencing the sample temperature include sample amount, positioning of sample, and whether or not a quartz sampling tube is employed. To achieve reproducibility, sample sizes for analytical pyrolysis should be in the lower microgram range. Large sample sizes (more than 100 μg) make controlled heat transfer and uniform heating difficult to achieve; secondary reactions are promoted, resulting in more complicated and less reproducible pyrograms. Small sample sizes allow formation of thin films and permit effective and rapid heat transfer from the heat source [43]. If analytical profiling is the goal, sample amounts applied to the tube are typically in the range of 10–100 μg. Acquisition of quality mass spectra for all peaks (including minor pyrolysis products) may require samples of several hundred micrograms.

With Curie-point pyrolysis, the active heating element is a ferromagnetic metal wire (0.5 mm in diameter). The sample (10–20 μg) may be coated on the wire by depositing drops (5 μl each) of suspension (2–4 μg/μl) and then dried while rotating the wire [11]. As mentioned already, placement of sample should be done reproducibly to reduce variability in pyrolysis results. Carbon disulfide has been recommended as a solvent for sampling into Curie-point wires as it evaporates easily and leaves no detectable residue [44]. Sampling factors influencing Curie-point techniques in Py-MS have been extensively studied [45–47]. The deposition of unpyrolyzed sample or nonvolatile residue onto the walls of the pyrolysis interface increases with the amount of sample loaded on Curie-point wires, with optimal loading between 1 and 20 μg [11,45].

Following loading of sample, the filament probe or Curie-point wire is inserted into a heated interface connected to the GC injection port or Py-MS vacuum. In the original design of filament pyrolysis systems from CDS Analytical, the ribbon or coil is an element in a Wheatstone bridge circuit. The bridge is set so that it is balanced at the setpoint temperature and a capacitor is discharged to rapidly heat the filament (up to 20°C/ms and 1000°C). Although resistively heated filament pyrolyzers are generally calibrated by the manufacturer, the actual temperature of the pyrolyzer element can be determined by appropriate calibration [48].

Curie-point pyrolysis involves placing the sample wire into a radio frequency field that induces eddy currents in the ferromagnetic material and causes a temperature rise. When the wire reaches the Curie-point temperature it becomes paramagnetic and stops inducting power. The temperature at which the wire stabilizes (the Curie-point) is a function of the type of metal. For example, the Curie-points of cobalt, iron, and nickel are 1128, 770, and 358°C, respectively. Wires made from alloys of these metals produce intermediate temperatures. For example, the commonly used nickel-iron wire has a Curie-point of 510°C [11]. Differences between filament and Curie-point pyrolyzers depend on the pyrolyzates examined and may be obscured by other instrumental differences, including the design of the transmission system of the detector.

Long-term reproducibility is affected by eventual deterioration of resistive filaments or sample wires. All components exposed to sample during pyrolysis (GC injection port liners and quartz sample tubes is used) often require acid cleaning, solvent washing, and oven drying. Active pyrolyzer elements (coils and ribbons in filament pyrolyzers) can be heated without sample to remove contamination (1000°C for 2 s is usually adequate). Curie-point wires are inexpensive enough to be discarded after use.

Preliminary experiments are usually required to choose pyrolysis conditions. The pyrolysis literature may guide the researcher, but optimal conditions for a particular sample must be determined empirically. The question of appropriate heating rates and final temperatures has been addressed by other authors [11,12]. Pyrolysis heating conditions interact with sample size and effects are usually system-dependent (i.e., resistive heating *versus* Curie-point, wires *versus* ribbons *versus* quartz tubes). With microorganisms, temperatures that produce pyrolyzates characteristic of the parent sample may range from 400 to 800°C or even 1000°C. Generally, lower pyrolysis temperatures induce less fragmentation and decrease the total amount of pyrolyzates produced. Higher temperatures ensure more complete fragmentation and thus might reduce the structural information obtained. Extremely low pyrolysis temperatures may not produce significant amounts of volatile fragments from intractable biomaterials and may lead to the accumulation of nonvolatile residues contaminating the analytical system. Pyrolysis at a variety of different temperatures may provide selective information about degradation patterns and thereby reveal different aspects of structure. A time profile of volatile products generated at different temperatures by linear programmed thermal degradation may also provide characteristic structural information [49].

C. Chromatographic Separation and Mass Spectrometric Detection of Pyrolysis Products from Microorganisms

Once the sample has been pyrolyzed, volatile fragments are swept from the heated pyrolysis/injection port by carrier gas into the GC or GC/MS system. In Py-MS, it is likewise desired to transfer pyrolysis products to the ionization source of the MS without appreciable degradation, condensation loss, or recombination. Designs of Curie-point Py-MS systems have incorporated glass reaction tubes, expansion chambers, heated walls, and positioning of the pyrolysis reactor directly in front of the ion source [11].

In Py-GC, if the column is kept relatively cold at the start of the run (e.g., as in a programmed temperature run), pyrolyzates will be focused at the head of the column until eluted by the increasing temperature. The amount of pyrolysis products transferred to the chromatographic column is dependent on whether a split injection or a direct injection without splitting is employed. For highest sensitivity, direct flow without splitting is preferred; good chromatography may dictate using a reasonable split ratio (e.g., 10–20 parts split flow, 1 part column flow).

In their profiling of oral streptococci, French and co-workers [50,51] used 1.5 m × 4 mm columns packed with a porous polymer (Chromosorb 104) combined with isothermal chromatographic conditions for fast repetitive analysis. Porous polymer columns are suitable for analysis of low-molecular-weight volatile pyrolyzates but provide only limited higher-molecular-weight information. Although some conventional packed columns have been used in pyrolysis studies of microorganisms, over the past decade GC applications have converted completely to the use of fused silica open tubular capillary columns. Because microbial pyrolyzates are often labile polar species, the column material and chromatographic column should be as inert as possible. Fused silica capillary columns coated with "bonded" phases satisfy these needs. Fused silica columns offer improved resolution, increased inertness, and better analytical precision than packed columns. Superior resolution per unit time available with capillary columns means that adequate resolution can often be achieved using short (5–10 m) columns. With optimized conditions, analysis times are usually less than 10–30 min. Engman et al. [35], in studying the pyrolysis of *flavobacteria*, employed fast temperature programs to shorten analysis times when using selected ion monitoring (SIM). Despite the resultant loss of resolution, capillary pyrograms were found to retain discriminating information. Another advantageous approach is to directly interface short (1–5 m) capillary columns to low-pressure MS ion sources. The decrease in analysis time, maintenance of resolution, and elution of labile components at reduced temperatures have been described by Trehy et al. [52] and applied to microbial pyrolysis by Snyder et al. [53].

Choice of chromatographic stationary phases for fused silica capillary columns can range from nonpolar (e.g., 100% methyl silicone) to polar (e.g., polyethylene glycol) phases. A particular stationary phase is employed, of course, to affect elution order of components being chromatographed. Figure 3A and 3B shows replicate pyrolysis-GC/MS analyses of *B. anthracis* strain VNR-1-D1. These pyrograms were obtained using a fused silica capillary column coated with a 1μ film thickness of DB-1701 (moderate polarity, 14% cyanopropylphenyl-86% methyl silicone). Figure 3C and 3D shows replicate pyrograms of the same strain on a different fused silica capillary column coated with a 0.25 μ film thickness of free fatty acid phase (FFAP, polar, polyethylene glycol-acid modified). Correlation of peaks between the two sets of chromatograms on different columns is extremely difficult without an additional dimension of information (e.g., mass spectra for all peaks). Nevertheless, replicate chromatograms on the same phase are reproducibly identical. The different stationary phases employed here "process" the mixture of pyrolysis products differently. Both stationary phases provide a characteristic and reproducible signature for the microorganism. In the same way, chromatographic systems (instrument plus column) with differing active metal or other sorption sites may preferentially filter polar or reactive components from the chromatogram.

Contamination in GC systems is a continual problem in analytical pyrolysis. Early reports [54,55] recognized that production of a residue from a sample during pyrolysis may be unavoidable, particularly with biological samples. Extraneous peaks that mysteriously appear in some pyrograms have determinate causes that, if understood, can be eliminated [56]. Contamination in pyrolysis is often due to carryover of higher-molecular-weight, polar, or less volatile material produced during pyrolysis that remains behind in the interface, connecting tubing, or other parts of the chromatographic system. Tarry products or smoke may

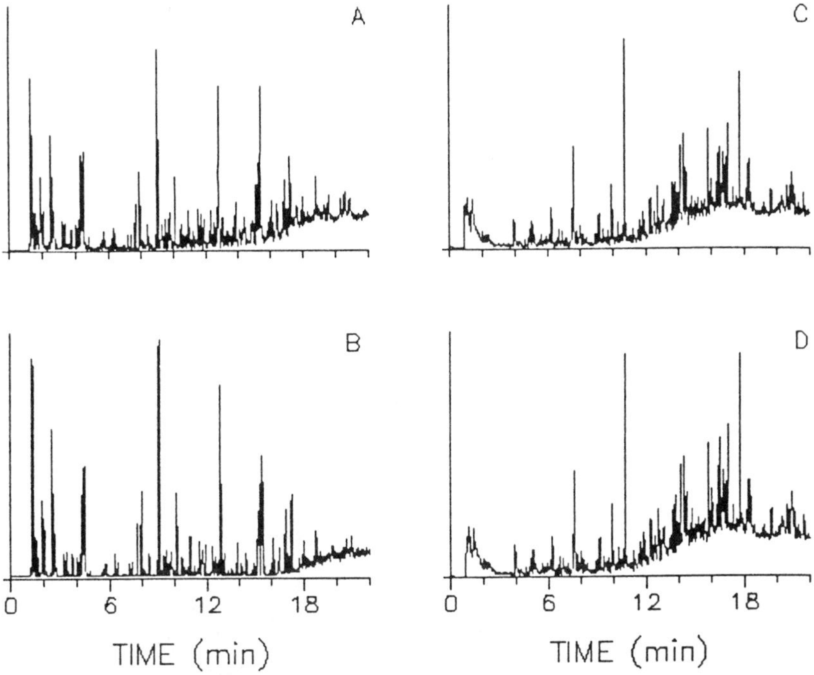

FIGURE 3 Pyrolysis of *B. anthracis* strain VNR-1-D1 at two different temperatures and analyzed by two different columns. (A, B) Replicate pyrograms produced by pyrolysis at 650°C with pyrolyzates separated on a 1.0 μm film thickness DB-1701 fused silica column (J & W Scientific). (C, D) Replicate pyrograms produced by pyrolysis at 800°C with pyrolyzates separated on a 0.25 μm film thickness FFAP fused silica column (Quadrex).

be produced by many samples and build-up inside the system after a period of use, confounding long-term reproducibility. During temperature programming, small amounts of the contaminating materials are flushed from the system. Possible troubleshooting solutions include investigating different pyrolysis temperatures and times to pyrolyze the sample more effectively and evaluating the pyrolysis elements for possible contamination. Wiping the inner surfaces of the pyrolysis interface with a cotton swab will reveal if residue accumulation is the problem. If residue is present, the pyrolysis interface and GC interface should be cleaned using various solvents, including dilute nitric acid, hexane, chloroform, and acetone in that order. Irwin [12] recommends overnight heating (600–700°C) of solvent-washed pyrolysis elements, heating in a water-saturated stream of hydrogen at 550°C, or chemical polishing followed by washing with acetone and heating. Direct flame heating should be avoided as it may oxidize metal, contaminating surfaces and changing heat transfer characteristics. Quartz sample tubes should be cleaned in dilute nitric acid at 100°C overnight, rinsed with methanol and acetone, then dried in an oven.

Sample sizes that are too large (approaching 1 mg) may contaminate the column. If residue from pyrolysis cannot be avoided, using less sample may reduce contamination and extend column life. However, contamination becomes even more obvious when small samples are involved. Occasionally with increased usage, contamination may be visibly detected on the first few cm of a column. A rather drastic but appropriate solution is to clip off the initial section of column. Removal of a section of contaminated column does not have a great effect on resolution and may restore performance to a degraded system. Crosslinked or bonded phase columns can sometimes be rejuvenated by rinsing with organic solvent such as methylene chloride and then reconditioning. Mitchell and Needleman [57] described a disposable precolumn with backflushing capability to remove contaminants from a pyrolysis-GC system. Sugimura and Tsuge [58] designed a splitter system for pyrolysis with capillary GC that may be used to reduce column contamination without decreasing chromatographic performance. More recently, Halket and Schulten [59] modified a Curie-point inlet system for capillary columns to permit better identification of all products, including less volatile products left behind in a quartz pyrolysis chamber.

In Py-GC or Py-MS systems, contamination may be caused by a cold spot that is trapping less volatile residue. To see if any temperature zones are set too low, temperature settings should be increased by small increments (without exceeding temperature limits of the column or instrument) and the results checked. Wrapping connecting tubing with heating tape or insulation may eliminate cold spots. Condensation losses due to cold spots can also severely limit the ability to observe higher-molecular-weight and less volatile components. Flexible fused silica columns can be inserted directly into the pyrolysis interface near the pyrolyzer element at one end and threaded straight through to the FID flame tip or the MS ion source at the other end. Contact with active surfaces is minimized and more complete transfer of the pyrolysis products from the pyrolysis interface to detector is achieved.

Similar considerations are important in Py-MS: cold spots and wall contact in the Py-MS vacuum systems can also affect transmission of pyrolyzates [11]. Py-MS produces a pyrogram of all compounds in the pyrolysis product mixtures superimposed in a single mass spectrum. For that reason, interpretation of Py-MS results from a complex sample can be more difficult than interpretation of a Py-GC/MS pyrogram, in which pyrolysis products are chromatographically separated before MS detection. As pointed out by Snyder et al. [53], Py-MS with relatively cold pyrolysis interface walls or with expansion chambers tend to provide only low mass range analysis (under m/z 200). When direct Py-MS is performed with the pyrolysis reactor close to the ion source so as to detect larger mass pyrolysis products (m/z 200–1000), the ion source tends to contaminate rather quickly, jeopardizing reproducibility.

Long-term reproducibility of pyrograms in Py-GC or Py-GC/MS depends on the lifetime of chromatographic columns. Fused silica columns coated with polar bonded phases (such as the DB-1701 and FFAP columns mentioned above) have been used in our laboratory for periods longer than one year without appreciable degradation.

In any pyrolysis work, a valuable habit to acquire is the analysis of a standard sample at regular intervals between other samples. When "ghost peaks" appear in a pyrogram, analysis of the standard sample as a quality check, coupled with one or more of the ideas described above, may help to resolve the contamination problem.

IV. SELECTED APPLICATIONS

A summary of experimental conditions for selected analytical pyrolysis studies of microorganisms is given in Table 1. The literature review included here is not intended to be comprehensive, but is designed to summarize selected recent applications of analytical pyrolysis to the analysis of microorganisms.

A. Discrimination of Gram-Positive and Gram-Negative Bacteria

Perhaps the simplest and most basic distinction made of microbial cells is that of gram type, and numerous studies have assessed the value of analytical pyrolysis for this task. Early systematic studies included those of Simmonds et al. [60,61], who correlated the formation of specific pyrolysis products with their site of origin in the microorganism. Pyrolysis products were assigned to protein, carbohydrate, nucleic acid, lipid, and porphyrin sources in the cell. Acetamide was identified in pyrograms of whole cells and it was speculated that acetamide was produced from the *N*-acetyl groups on the glycan backbone of peptidoglycan (see Fig. 2). Model compounds with chemical structures similar to substructures of PG were pyrolyzed by Hudson et al. [62] and by Eudy et al. [63]. Only very low amounts of acetamide were found in pyrograms of glucosamine, muramic acid, and trialanine. Acetylated compounds such as *N*-acetyl glucosamine, *N*-acetyl muramic acid, and muramyl dipeptide produced much larger amounts of acetamide when pyrolyzed. It was not suggested that PG is the only source of acetamide in microbes, only that it is a major source. Levels of acetamide were greater in pyrograms of gram-positive bacteria than in gram-negative bacteria [18,62,63]. These results correlate well with the higher levels of PG found in gram-positive cells [18,64]. A second pyrolysis product, propionamide, was also seen to originate in part from

TABLE 1 Summary of Selected Experimental Conditions for Pyrolysis-GC and Pyrolysis-GC/MS of Microorganisms.

Organism/ components	Pyrolysis conditions	Stationary phase	Column and material	Detector	Ref.
Bacillus species	Curie-point 510°C	—	—	VG Pyromass 8–80 Py-MS	80
Bacilli	CDS Pyroprobe 650°C, 10 s fast ramp	DB-1701	30 m × 0.329 mm i.d., 1 μm film fused silica	Hewlett-Packard MSD EI GC/MS	83
	CDS Pyroprobe 800°C, 5 s fast ramp	FFAP	25 m × 0.25 mm i.d., 0.25 μm film fused silica	Hewlett-Packard MSD EI GC/MS	36
B. cereus, B. globigii spores, *B. subtilis,* MS-2 aldolase, *E. coli, Secale cerale* coliphage, fog oil, diesel and wood smoke, pollen, Type X, dry yeast	Curie-point 610°C	—	—	Extrel ELQ400 Extrel EL-400 triple quadrupole MS, EI	79 82

Table 1 Continued

Organism/components	Pyrolysis conditions	Stationary phase	Column and material	Detector	Ref.
B. cereus, B. licheniformis B. subtilis, E. coli, S. aureus	Curie-point 610°C	—	—	Extrel EL-400 triple quadrupole MS, EI	97
B. subtilis, E. coli, M. tuberculosis	Curie-point 358, 510, and 610°C	BP-1	12 m × 0.22 mm i.d., 0.15 μm film fused silica	Perkin-Elmer 8500 GC, Finnigan 700 MAT ion trap MS	93
B. cereus, B. licheniformis, B. subtilis, B. thuringiensis, E. coli, S. aureus	Curie-point 610°C	007-1 Quadrex methyl silicone	15 m × 0.250 mm i.d., 100%	Extrel EL-400 triple quadrupole MS, EI Varian 4000 GC	98
B. anthracis, B. cereus, B. licheniformis, B. thuringiensis, B.	CDS Pyroprobe 1000°C, 20 s fast ramp	BP-1	5 m × 0.25 μm film, capillary	HP 5890 GC and Finnigan-MAT 700	95

subtilis, E. coli, P. fluorescens, S. albus, S. aureus				ion trap MS	
B. cereus, B. subtilis, E. coli	Curie-point 510°C	methyl silicone	4 mm × 0.25 mm i.d., fused silica	Extrel triple quadrupole MS, CI	81
Bacteroides gingivalis	Curie-point 358°C, 510°C	—	—	FOM Institute quadrupole MS	100
Biomaterials	Curie-point 500°C	DB-1	30 m × 0.32 mm i.d., 0.25 μm film, fused silica	Varian 3700 GC Finnigan MAT 212 MS, EI/FID, FID	59
E. coli HB101, HB101 (pSLM204)	Curie-point 530 °C	—	—	Horizon Instruments PYMS-200X quadrupole MS, EI	102
Flavobacterium	CDS Pyroprobe 900°C, 10 s fast ramp	methyl-silicone or SE-52	20 m × 0.21 mm i.d. fused silica or 8 m glass capillary	Hewlett-Packard 5985 EI GC/MS	35

TABLE 1 Continued

Organism/ components	Pyrolysis conditions	Stationary phase	Column and material	Detector	Ref.
Gram-positive, gram-negative bacteria, and cell walls	CDS Pyroprobe 800°C fast ramp	Carbowax 20M or Superox-4	20 m × 0.25 mm i.d. glass capillary or 25 m × 0.2 mm i.d. fused silica	Hewlett-Packard 5992 EI GC/MS and Finnigan EI GC/MS	62 63 64
Capsular polysaccharide from *Klebsiella*	CDS Pyroprobe 650°C, 10 s	DB-1701	30 m × 0.329 mm i.d. 1 μm film fused silica	Finnigan MAT GC/ MS, EI/CI	87
Lipids, lipopolysaccharides, fatty acids	Direct CI 570°C	—	—	Scientific Res. Instruments CI MS	99
Listeria monocytogenes	Curie-point 358°C	—	—	Spectrel quadrupole MS	88

Mycobacteria/lipids	Curie-point 610°C	SE-30	5 m × 0.32 mm i.d. fused silica	Finnigan ion trap	94
		BP-1	5 m, 0.25 1 film	GC/MS	
Mycobacteria/methyl mycolates	Curie-point 358°C	OV-1	3% (w/w) packed column 1 m × 3 mm i.d.	Hitachi EI GC/MS	90
Methylated microbial fatty acids	Curie-point 510°C	DB-5	7.2 m × 0.2 mm i.d. fused silica 0.11 μm film	Hewlett-Packard MSD, EI GC/MS, FID	92
Penicillium italicum/ polysaccharides	DCI pyrolysis	—	—	Finnigan MAT HSQ-30	105
Pseudomonas and *Xanthomonas* bacteria	Curie-point 510°C	—	—	Extrel quadrupole MS	101
Streptococci/glucitol phosphate	CDS Pyroprobe 800°C fast ramp	FFAP	25 m × 0.25 mm i.d. fused silica EI GC/MS	Hewlett-Packard MSD	75, 76 37, 42 18

the lactyl-peptide bridge region of PG [63]. This pyrolysis product is not produced in sufficient amount, and probably has other sources of origin in the microbial cell, so as to not be valuable for differentiating gram type [18].

Furfuryl alcohol has been suggested to derive in minor amounts from carbohydrates and in major amounts from RNA and DNA in microorganisms [60,61,63]. As gram-negative bacteria have relatively thin cell walls compared to gram-positive bacteria, the higher levels of furfuryl alcohol in pyrograms of gram-negative bacteria may be due to intracellular DNA being more readily released from gram-negative bacteria [18]. Although not directly related to gram typing, a pyrolysis-GC/MS method for measuring DNA content of cultured cells was recently compared to results from an established colorimetric method based on reaction with diphenylamine [65]. The single step procedure used cell samples containing 0.1–25 μg of DNA and had a detection limit of about 100 ng DNA. Furfuryl alcohol from DNA was also found to be the source peak in pyrograms of virally transformed mammalian cells that could discriminate them from their normal counterparts [66,67]. This was not surprising as DNA levels are known to be higher in cancer cells.

Figure 4 shows relative amounts of acetamide and furfuryl alcohol measured in the GC/MS pyrograms of a group of microorganisms [18]. Replicate pyrolysis results are connected by lines to indicate their reproducibility. Fungal samples included in this study as a control produced low levels of both pyrolysis products. Gram-positive bacteria cluster towards higher levels of acetamide and moderate levels of furfuryl alcohol. Gram-negative bacteria produced higher levels of furfuryl alcohol and moderate levels of acetamide. Diagonal lines (with no statistical significance) can be drawn separating the gram-positive and gram-negative groups. The data from some organisms are close to the dividing lines. Obviously, acetamide and furfuryl alcohol have other sources within cells and are not generally useful as specific chemical markers for bacterial PG. Morgan et al. [18] concluded that it remains to be seen if individual bacteria can be gram typed by analytical pyrolysis.

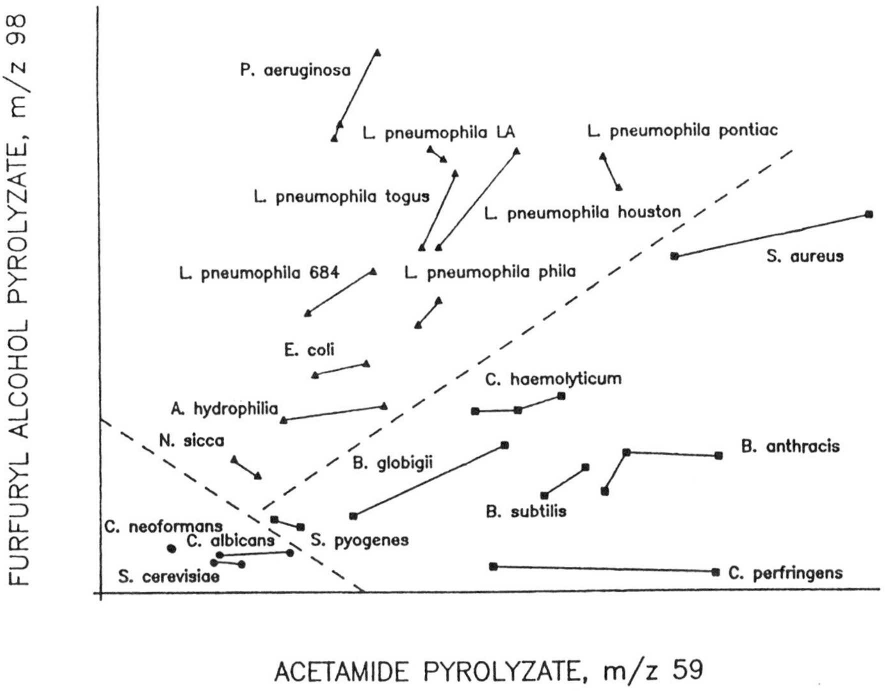

Figure 4 Plot of relative amounts of acetamide and furfuryl alcohol in a diverse group of microorganisms. Amounts are measured as the integrated reconstructed ion intensity (m/z 59 for acetamide, m/z 98 for furfuryl alcohol) at the appropriate retention time as a percentage of the total ion intensity. Gram-positive bacteria are represented by squares, gram-negative bacteria by triangles, and fungi by circles. (Reproduced with permission from Ref. 18. Copyright 1990, Plenum Publishing.)

Adkins et al. [49] reported ions at m/z 166 in negative ion mass spectra following pyrolysis of *N*-acetylglucosamine, chitin, and *B. subtilis* peptidoglycan. It has been speculated that larger pyrolysis products more specific for bacterial PG might be generated by pyrolysis, but a comprehensive study with diverse gram-positive and gram-negative bacteria has not validated this possibility. However, recent pyrolysis results from Voorhees et al. [68] indicate that gram typing may be more readily demonstrated with the higher selectivity of the triple quadrupole MS.

B. Analysis of Gram-Positive Organisms

Streptococci have been popular targets for analytical pyrolysis [18,37,42,69–76], perhaps because their chemical composition has been well-characterized by other techniques. One of the first papers describing additivity of pyrograms of individual components in a complex microbial pyrogram was published by Huis In't Veld et al. [69]. The pyrogram of a streptococcal strain containing a polysaccharide cell wall antigen was the sum of the pyrogram of its streptococcal mutant lacking this antigen and the pyrogram of the isolated antigen.

In a study involving 14 different strains, Smith et al. [75] found that group B streptococci (*Streptococcus agalactiae*) could be differentiated from group A streptococci by a single peak. Electron impact (EI) mass spectra of model compounds, including group B polysaccharide antigen and glucitol, suggested that the distinctive peak in the pyrogram of group B organisms was derived from glucitol phosphate residues in the group-specific polysaccharide. This carbohydrate-derived chemical marker for group B streptococci was identified by EI and methane CI MS as a dehydration product of glucitol-6-phosphate, dianhydroglucitol [18]. Figure 5 shows this differentiation using selected ion monitoring to focus on the characteristic m/z 86 ion of the dianhydroglucitol marker. Later studies using 29 different strains differentiated groups A, B, C, F, and G streptococci from one another [37,42,76]. Group B pyrograms were clearly distinguishable from the other groups, and samples from groups A and F were usually classified correctly [37]. That group C and G strains were indistinguishable was not surprising because no commercial available system was capable of distinguishing these groups. The discrimination by pyrolysis was judged as good as that achievable by any technique except serogrouping.

Identification of *Staphylococci* tends to be slow and irre-

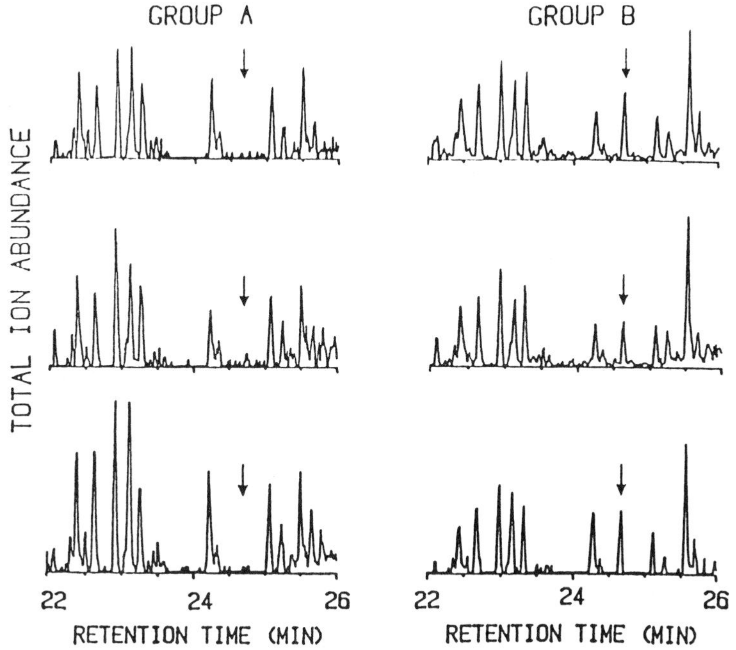

FIGURE 5 Discrimination of group A and group B streptococci shown in an expanded view of total ion abundance pyrograms in the region from 22–26 min. Pyrograms from three different samples of group A and group B isolates are shown. The peak marked with an arrow is dianhydroglucitol. Pyrolysis conditions: CDS model 120 Pyroprobe, 80 μg sample, pyrolysis settings at 800°C and 75°C/ms. Chromatographic conditions: 25 m $\times$ 0.24 mm FFAP fused silica capillary, Hewlett-Packard model 5880A GC coupled to a model 5970 mass selective detector. (Reproduced with permission from Ref. 75. Copyright 1987 American Chemical Society.)

producible and a commercially available system (APIStaph) only reduces identification time to 24 h [77]. Hindmarch and Magee [78] analyzed a total of 451 strains of staphylococci from a clinical investigation by Curie-point pyrolysis-GC. Samples were pyrolyzed at 610°C and packed GC columns were used to acquire pyrograms. After discarding some results from anomalous strains, discrimination by pyrolysis for only 60 of 415 strains were in disagreement with biochemical testing. At least a third of the misidentified strains were atypical strains.

Samples of a single *S. aureus* strain were pyrolyzed in an air atmosphere tube furnace and in a vacuum Curie-point pyrolyzer connected to a triple quadrupole mass spectrometer [79]. Pyrograms of *S. aureus* were distinguishable from those of *E. coli* and *Bacillus* strains. Peaks related to fatty acids were identified in the mass pyrograms.

Curie-point pyrolysis MS was applied to the identification of selected *Bacillus* species by Shute et al. [80]. Fifty-three strains of *B. subtilis*, *B. pumilus*, *B. licheniformis*, and *B. amyloliquefaciens* in both a sporulating and nonsporulating state were pyrolyzed at 510°C using a PyroMass 8-80 instrument (VG Gas Analysis, Middlewich, Cheshire, UK). The four strains could be differentiated by Py-MS from nonsporulated cultures. With sporulated cultures, the pyrograms were more similar and only *B. licheniformis* could be completely differentiated.

Adkins et al. [49] presented temperature-resolved pyrolysis-mass spectra of cell fractions from gram-positive bacteria of the genus *Bacillus*. Culture conditions were found to play a major role in determining exact pyrolysis profiles. Figure 6 shows two pyrolysis mass spectra of *B. subtilis* samples grown under different conditions. Pyrolysis products were identified from amino sugars in peptidoglycan as well from teichoic acids present in the *bacilli.* In one of the first pyrolysis-triple quadrupole MS analyses of bacteria, Voorhees et al. [81] compared Curie-point pyrograms of *B. cereus*, *B. subtilis*, and *E. coli*. Parent ion scans of daughter ions selected by pattern recognition were found to give the best identification of the three species. Morgan et al. [17] reported that dipicolinic acid from sporulating organisms produces pyridine upon pyrolysis and found a pyridine peak to differentiate spore-forming and vegetative strains of *B. anthracis*. Voorhees et al. [82] have also recently used pyrolysis-tandem MS to analyze samples of *B. subtilis*, *B. globigii*, *E. coli*, and other possible interferents (fog oil, diesel smoke, dry yeast,

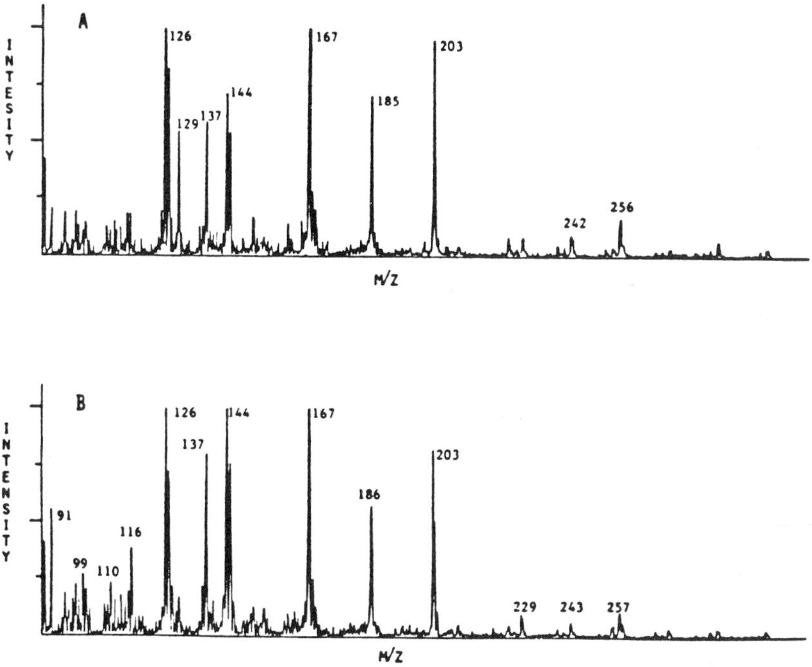

FIGURE 6 Pyrolysis mass spectra of *B. subtilis* grown (A) on minimal media, and (B) in penassay broth. (Reproduced with permission from Ref. 49. Copyright 1984, Elsevier.)

and pollen). Four useful markers for biological substances were identified: adenine from nucleic acids; diketopiperazine and indole from proteins; and pyridine from picolinic acid in sporulating bacteria.

A pyrolysis product of galactose was identified in *B. anthracis* by Watt et al. [83]. The anhydrosugar 1,6-anhydro-galactopyranose was found in pyrograms of *B. anthracis* and absent in pyrograms of *B. cereus*. The presence of galactose in *B. anthracis* was confirmed by an alditol acetate derivatization GC/MS method. The results support the assertions of Helleur et al. [84–87] that analytical pyrolysis can be used for rapid carbohydrate profiling of complex biological samples such as microorganisms.

Samples of the gram-positive cocobacillus *Listeria monocytogenes* from infected patients and food sources were analyzed by Py-MS [88]. Samples within the serogroup could be differentiated to the extent that the strain obtained from the food was ruled out as a possible source for human infections. The independence of Py-MS from the labor intensive use of reagent probes for classification of these organisms was noted by these researchers.

C. Analysis of Gram-Negative Organisms

During the 1980s, organisms classified in the family Legionellaceae grew rapidly as new serogroups of Legionella-like organisms were identified. Pyrolysis-GC/MS with selected ion monitoring was used to differentiate 21 strains of *Legionella* organisms [6]. Replicate samples (100 μg) were pyrolyzed at 800°C using a model 120 Pyroprobe (CDS Analytical) and 15 pyrolyzates were monitored by SIM GC/MS. The pyrograms of *Legionella pneumophila*, *Tatlockia micdadei*, and *Fluoribacter* could be distinguished from one another. In a Py-MS study of *Legionella* species, Kajioka and Tang [89] suggested that differences in ion profiles could be used to distinguish the various species. Samples were pyrolyzed in a Curie-point pyrolyzer at 358°C and analyzed by a quadrupole mass spectrometer. Specific and reproducible pyrograms providing distinctive fingerprints for *Legionella* strains were obtained by Py-MS.

A pyrolysis fragment, 2-butenoic acid, from the lipid substance poly-β-hydroxybutyrate present in microbial samples was identified by Watt et al. [36]. The use of this pyrolysis product as a chemical marker was validated by correlating results from an extraction, hydrolysis, and derivatization GC/MS method, by growth trials that profiled PHB content as function of culture age, and by comparative pyrolysis studies of a diverse group of organisms. PHB was shown to be common at moderate to high levels in the family Legionellaceae. Only background to low levels of PHB were in organisms of the family Enterobacteriaceae. Figure 7 compares total ion abundance GC/MS pyrograms and reconstructed ion m/z 86 pyrograms for two organisms, *L. pneumophila* and *P. vulagaris*. In common with many other bacilli, *B. anthracis* was also found to contain PHB.

The bacterium *Klebsiella* is known to possess capsular polysaccharides. Helleur [87] studied isolated capsular polysaccharides from *Klebsiella* by pyrolysis-GC/MS. The polysaccharide is composed of repeating three glucopyranosyl units, one galactopyranosyl unit, one galactofuranosyl unit, one rhamnopyranosyl unit, and one glucouronosyl unit. Specific anhydrosugars pro-

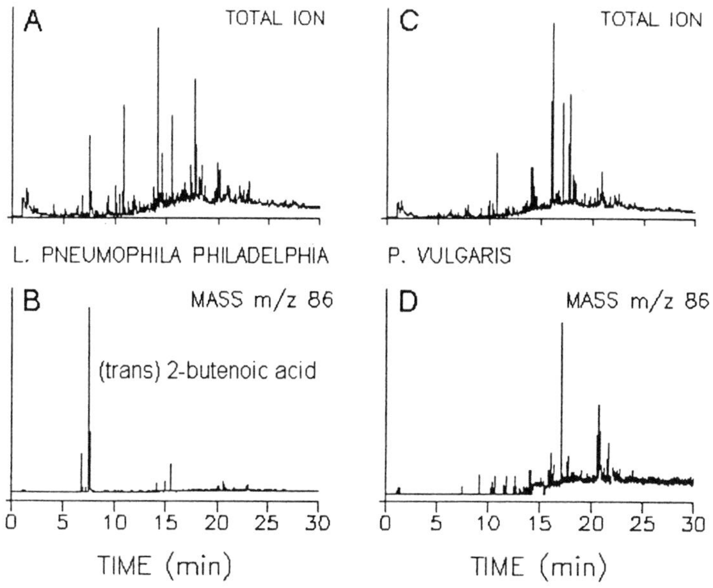

FIGURE 7 (A) Total ion abundance pyrogram of *L. pneumophila* Philadelphia. (B) Reconstructed ion pyrogram for ion mass m/z 86 of *L. pneumophila* Philadelphia. Peak 3 is *trans*-2-butenoic acid. (C) Total ion abundance pyrogram of *P. vulgaris*. (D) Reconstructed ion pyrogram for ion mass m/z 86 of *P. vulgaris* (Adapted with permission from Ref. 36.)

duced upon pyrolysis for each type of saccharide unit were identified and quantified (Fig. 8).

Pyrolysis-GC/MS of 32 species of mycobacteria for classification based upon the methyl ester fatty acid profiles was performed by Kusaka et al. [90]. The mycobacteria were successfully classified into four groups: 22-group, C_{20} to C_{24} fatty acids with C_{22} predominating; 24-group, C_{22} to C_{24} fatty acids with C_{24} predominating; 24′-group, C_{22} to C_{26} fatty acids with C_{24}

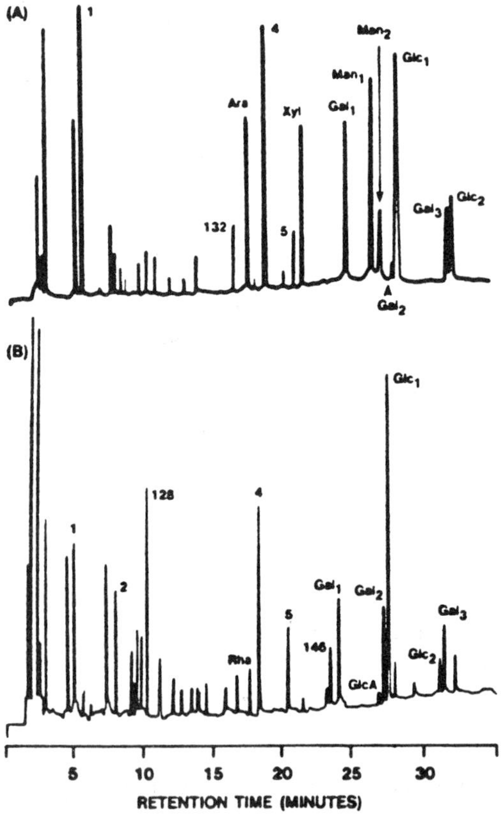

FIGURE 8 Capsular polysaccharides identified by pyrolysis GC/MS from Klebsiella K41 strains. (Reproduced with permission from Ref. 87. Copyright 1987, Elsevier.)

predominating; and 26-group, C_{22} to C_{26} fatty acids with C_{26} predominating.

Curie-point pyrolysis coupled with short-column GC interfaced to an ion trap mass spectrometer was employed by Snyder et al. [53,91] to identify lipid components of microorganisms. A portion of the lipid component was further distinguished as dehydrated mono- and diacylglycerides. Discrimination of diverse microorganisms (four different bacilli, *S. aureus*, *E. coli*, *L. pneumophila*) was performed by visual comparison of total ion chromatograms and selected reconstructed ion chromatograms.

Holzer et al. [92] employed Curie-point pyrolysis to analyze microbial fatty acid by in situ methylation with trimethylanilinium hydroxide (TMAH). The fatty acid methyl ester profiles produced during pyrolysis of the whole cell agreed well with the analysis of lipid extracts from the same microorganism. Dworzanski et al. [93] also used pyrolytic methylation-GC to profile fatty acids in whole cells of *E. coli*, *Mycobacterium tuberculosis*, and *B. subtilis*. Curie-point pyrolysis coupled to GC/ion trap MS was used by Snyder et al. [94] to characterize 14 strains representing nine bacterial species for their lipid biomarker content. A quartz-tube Pyroprobe from CDS Analytical was used with GC/ion trap MS by Smith and Snyder [95] to analyze lipid and fatty acid components in a variety of bacteria with a total analysis time under 10 min. Snyder et al. [96] also used 13 strains of eight bacterial species to demonstrate the use of Curie-point wire and quartz-tube pyrolysis coupled to short column GC/MS for the detection of biological warfare agents. Lipid and nucleic acid components were the primary target for discrimination purposes. Two articles by DeLuca et al. [97,98] studied Curie-point pyrolysis-tandem MS for direct analysis of bacterial fatty acids. Py-MS results correlated well with previous Py-GC/MS data on the same bacteria. Interestingly, better classification results were obtained when just Py-MS fatty acid profiles were employed rather than total ion spectra.

Adkins et al. [99] also produced temperature-resolved pyrolysis results for cell envelope fractions and whole cells of *Salmonella typhimurium* and for several *Salmonella* LPS samples.

Model compounds (fatty acids, phospholipids, cholines, etc.) for LPS structures were also pyrolyzed using linear-programmed thermal degradation strategies. Common ions were found in the Py-MS fingerprints of cell wall components and the model compounds. Boon et al. [100] applied Py-MS to the analysis of samples from *Bacteroides gingivalis* isolated from dental patients. Pyrolysis mass spectra were dominated by volatile components but clear differences were detected between related *Bacteroides* strains. A rule-building expert system was shown to be able to classify pyrolysis mass spectra from gram-negative *Pseudomonas* samples versus samples containing *Xanthomonas*, a plant pathogen [101]. Goodacre et al. [102] demonstrated the use of Py-MS to detect the fimbrial adhesive antigen F41 from *E. coli* and to discriminate between bacteria which differ genotypically in that regard.

D. Analysis of Fungi

Pyrolysis-GC fingerprints were obtained for fungal spores by Papa et al. [103] using packed column chromatography with Carbowax 20M on 80–100 mesh Supelcoport. Papa et al. [104] also characterized several fungi, including *Agaricales Boletus*, *Russula*, *Aminata Lepiota*, *Agaricus*, and *Lycoperdon*. Pyrolysis coupled to capillary GC/MS was used to identify discriminating components. Among other results, it was noted that pyrograms of *B. calopus* and *B. bovinus* contained saturated and unsaturated aliphatic hydrocarbons and toluene. *B. bovinus* pyrograms contained esadecanoic acid.

Pyrolysis-direct CI MS was employed by Tas et al. [105] to detect ions related to polysaccharides in pyrograms of fungal *Penicillium italicum* strains. A series of Py-MS ions at m/z 167, 185, 187, and 210 were identified as originating from *N*-acetylglucosamine residues, present at higher levels in the strain cultured in the presence of imazalil fungicide. Tas et al. [106] also used pyrolysis-direct CI MS to investigate differences between *Candida albicans* and *Ophiostoma ulmi* fungi. Hexoses, deoxyhexose, *N*-acetylglucosamine from chitin, and sterols in treated yeast samples could be detected in the pyrograms.

V. CONCLUSIONS

In his 1982 text on analytical pyrolysis, Irwin [12] described the ideal taxonomic method for the analysis of microorganisms: universally applicable to diverse microorganisms, capable of differentiating groups of organisms, reproducible independent of operator and system, rapid, sensitive, and capable of both being automated and interpreted in biochemical or chemical terms. At the time, Irwin also noted that "no comprehensive taxonomic scheme based on pyrograms of mass pyrograms had been proposed."

The identification of pyrolysis products and their sources in intact organisms was not widespread in pyrolysis studies until the last decade. Classification and differentiation of organisms were often based on the relative peak heights of one or more peaks at a given retention time in the pyrogram. This simplistic approach can lead to erroneous conclusions when the chemical identities of pyrolysis products and their origin are unknown.

When pyrolysis is applied to a microbial sample, a complex mixture of thermal degradation products is produced. Analytical pyrolysis is often performed in a fingerprinting mode using sophisticated pattern recognition methods. However, if the chemical basis of pattern differences is not defined, pyrograms are so complex that minor variations in instrumental conditions causes sufficient changes in patterns that making comparisons over extended time and between laboratories is a formidable task.

Ideally, pyrolysis products are preferred that retain as much of the structural integrity and chemical uniqueness of the original monomer so that their origin can be identified. Simple scission fragments may not retain as much chemical uniqueness of the parent structure as products such as those from dehydration, simple rearrangements, or generated by newer soft ionization methods and detected directly by MS. Many pyrolysis products are common organic compounds that could potentially be derived from multiple sources. Some pyrolysis products, although their origin may be well defined, may not be useful as chemical markers for bacterial discrimination.

By identifying specific chemicals in pyrograms as originating from taxonomically relevant microbial structures, discrimination may be based on well-defined features, simplifying this process dramatically. Invariant chemical features of the organism will provide effective discrimination if distinctive pyrolysis products can be generated under a wide range of experimental conditions. Furthermore, the discrimination thus achieved should be reproducible between different instruments and different laboratories. While quantitative amounts of pyrolyzates generated may vary, the absence or presence of pyrolysis products due to a true chemotaxonomic marker should not depend dramatically on the choice of pyrolysis system.

Appropriate groups of organisms for chemotaxonomic studies involving pyrolysis can often be chosen on the basis of a priori information concerning microbial chemical structures. Interpretation is simplified by selecting groups of organisms, or cell fractions, that differ in defined structural characteristics. Analysis of bacteria possessing unrelated differences may not permit significance of a particular pyrolysis product to be evaluated. Organisms not containing particular known chemical structures act as a blank indicating background levels produced from other sources. Analysis of multiple strains also confirms the consistency of correlations established among results from different organisms.

Automation of sampling is more readily accomplished using Py-MS than Py-GC/MS; however, better and automated sampling methods are needed for all forms of analytical pyrolysis. Absence of standard reference materials for many microbial components makes peak validation difficult and is a hindrance to systematic studies. Improved high-resolution capillary GC with inert polar phases capable of separating the complex mixture of pyrolysis products from microbial samples are needed. A new generation of MS techniques are evolving and being applied to microbiological problems: laser and plasma desorption, fast-atom bombardment, new chemical ionization approaches, and tandem MS. Many of these advances in MS detection will find their way into pyrolysis applications. Multiple detectors coupled in "hyphenated" modes will allow more thorough characterization of the microbial pyrolysis product mixture. Finally, advances in small computers are also driving data analysis techniques closer to the analytical instrument. The future may bring us closer to Irwin's ideal taxonomic method for the analysis of microorganisms.

REFERENCES

1. R. E. Buchanan and N. E. Gibbons, eds., *Bergey's Manual of Determinative Bacteriology*, 8th ed., Williams and Wilkins, Baltimore (1974).
2. C. S. Cummins, in *Analytical Microbiology Methods: Chromatography and Mass Spectrometry* (A. Fox, S. L. Morgan, L. Larsson, and G. Odham, eds.), Plenum Press, New York, pp. 53–57 (1990).
3. S. L. Morgan, A. Fox, and J. Gilbart, *J. Microbiol. Methods*, *9*: 57–69 (1989).
4. A. Fox, J. Gilbart, and S. L. Morgan, in *Analytical Microbiology Methods: Chromatography and Mass Spectrometry* (A. Fox, S. L. Morgan, L. Larsson, and G. Odham, eds.), Plenum Press, New York, pp. 1–17 (1990).
5. W. K. Joklik, H. P. Willet, and D. B. Amos, eds., *Zinsser Microbiology* 17th ed., Appleton-Century-Crofts, New York, (1980).
6. A. Fox and S. L. Morgan, in *Instrumental Methods for Rapid Microbiological Analysis* (W. H. Nelson, ed.), VCH Publishers Inc., Deerfield Beach, FL, pp. 135–164 (1985).
7. P. D. Zemany, *Anal. Chem.*, *24*: 709 (1952).
8. V. I. Oyama, *Nature (London)*, *200*: 1058 (1963).
9. E. Reiner, *Nature*, *206*: 1272 (1965).
10. C. S. Gutteridge and J. R. Norris, *J. Appl. Bacteriol.*, *47*:5 (1979).
11. H. L. C. Meuzelaar, J. Haverkamp, and F. D. Hileman, *Pyrolysis Mass Spectrometry of Recent and Fossil Biomaterials*, Elsevier, Amsterdam (1982).
12. W. J. Irwin, *Analytical Pyrolysis: A Comprehensive Guide*, Marcel Dekker, New York (1982).

13. G. Wieten, H. L. C. Meuzelaar, and J. Haverkamp, in *Gas Chromatography/Mass Spectrometry Applications in Microbiology* (G. Odham, L. Larsson, P.-A. Maardh, eds.), Plenum, New York, pp. 335–380 (1984).
14. F. L. Bayer and S. L. Morgan, in *Pyrolysis and GC in Polymer Analysis* (E. J. Levy and S. Liebman, eds.), Marcel Dekker, New York, pp. 277–337 (1985).
15. T. P. Wampler, *J. Anal. Appl. Pyrol.*, *16*: 291–322 (1989).
16. G. A. Eiceman, W. Windig, and A. P. Snyder, in *Gas Chromatography: Biochemical, Biomedical, and Clinical Applications* (R. E. Clement, ed.), John Wiley & Sons, New York, pp. 327–347 (1990).
17. S. L. Morgan, A. Fox, J. C. Rogers, and B. E. Watt, in *Modern Techniques for Rapid Microbiological Analysis* (W. H. Nelson, ed.), VCH Publishers, Inc., New York, pp. 1–18 (1991).
18. S. L. Morgan, B. E. Watt, K. Ueda, and A. Fox, in *Analytical Microbiology Methods: Chromatography and Mass Spectrometry* (A. Fox, S. L. Morgan, L. Larsson, and G. Odham, eds.), Plenum Press, New York, pp. 179–200 (1990).
19. J. Postgate, *Microbes and Man*, 3rd ed., Cambridge University Press, Cambridge (1992).
20. J. D. Watson, N. H. Hopkins, J. W. Roberts, J. A. Steitz, and A. M. Weiner, *Molecular Biology of the Gene*, 4th ed., Benjamin/ Cummings Company, Inc., Menlo Park, CA, p. 101 (1987).
21. H. J. Rogers, *Bacterial Cell Structure*, Van Nostrand Reinhold Co. Ltd., Wokingham, Berkshire (1983).
22. H. J. Rogers, H. R. Perkins, and J. B. Ward, *Microbial Cell Walls and Membranes*, Chapman and Hall, London (1980).
23. K. H. Schliefer and O. Kandler, *Bacteriol. Rev.*, *36*: 407 (1972).
24. K. Ueda, S. L. Morgan, A. Fox, A. Sonesson, L. Larsson, and G. Odham, *Anal. Chem.*, *61*: 265–270, (1989).
25. A. Fox, K. Ueda, and S. L. Morgan, in *Analytical Microbiology Methods: Chromatography and Mass Spectrometry* (A. Fox, S. L. Morgan, L. Larsson, and G. Odham, eds.), Plenum Press, New York, pp. 89–99 (1990).
26. J. Baddiley, in *Essays in Biochemistry*, Vol. 8 (P. N. Campbell and F. Dickens, eds.), Academic Press, London, pp. 35–78 (1972).

27. O. Luderitz, O. Westphal, A. M. Staub, and H. Nikaido, in *Microbial Endotoxins*, Vol. 4 (G. Weingbaum, S. Kadis, and S. J. Ayl, eds.), Academic Press, London, pp. 145–223 (1971).
28. C. W. Moss, in *Analytical Microbiology Methods: Chromatography and Mass Spectrometry* (A. Fox, S. L. Morgan, L. Larsson, and G. Odham, eds.), Plenum Press, New York, pp. 59–69 (1990).
29. V. Holdeman, E. P. Cato, and W. C. Moore, *Anaerobe Laboratory Manual*, 4th ed., Virginia Polytechnic Institute and State University, Anaerobe Laboratory, Blacksburg (1977).
30. L. Miller and T. Berger, Bacteria identification by gas chromatography of whole cell fatty acids, Hewlett-Packard Application Note, *228*: 41, (1985).
31. A. Fox, S. L. Morgan, and J. Gilbart, in *Analysis of Carbohydrates by GLC and MS* (C. J. Bierman and G. McGinnis, eds.), CRC Press, Boca Raton, FL (1989).
32. D. G. Pritchard, B. M. Gray, and H. C. Dillon, *Arch. Biochem. Biophys.*, *235*: 385, (1984).
33. A. Fox, J. C. Rogers, J. Gilbart, S. L. Morgan, C. H. Davis, S. Knight, and P. B. Wyrick, *Infect. Immun.* *58*: 835–837 (1990).
34. C. S. Gutteridge and J. R. Norris, *Appl. Environm. Microbiol.*, *40*: 462 (1980).
35. H. Engman, H. T. Mayfield, T. Mar, and W. Bertsch, *J. Anal. Appl. Pyrol.*, *6*: 137 (1984).
36. B. E. Watt, S. L. Morgan, and A. Fox, *J. Anal. Appl. Pyrol.*, *20*: 237–250 (1991).
37. C. S. Smith, S. L. Morgan, C. D. Parks, A. Fox, *J. Anal. Appl. Pyrolysis*, *18*: 97–115 (1990).
38. *NIH Biohazards Safety* Guide, U.S. Department of Health, Education, and Welfare, Washington, D.C. (1974).
39. Occupational Safety and Health Administration, U.S. Department of Labor, Occupational safety and health standards for general industry (29 CFR Part 1910), Commerce Clearing House, Inc., Chicago, IL.
40. W. Windig, *J. Anal. Appl. Pyrol.*, *17*: 283–289 (1990).
41. G. Montaudo, *J. Anal. Appl. Pyrol.*, *13*: 1–7 (1988).
42. J. Gilbart, A. Fox, and S. L. Morgan, *Eur. J. Clin. Microbiol.*, *6*: 715–723 (1987).

43. T. P. Wampler and E. J. Levy, *J. Anal. Appl. Pyrol.*, *12*: 75–82 (1987).
44. H. L. C. Meuzelaar and R. A. in't Veld, *J. Chromatogr. Sci.*, *10*: 213–216 (1972).
45. W. Windig, P. G. Kistemaker, J. Haverkamp, and H. L. C. Meuzelaar, *J. Anal. Appl. Pyrol.*, *1*: 39–52 (1979).
46. W. Windig, P. G. Kistemaker, J. Haverkamp, and H. L. C. Meuzelaar, *J. Anal. Appl. Pyrol.*, *2*: 7–18 (1980).
47. A. van der Kaaden, R. Hoogerbrugge, and P. G. Kistemaker, *J. Anal. Appl. Pyrol.*, *9*: 267 (1986).
48. G. Wells, K. J. Voorhees, and J. H. Futrell, *Anal. Chem.*, *52*: 1782 (1980).
49. J. A. Adkins, T. H. Risby, J. J. Scocca, R. E. Yasbin, and J. W. Ezzell, *J. Anal. Appl. Pyrol.*, *7*: 15–33, (1984).
50. G. L. French, I. Phillips, and S. Chin, *J. Gen. Microbiol.*, *125*: 347 (1981).
51. G. L. French, H. Talsania, and I. Philips, *Med. Microbiol.*, *29*: 19 (1989).
52. M. L. Trehy, R. A. Yost, and J. G. Dorsey, *Anal. Chem.*, *58*: 14 (1986).
53. A. P. Snyder, W. H. McClennen, and H. L. C. Meuzelaar, in *Analytical Microbiology Methods: Chromatography and Mass Spectrometry* (A. Fox, S. L. Morgan, L. Larsson, and G. Odham, eds.), Plenum Press, New York, pp. 201–217 (1990).
54. P. A. Quinn, *J. Chromatogr. Sci.*, *12*: 796–806 (1974).
55. H. L. C. Meuzelaar, H. G. Ficke, H. C. den Harink, *J. Chromatogr. Sci.*, *13*: 12–17 (1975).
56. J. E. Purcell, H. D. Downs, and L. S. Ettre, *Chromatographia*, *8*: 605–616 (1975).
57. A. Mitchell and M. Needleman, *Anal. Chem.*, *50*: 668–669 (1978).
58. Y. Sugimura and S. Tsuge, *Anal. Chem.*, *50*: 1968–1972 (1978).
59. J. M. Halket and H.-R. Schulten, *J. High Resolut. Chromatogr. Chromatogr. Commun.*, *9*: 596–597 (1986).
60. P. G. Simmonds, *Appl. Microbiol.*, *20*: 567 (1970).
61. E. E. Medley, P. G. Simmonds, and S. L. Mannatt, *Biomed. Mass Spectrum.*, *2*: 261, (1975).

62. J. R. Hudson, S. L. Morgan, and A. Fox, *Anal. Biochem.*, *120*: 59–65 (1982).
63. L. W. Eudy, M. D. Walla, J. R. Hudson, S. L. Morgan, and A. Fox, *J. Anal. Appl. Pyrol.*, *7*: 231–247 (1985).
64. L. W. Eudy, M. D. Walla, S. L. Morgan, and A. Fox, *Analyst*, *110*: 381–385 (1985).
65. R. S. Sahota, S. L. Morgan, and K. E. Creek, *J. Anal. Appl. Pyrol.*, *24*: 107–122 (1992).
66. R. S. Sahota and S. L. Morgan, *Anal. Chem.*, *64*: 2383–2392 (1992).
67. R. S. Sahota, S. L. Morgan, K. E. Creek, and L. Pirisi, unpublished manuscript (1994).
68. K. J. Voorhees, Department of Chemistry & Geochemistry, Colorado School of Mines, Golden, CO, pers. commun. (1994).
69. J. H. J. Huis In't Veld, H. L. C. Meuzelaar, and A. Tom, *Appl. Microbiol.*, *26*: 92–97 (1973).
70. M. V. Stack, H. D. Donoghue, J. E. Tyler, and M. Marshall, in *Analytical Pyrolysis* (C. E. R. Jones and C. A. Cramers, eds.), Elsevier, Amsterdam, p. 57 (1977).
71. M. V. Stack, H. D. Donoghue, and J. E. Tyler, *Appl. Environ. Microbiol.*, *35*: 45 (1980).
72. G. L. French, I. Phillips, and S. Chin, *J. Gen. Microbiol.*, *125*: 347 (1981).
73. M. V. Stack, H. D. Donoghue, and J. E. Tyler, *J. Anal. Appl. Pyrol.*, *3*: 221 (1981/1982).
74. G. L. French, H., Talsania, and I. Phillips, *Med. Microbiol.*, *29*: 19 (1989).
75. C. S. Smith, S. L. Morgan, C. D. Parks, A. Fox, and D. G. Pritchard, *Anal. Chem.*, *59*: 1410–1413 (1987).
76. A. Fox, J. Gilbart, B. Chritensson, and S. L. Morgan, in *Rapid Methods and Automation in Microbiology and Immunology* (A. Balows, R. C. Titon, and A. Turano, eds.), Brixia Academic Press, Brescia, Italy, pp. 379–388 (1989).
77. Y. Brun, J. Fleurette, and F. Forey, *J. Clin. Microbiol.*, *8*: 503–508 (1978).
78. J. M. Hindmarch and J. T. Magee, *J. Anal. Appl. Pyrol.*, *11*: 527–538 (1987).

79. S. J. DeLuca and K. J. Voorhees, *J. Anal. Appl. Pyrol.*, *24*: 211–225 (1993).
80. L. A. Shute, C. S. Gutteridge, J. R. Norris, and R. C. W. Berkeley, *J. Gen. Microbiol.*, *130*: 343–355 (1984).
81. K. J. Voorhees, S. L. Durfee, J. R. Holtzclaw, C. G. Enke, and M. R. Bauer, *J. Anal. Appl. Pyrol.*, *14*: 7 (1988).
82. K. J. Voorhees, S. J. DeLuca, and A. Noguerola, *J. Anal. Appl. Pyrol.*, *24*: 1–21 (1992).
83. B. E. Watt, S. L. Morgan, K. Fox, and A. Fox, *Anal. Chem.*, submitted (1994).
84. D. R. Budgell, E. R. Hayes, and R. J. Helleur, *Anal. Chim. Acta.*, *192*: 243 (1987).
85. R. J. Helleur, E. R. Hayes, J. S. Craigie, and J. L. MacLachlan, *J. Anal. Appl. Pyrol.*, *8*: 349 (1985).
86. R. J. Helleur, D. R. Budgell, and E. R. Hayes, *Anal. Chim. Acta*, *192*: 367 (1987).
87. R. J. Helleur, *J. Anal. Appl. Pyrol.*, *11*: 297–311 (1987).
88. R. Kajioka and M. A. Noble, *J. Anal. Appl. Pyrol.*, *22*: 29–38 (1991).
89. R. Kajioka and P. W. Tang, *J. Anal. Appl. Pyrol.*, *6*: 59–68 (1984).
90. T. Kusaka, and T. Mori, *J. Gen. Microbiol.*, *132*: 3403 (1986).
91. A. P. Snyder, W. H. McClennen, J. P. Dworzanski, and H. L. C. Meuzelaar, *Anal. Chem.*, *62*: 2565–2573 (1990).
92. G. Holzer, T. F. Bourne, and W. Bertsch, *J. Chromatogr.*, *468*: 181 (1989).
93. J. P. Dworzanski, L. Berwald, and H. L. C. Meuzelaar, *Appl. Environ. Microbiol.*, *56*: 1717–1724 (1990).
94. A. P. Snyder, W. H. McClennen, J. P. Dworzanski, and H. L. C. Meuzelaar, *Anal. Chem.*, 2565–2573 (1990).
95. P. B. Smith and A. P. Snyder, *J. Anal. Appl. Pyrol.*, *24*: 23–38 (1992).
96. A. P. Snyder, P. B. W. Smith, J. P. Dworzanski, and H. L. C. Meuzelaar, in *Mass Spectrometry for the Characterization of Microorganisms* (C. Fenselau, ed.), ACS Symp. Ser. 541, American Chemical Society, Washington, D.C., pp. 63–84 (1994).

97. S. J. Deluca, E. W. Sarver, P. De B. Harrington, and K. J. Voorhees, *Anal. Chem. 62*: 1465–1472 (1990).
98. S. J. DeLuca, E. W. Sarver, and K. J. Voorhees, *J. Anal. Appl. Pyrol.*, *23*: 1–14 (1992).
99. J. A. Adkins, T. H. Risby, J. J. Scocca, R. E. Yasbin, and J. E. Ezzell, *J. Anal. Appl. Pyrol.*, *7*: 35–51 (1984).
100. J. J. Boon, A. Tom, B. Brandt, G. B. Eijkel, P. G. Kistemaker, F. J. W. Notten, and F. H. M. Mikx, *Anal. Chim. Acta*, *163*: 193–205 (1984).
101. P. B. Harrington, T. E. Street, K. J. Voorhees, F. R. Brozolo, and R. W. Odom, *Anal. Chem.*, *61*: 715–719 (1989).
102. R. Goodacre, R. C. W. Berkeley, and J. E. Beringer, *J. Anal. Appl. Pyrol.*, *22*: 19–28 (1991).
103. G. Papa, P. Balbi, and G. Audisio, *J. Anal. Appl. Pyrol.*, *11*: 539–548 (1987).
104. G. Papa, P. Balbi, and G. Audisio, *J. Anal. Appl. Pyrol.*, *15*: 137 (1989).
105. A. C. Tas, A. Kerkenaar, G. F. LaVos, and J. Van der Greef, *J. Anal. Appl. Pyrol.*, *15*: 55–70 (1989).
106. A. C. Tas, H. B. Bastiaanse, J. Van der Greef, and A. Kerkenaar, *J. Anal. Appl. Pyrol.*, *14*: 309–321 (1989).

10

Analytical Pyrolysis of Polar Macromolecules

John W. Washall
CDS Analytical Inc., Oxford, Pennsylvania

I. INTRODUCTION

There is a wide range of polymers that are polar in their chemical make-up; frequently they are macromolecules formed by condensation reactions or through other catalyzed means. Analytical pyrolysis can be a useful method for the characterization of such polymers, both from a qualitative and quantitative standpoint.

When heated sufficiently to cause bond dissociation, these polymers degrade to produce a pyrolyzate that is individually indicative of the particular polymer. The result in pyrolysis-gas chromatography (Py-GC) is a fingerprint chromatogram which qualitatively identifies the macromolecule.

Quantitative information can also be obtained in the analysis of copolymeric systems. Many polar macromolecular systems produce high yields of monomers that may be used in quantitatively determining monomer composition in the polymer. This type of information can be valuable not only in terms of characterizing the polymer, but also in performing degradation studies.

The macromolecules discussed in this chapter are primarily synthetic polymers such as poly(methacrylates), poly(esters), poly(amides), and poly(urethanes). There will also be a discussion on the analysis of surfactants by Py-GC.

II. SYNTHETIC POLYMERS

A. Poly(methacrylates)

Poly(methacrylates) fit into a class of polymers that tend to undergo depolymerization under pyrolysis conditions. Poly-(methyl methacrylate), PMMA, when pyrolyzed, yields primarily methyl methacrylate monomer. This process begins by initial cleavage of the skeletal backbone of the polymer, forming two free radical ends. Subsequent beta scissions produce an unwinding effect as sequential monomer units are formed. Once initiated, this process proceeds down the entire length of the polymer. This process is known as "unzipping." Figure 1 shows a typical pyrogram of poly(methyl methacrylate) at 600°C. From this chromatogram, it is clearly evident how extensively the depolymerization process works.

Figure 2 shows the mechanistic description of depolymerization initiation and product formation for methacrylate polymers. From the literature concerning monomer yields for poly (methyl methacrylate), typical values are 92–98% recovery of methyl methacrylate monomer.

These recovery values are fairly consistent regardless of pyrolysis temperature and heating rate. It seems that in the case of PMMA, as long as there is sufficient energy delivered to the sample in order to break a C-C bond, then depolymerization occurs readily.

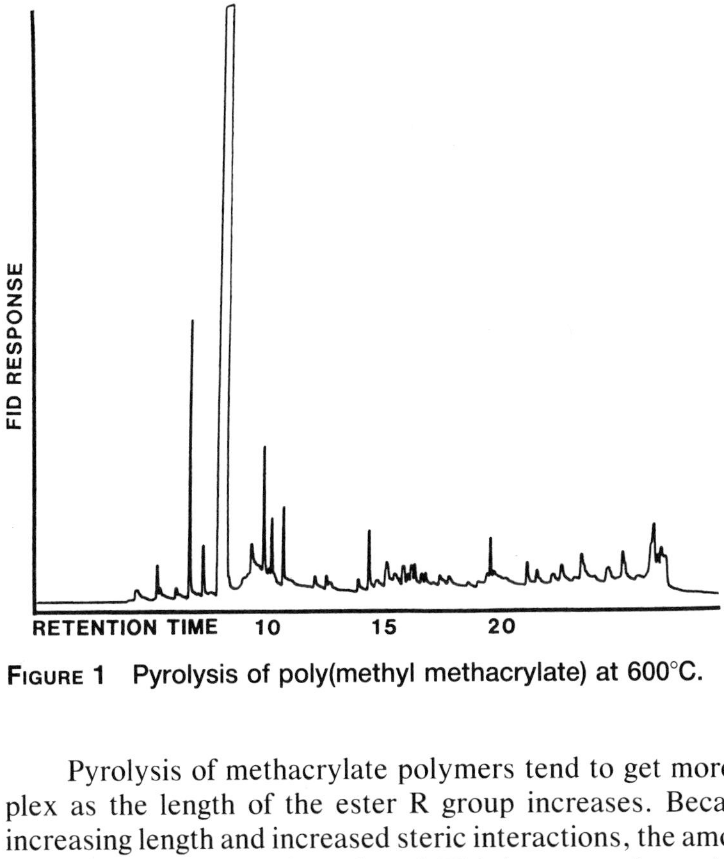

FIGURE 1 Pyrolysis of poly(methyl methacrylate) at 600°C.

Pyrolysis of methacrylate polymers tend to get more complex as the length of the ester R group increases. Because of increasing length and increased steric interactions, the amount of depolymerization may be reduced. This increases the probability that products other than, or in addition to, the monomer will be produced. For instance, the pyrogram of poly(butyl methacrylate), PBMA, is shown in Figure 3. The monomer yield for PBMA is in the 93–95% range. Minor products include 1-butanol, butyl acrylate, 1-propene, 2-methyl-1-propene, and butanal. Among these minor products, 1-butanol is the most abundant and is formed as a result of cleavage of the etherial C-O bond of the ester group.

The last methacrylate polymer example is poly(lauryl methacrylate), PLMA. This polymer is interesting because it contains an extremely long R group. Having such a long hydrocarbon

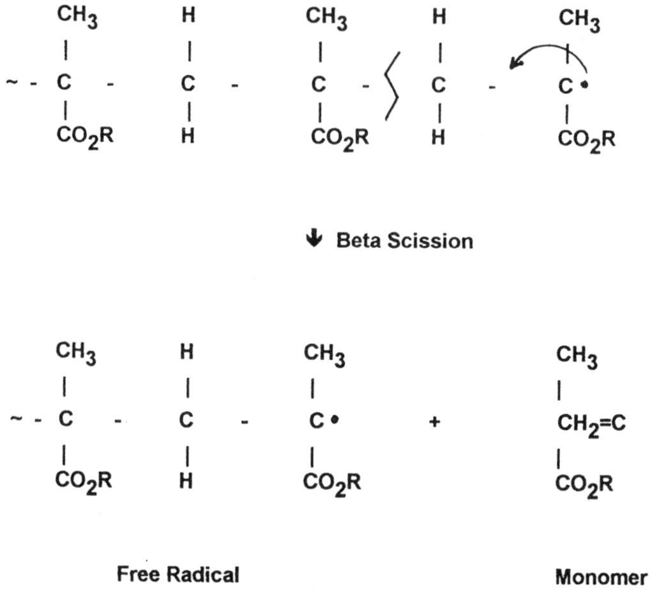

FIGURE 2 Depolymerization mechanism for poly(methyacrylates).

chain dangling from the polymer backbone makes it much easier to cleave. The result is a pyrolyzate composed not only of lauryl methacrylate monomer, but also many unsaturated hydrocarbons as the result of random scission reactions in the chain of the C_{12} group. Figure 4 shows an example of PLMA pyrolyzed at 600°C. There is quite a noticeable repeating series of peaks corresponding to the various monounsaturated hydrocarbons. In this particular example, the yield for lauryl methacrylate was 68%. This value is much lower than those obtained for PMMA and PBMA, which may be attributed to the length of the R group.

The qualitative aspect of methacrylate polymer pyrolysis as well as the quantitative determination of recovered monomer have been described. What remains to be discussed is a quantitative assessment of monomer composition in a copolymer sample. For quality control purposes, it is often helpful to know the composition of a copolymer product. Such determinations can be correlated to physical properties, temperature stability, etc. One

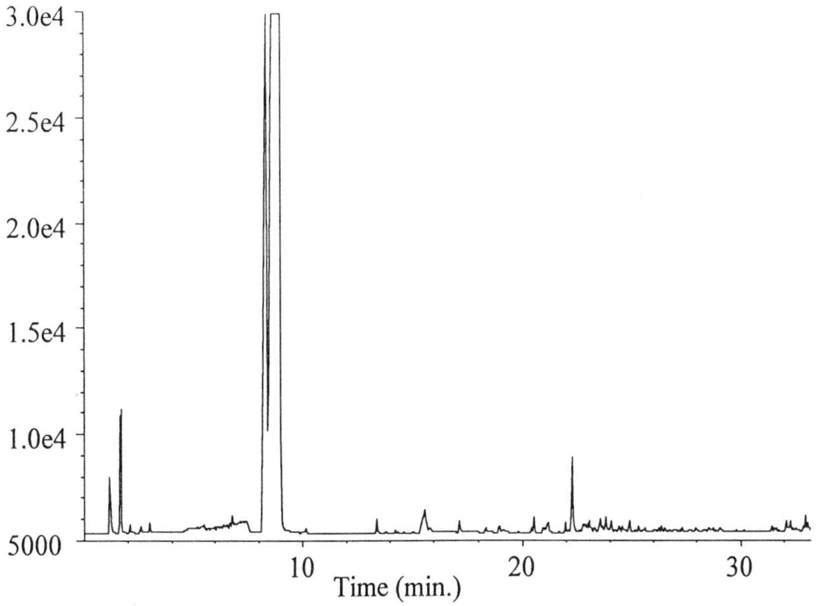

FIGURE 3 Pyrolysis of poly(butyl methacrylate) at 600°C.

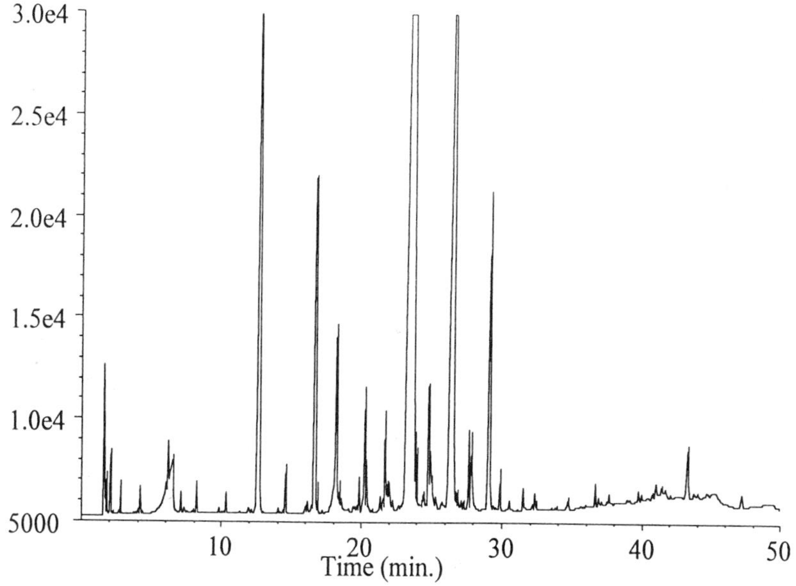

FIGURE 4 Pyrolysis of poly(lauryl methacrylate) at 600°C.

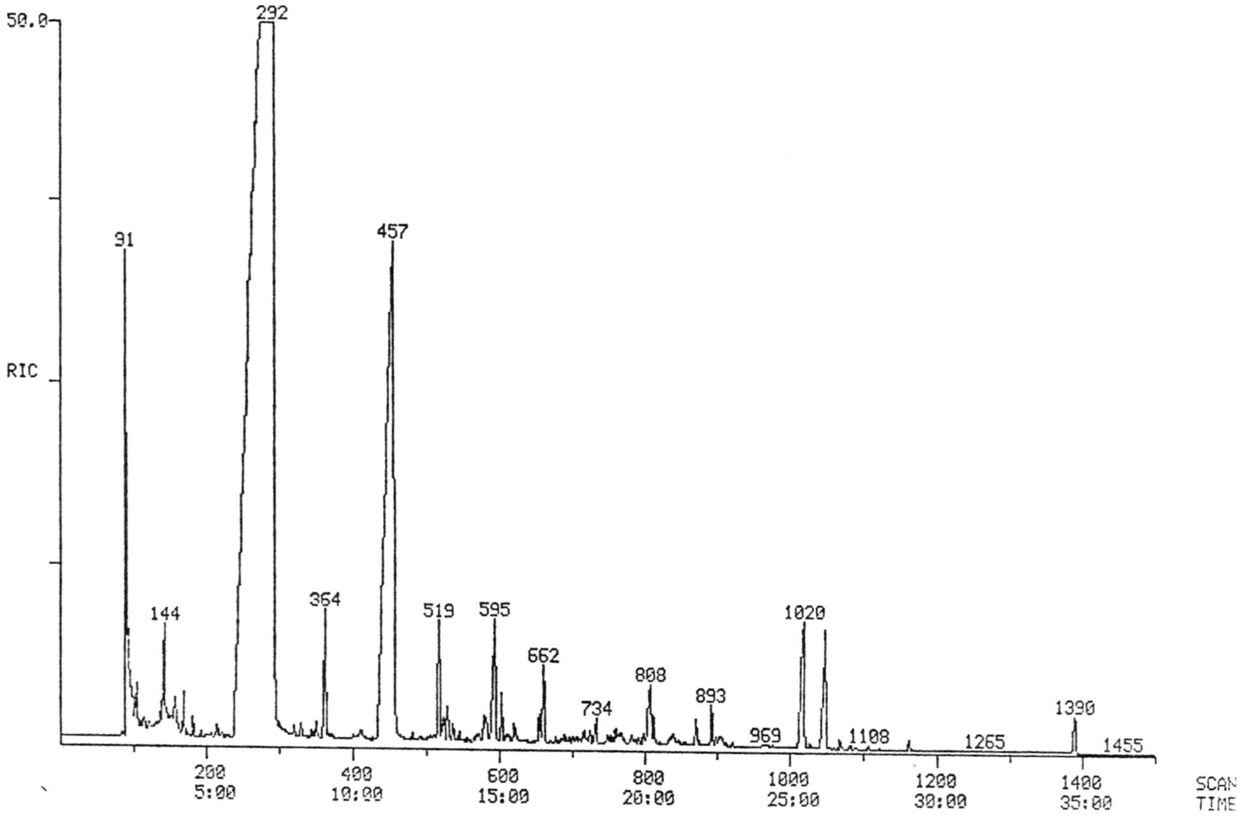

FIGURE 5 Pyrolysis of copolymer (30% methyl methacrylate, 70% butyl acrylate) at 600°C.

copolymer system, which has been studied extensively, is poly-(methyl methacrylate/butyl acrylate). A pyrogram of this copolymer is shown in Figure 5. The two primary peaks in this chromatogram correspond to the respective monomers. A series of such copolymers with varying concentrations of monomer can be pyrolyzed in order to obtain a calibration curve. A typical calibration curve is shown in Figure 6, plotting the area ratio of the methyl methacrylate peak to the butanol peak versus concentration of methyl methacrylate. With proper calibration of the pyrolyzer, it is possible to obtain linear quantitative relationships for most polymeric systems.

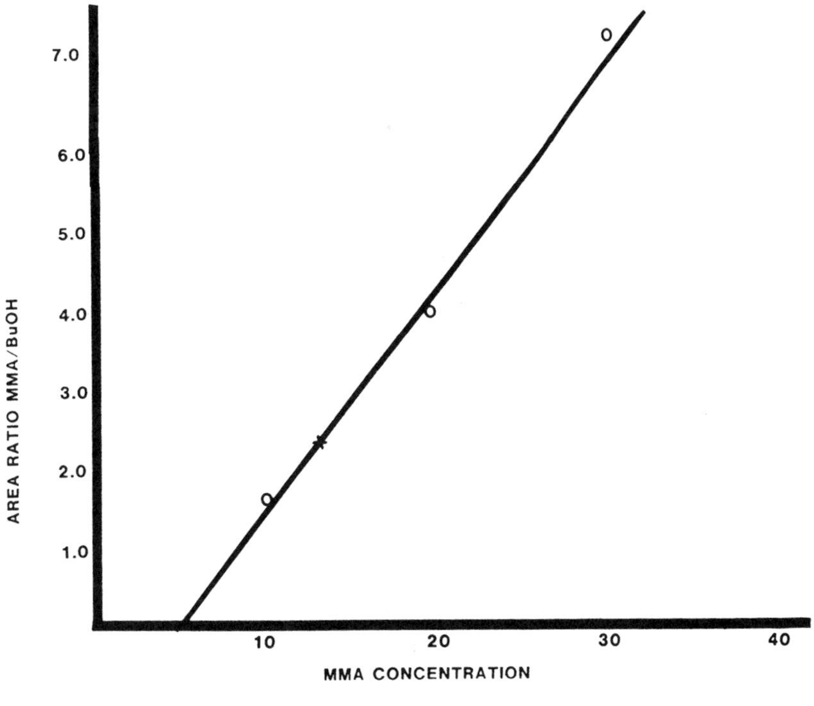

FIGURE 6 Graph of peak area ratio of methyl methacrylate monomer (from PMMA) to butanol (from PBA) vs. the concentration of methyl methacrylate in the copolymer.

B. Polyethers

Pyrolysis of polyethers shows some interesting thermal degradation mechanisms. Polyethers can be made from a variety of monomers, a common example being polyethyleneglycol terephthalate. This material produces five primary products upon pyrolysis, which are shown in Figure 7 [1]. Benzene is by far the most abundant product from the pyrolysis of polyethyleneglycol terephthalate. The reaction mechanism for the formation of benzene proceeds first through an intramolecular rearrangement involving the carbonyl oxygen with a $-CH_2$ hydrogen atom (beta to the etherial oxygen) in a 1,5-hydrogen shift. Decarboxylation and subsequent random scission account for the formation of the very stable benzene product.

C. Polyesters

In the pyrolysis of polar macromolecules, stability of the final products plays a crucial role in determining the relative product abundances. With polyesters, the polymer generally unzips to give the original monomer unit. However, when ester copolymers are pyrolyzed, a number of other reactions are likely. For instance, a polyester prepared by the reaction of chlorinated norbornene dicarboxylic acid, maleic anhydride, and 1,2-propa-

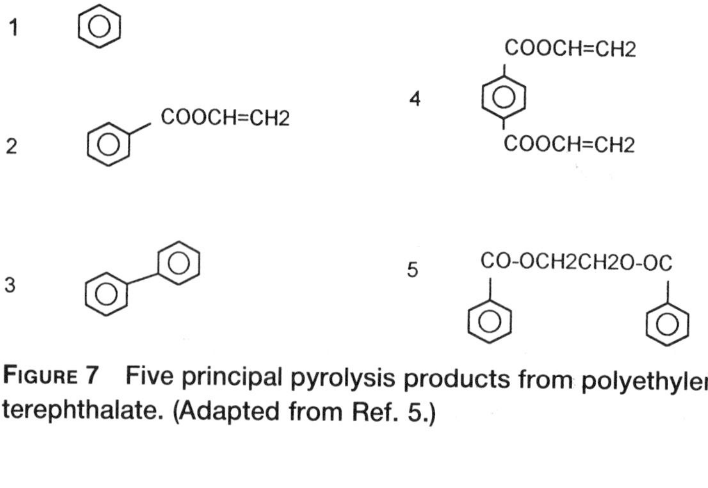

Figure 7 Five principal pyrolysis products from polyethyleneglycol terephthalate. (Adapted from Ref. 5.)

nediol in vacuum produces an unsaturated polyester. Pyrograms of this material [2] show the following major pyrolysis products: 1,2,3,4,5-pentachlorocyclopenta-1,3-diene and 1,2,3,4,5,5-hexachlorocyclopenta-1,3-diene. The monomer yield in this case was very small. Product formation in this case takes place via retro-Diels Alder reactions because of the presence of various polychloro-5-norbornene-dicarboxylic acid units in the backbone of the polymer.

In the case of styrene-cured unsaturated polyesters, these polymers showed increased quantities of toluene, ethylbenzene, and alpha-methylstyrene in their pyrograms. Interestingly, dimer formation was not seen under these circumstances. It is believed that chlorine radicals participate to prohibit dimer formation through transfer reactions. The result of this reaction is the formation of phenylalkyl chlorides [3].

Structural determination often hinges on identification of higher-molecular-weight pyrolysis fragments. For this reason, it may be necessary to choose a pyrolysis temperature that gives a product distribution that favors formation of dimeric, trimeric, and higher units. This provides information on the arrangement of the various monomer units, and will also provide an insight as to whether a copolymer is random or a block copolymer. For example, a styrene-glycidyl methacrylate copolymer shows monomer, dimer, and codimer fragments in the pyrogram [4].

Many acrylic copolymers are currently used in the textile industry as binders for nonwoven fabrics. The purpose of these fibers is to stabilize the material. In many instances, these copolymers are used in conjunction with amino resins. Casanovas and Rovira [4] have done a study of methyl methacrylate-ethyl acrylate-*N*-methylol-acrylamide by pyrolysis-gas chromatography using mass spectrometry as the detector. Among the products identified were methane, ethylene, propene, isobutene, methanol, propionaldehyde, ethanol, ethyl acetate, methyl acrylate, methyl isobutyrate, ethyl acrylate, methyl methacrylate *n*-propyl acrylate, and ethyl methacrylate. In this sample, clearly monomer reversion is the primary degradation process occurring; however, several other degradation mechanisms are at work. When the sample contains an amino resin in the mixture, acrylonitrile is observed in the pyrogram. Another effect of the amino presence was a marked increase in the amount of methanol detected. Other products detected were methoxyhydrazine, methyl isocyanate, and methyl isocyanide.

Thus far we have looked at a number of oxygenated polymers and copolymers. We have seen that depolymerization is the primary degradation mechanism for many acrylate polymers and polyesters. This mechanism is affected by the nature of the monomer unit; for instance, with methacrylate polymers, the product distribution varies as the length of the R group increases. The result is that random scission begins to be more of a factor as the alkyl chain gets longer. Now we will take a look at macromolecules that contain a nitrogen heteroatom.

Before doing so, however, it might be helpful to look at smaller molecules in order to understand the chemistry surrounding a central nitrogen atom. The first compounds that we will discuss are amino acids. These compounds are relatively simple chemically and offer some insight as to the pyrolytic behavior surrounding the nitrogen atom. Many alpha-amino acids, when exposed to pyrolysis conditions, undergo initial decarboxylation to form the amine fragment. With aliphatic amino acids this results in the amine as the primary pyrolysis product in the chromatogram. This, however, is not at all surprising; CO_2 is highly stable and electronically readily available in the amino acid. There are some bimolecular processes that can occur to give the formation of a nitrile upon pyrolysis of amino acids, these have been widely characterized in the literature.

Other products that form from the pyrolysis of amino acids are aldehydes, which contain one less carbon atom than the parent amino acid. This process occurs via an SNi deamination mechanism. Another process which can occur is side-chain stripping involving chain homolysis. The result is the production of various saturated and unsaturated compounds [5].

With materials that contain nitrogen heteroatoms, the pyrolysis conditions are of extreme importance. Because of the thermally labile nature of many nitrogen-containing compounds, the thermal stability of the final product is paramount. Sample size can also dramatically affect the degradation mechanism and product formation. The larger the sample size, the greater the corresponding thermal gradient within the sample. Large sample sizes tend to push the product distribution in the direction of secondary degradation processes.

Pyrolysis of phenylalanine, for example, can proceed through several different pathways. Figure 8 shows the pyrograms of phenylalanine and tyrosine. In the case of phenylalanine, a primary degradation product is toluene. This product results from the cleavage of the carbon-carbon bond beta to the amine nitrogen atom. There is the need for a hydrogen radical migration to form the final product. Likewise, cleavage of the carbon-nitrogen bond leads to the formation of styrene. Ethylbenzene is another product of the pyrolysis of phenylalanine. Bibenzyl is also a major pyrolysis product of phenylalanine resulting from the reaction of two toluyl radicals [6].

As mentioned previously, the nature of the pyrolysis products with nitrogen-containing macromolecules can be highly dependent on many analytical conditions, such as sample size, pyrolysis temperature, and inertness of the pyrolysis system. Many nitrogen-containing compounds are thermally labile and thus are sensitive to temperature and metal surfaces. Figure 8 shows the result of pyrolyzing phenylalanine using a glass-lined analytical system. Irwin [5] reports on the differences resulting from the presence of stainless steel, quartz, and pyrex in the same analysis. Although the pyrolysis products remain the same in all three experiments, the product distribution does change. For instance, in a stainless system, the amount of benzene produced is much higher than that of the pyrex and quartz systems. Quartz, which is the most inert, has the highest level of styrene and ethylbenzene. Pyrolysis at 500°C provides the largest yield of Ph-CH_2-CH_2-CN. As the temperature increases, the amount of the nitrile diminishes [6].

Understanding the product formation from the pyrolysis of amino acids is the first step in comprehending pyrolysis behavior for more complex systems, such as proteins and nucleic acids,

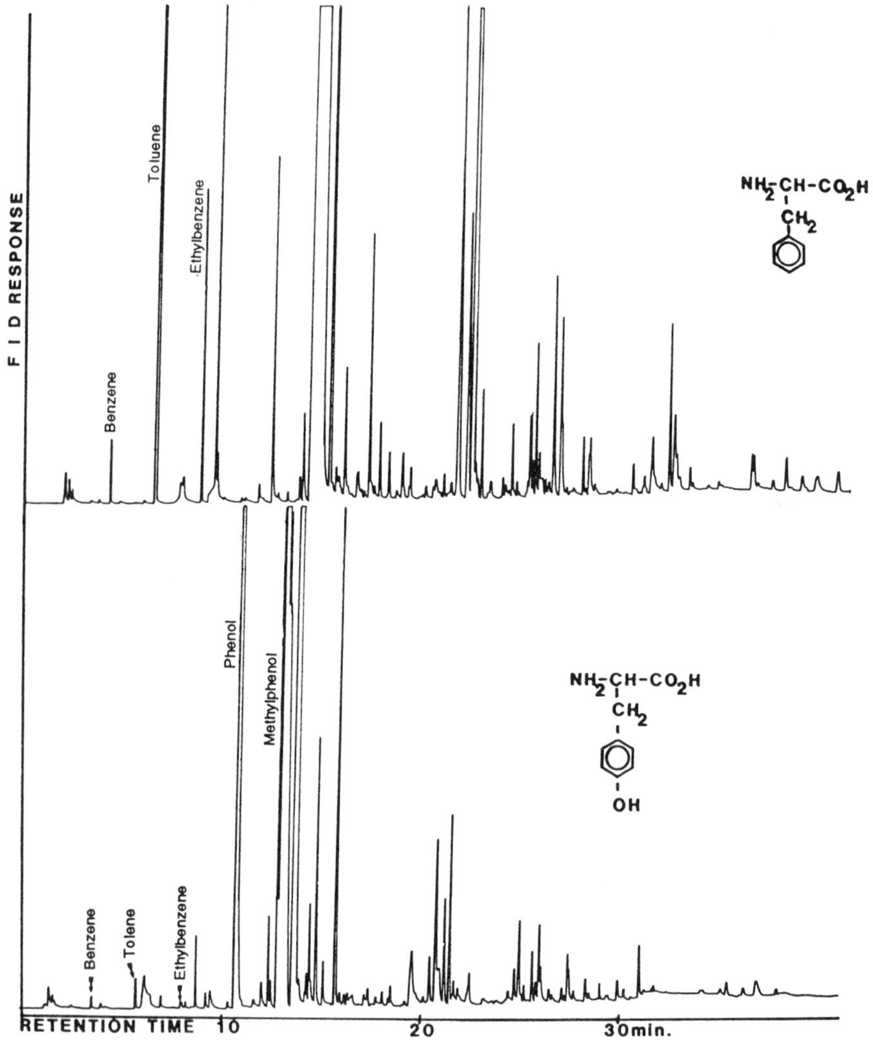

Figure 8 Pyrograms of phenylalanine (top) and tyrosine (bottom) (glass-lined system at 700°C for 10 s).

which comprise a large quantity of the material in biological systems.

III. SURFACTANTS

Surfactants serve a wide variety of consumer and industrial uses. There are three basic types: cationic, anionic, and nonionic surfactants. For our purposes, only cationic surfactants will be discussed, although pyrolysis is appropriate in the investigation of anionic and nonionic surfactants as well.

Cationic surfactants generally contain a central heteroatom with four side groups extending from the central atom. In the case of alkyl ammonium halide surfactants, there are typically three small alkyl or aromatic groups bonded to the central nitrogen atom. The cleansing properties arise from the presence of the fourth group, which is a long alkyl chain. The length of this alkyl chain varies from C_{12} to C_{20} in length. The alkyl chain, which is derived from the reaction of the amine with a fatty acid, is crucial to the formation of the micelle. Micelle formation is the mechanism by which surfactants get their cleansing properties. In commercial products, there are usually mixtures of various alkyl ammonium surfactants. For instance, several commercially available cleaning products contain the following surfactants: dodecyl dimethyl-, tetradecyl dimethyl-, hexadecyl dimethyl-, and octadecyl dimethylbenzyl ammonium chlorides. To understand how pyrolysis can be used to analyze these materials, the pyrolysis products of the pure components must be examined. For a generic surfactant of the chemical structure:

$$(CH_3)_2\ Ph\text{-}N\text{-}R$$

the following scenarios are possible. If bond energies alone are used to deduce where bond cleavage will occur, the natural choice for initial homolytic cleavage would be the R-N bond [7]. The dissociation energy for this bond is 78 kcal/mole. This can be compared to the bond dissociation energy of a typical carbon-carbon single bond of kcal/mole. If one looks at the pyrogram of hexadecyltrimethyl ammonium chloride (Fig. 9), it becomes

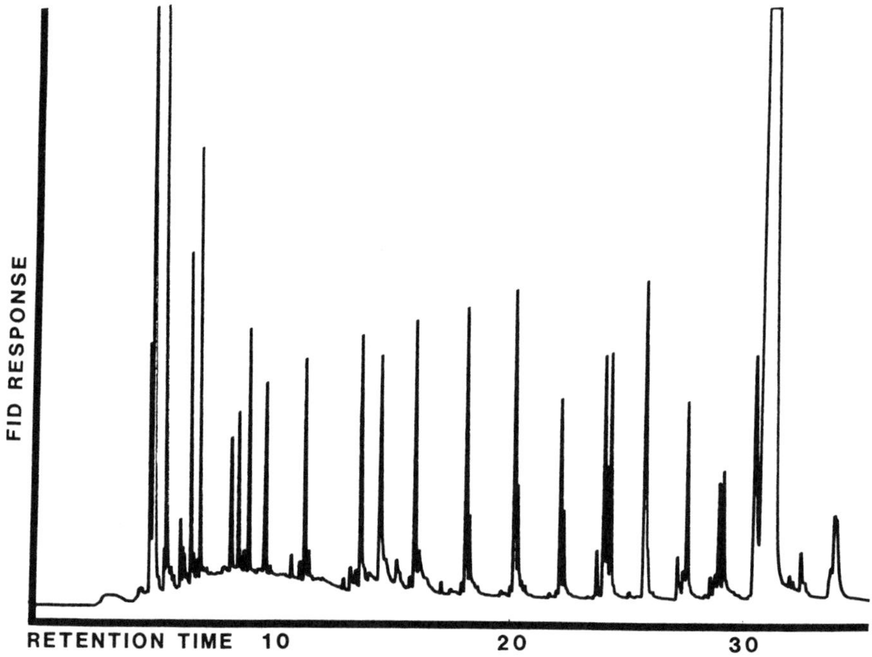

Figure 9 Pyrolysis of hexadecyltrimethyl ammonium chloride at 750°C.

obvious that the two major pyrolysis products are hexadecene and trimethyl amine. This is the result of cleavage of the long alkyl chain at the nitrogen atom. Hydrogen radical transfer from the alkyl radical to the amine radical completes the product formation. It can also be noticed that there are several other minor pyrolysis products that correspond to unsaturated straight-chain hydrocarbons. These result from random scission of the alkyl radical producing monoolefins from C_{15} and lower. This process is very similar to what happens when polyethylene is pyrolyzed and is in agreement with pyrolysis theory.

Similarly, when octadecyl dimethylbenzyl ammonium chloride is pyrolyzed (Fig. 10), the major products are octadecene and dimethylbenzyl ammonium chloride. Other products include

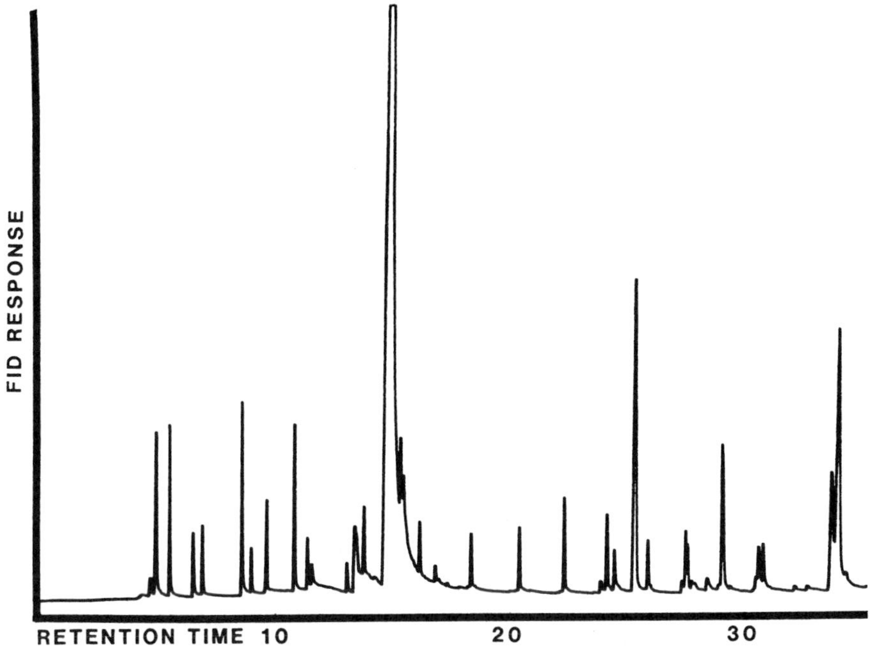

Figure 10 Pyrolysis of an octadecyldimethylbenzyl ammonium chloride at 750°C.

benzene, toluene, dimethyl amine, and unsaturated hydrocarbons with carbon numbers less than C_{18}.

The utility of pyrolysis for the analysis of surfactants can be shown by looking at aqueous solutions of these surfactants in dilute quantities. Analytical pyrolysis is frequently thought of as a technique that does not work well with trace quantities. However, dilute solutions of surfactants can be analyzed by applying the aqueous solution to the platinum ribbon of a filament pyrolyzer and allowing the water to evaporate. This procedure can be repeated to allow for the sensitivity needed. In Figure 11, the analysis of a commercial surfactant at the low ppm level was analyzed by pyrolysis at 700°C. Among the products seen are benzene, toluene (the peaks at 5 and 8 min, respectively), and

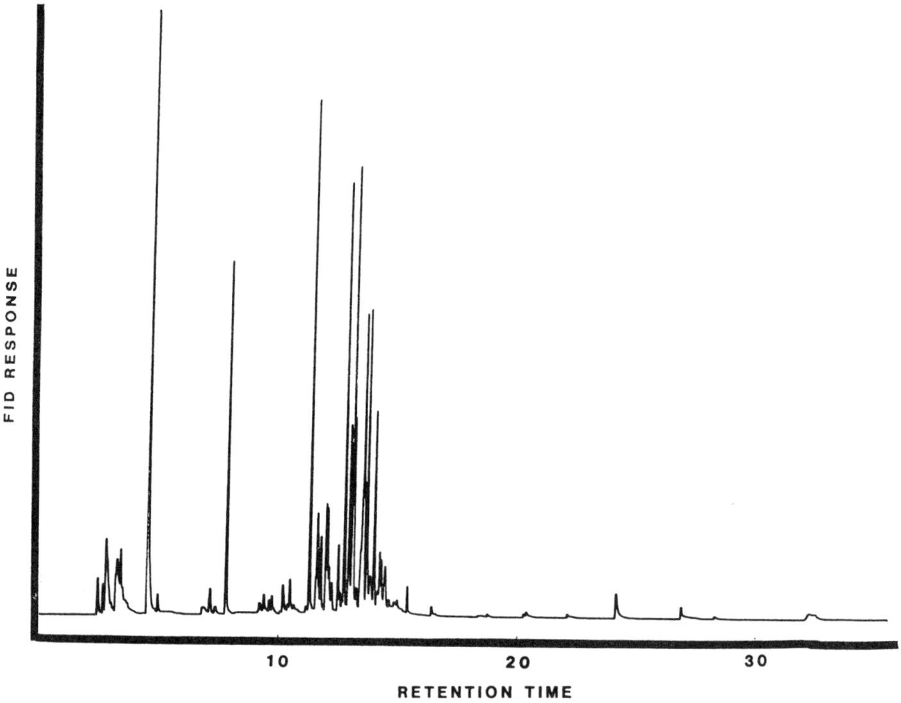

FIGURE 11 Pyrolysis of the residue of a 40 ppm solution in water of a quaternary ammonium chloride surfactant (800°C for 10 s).

many olefinic hydrocarbons. This exercise brings about the prospect of using pyrolysis as a tool for analyzing sediments, high-molecular-weight organics, pesticides, and semivolatiles by pyrolysis-gas chromatography. Such work is currently underway by several research groups.

IV. SULFUR-CONTAINING MACROMOLECULES

Sulfonamides are known for their ability to form antibacterial drugs, and the derivatives of 4-aminobenzenesulfonamide serve a number of medical applications. Analytical pyrolysis of these compounds proceeds in a similar manner to that of amino acids.

Pyrolysis generally occurs at the sulfonamido group, resulting in the liberation of sulfur dioxide. The other products will include aniline and a heterocyclic aromatic amine. The sulfonamide is generally characterized by the heterocyclic amine since aniline is always a product with these compounds [8]. For example, in the pyrolysis of sulfadiazine, pyrolysis occurs at the sulfonamido nitrogen atom to form an unstable $-SO_2$ radical, which readily converts to sulfur dioxide. The other remaining products are aniline and 2-aminopyrimidine. The same process holds for the sulfonamides sulfamerazine and sulfadimidine. The only difference is that the heterocyclic products are 2-amino-4-methylpyrimidine and 2-amino-4,6-dimethylpyrimidine, respectively [8].

These are very simple sulfur compounds in terms of pyrolytic behavior. The problem gets more sophisticated when the macromolecule is a polymer that may be crosslinked and have sulfur bridges. Many attempts have been made to address the topic of sulfur linkages in vulcanized crosslinked rubbers, with varying degrees of success. The problem that arises is that the sulfur bridges generally make up only a small portion of the polymer. This leads to difficulty because of sensitivity levels and the positive identification of pyrolysis fragments which can be attributed directly to the sulfur bridges. The chromatogram produced is often quite complex, not only because of the nature of the rubber itself, but because of the sulfur bridging. The complexity of this situation may make it necessary to use statistical modeling of known crosslinked rubbers to be able to perform adequate quantitative analysis of the degree of crosslinking for these polymers.

V. SUMMARY

The pyrolysis of polar macromolecules tends to be more complex from a mechanistic standpoint than nonpolar materials, since many of the products formed go through a cyclic intermediate. However, compounds such as polymethacrylates generally decompose via a depolymerization mechanism which produces mostly monomer. Analytical pyrolysis can be used to extract a great deal of information concerning nonvolatile samples; for instance, sample identification through fingerprinting, quantitative measurement of copolymers, and structural information related to branching and stereochemistry.

REFERENCES

1. Sugimura, Y. and Tsuge, S., *J. Chrom. Sci.*, *17*: 269–272 (1979).
2. Irzl, G. H., Vijayakumar, C. T., Fink, J. K., and Lederer, K., *J. Appl. Anal. Pyrol.*, *11*: 277–286 (1987).
3. Irzl, G. H., Vijayakumar, C. T., and Lederer, K., *J. Appl. Anal. Pyrol.*, *13*: 305–317 (1988).
4. Casanovas, A. M., and Rovira, X., *J. Appl. Anal. Pyrol.*, *11*: 227–232 (1987).
5. W. I. Irwin, *Analytical Pyrolysis, A Comprehensive Guide,* Marcel Dekker, New York (1982).
6. Patterson, J. M., Haidar, N. F., Papadopoulos, E. P., and Smith, W. T., *J. Org. Chem.*, *38*: 663–666 (1973).
7. Abraham, S. J. and Criddle, W. J., *J. Appl. Anal. Pyrol.*, *7:* 337–349.
8. Irwin, W. J. and Slack, J. A., *Analytical Pyrolysis*, 107–116, Elsevier, Amsterdam (1977).

11

Index of Sample Pyrograms

THOMAS P. WAMPLER
CDS Analytical, Inc., Oxford, Pennsylvania

Each analyst will optimize the chromatography of his pyrolysis-GC system to provide the most pertinent information about the specific sample being analyzed. It is helpful however, especially when first starting out, to have examples of typical analyses, giving some idea of what to expect in a pyrogram. This chapter includes capillary GC pyrograms of many different materials, all performed on readily available and typical columns. Although an analyst's samples will almost certainly be different in some way from the example materials shown here, it may be helpful to review results obtained on similar samples as a starting point in developing a specific pyrolysis method.

The examples are grouped roughly according to sample material type.

Pyrogram	Sample Material	Setpoint (°C)
Synthetic polymers		
S-1	Kraton 1107, Kraton is a copolymer of styrene and isoprene	800
S-2	Polyester shirt thread	750
S-3	Polychloroprene	750
S-4	Polyethyl methacrylate	600
S-5	Polymethyl methacrylate	600
S-6	Polystyrene	750
S-7	Polyvinyl chloride	600
Polyolefins		
O-1	Polyethylene	725
O-2	Polypropylene, isotactic	700
O-3	Polyisobutylene	800
O-4	Polybutadiene	800
O-5	Polypropylene-1-butene copolymer (butene content 47%)	750
O-6	Polyethylene in air	650
O-7	Polyethylene in air	750
O-8	Polypropylene in air, heating rate 100°/s	800
O-9	Polyisoprene	750
O-10	Poly 1-butene	600
Nylons		
N-1	Nylon 11	900
N-2	Nylon 12	800
N-3	Nylon 6/T	800
N-4	Nylon 6/6	800
N-5	Nylon 6/9	800
N-6	Nylon 6/10	850
N-7	Nylon 6/12	800
Biological and natural materials		
B-1	Amber	650
B-2	Baltic amber, peak marked "S" is succinic acid	650

Pyrogram	Sample Material	Setpoint (°C)
B-3	Chitin from crab shells	450
B-4	*E. coli*	650
B-5	Gelatin	750
B-6	Starch	650
B-7	Phenylalanine	700
B-8	Tyrosine	700
B-9	Animal glue from a 1500-year-old Egyptian artifact	500
B-10	Human hair	750
B-11	Lamb's wool (raw)	750
B-12	Cotton thread	750
B-13	Human finger nail	750
B-14	Kerogen	800
B-15	Dried linseed oil	700
B-16	Oak leaf	750
B-17	Straw	750
B-18	Natural rubber: polyisoprene, cis configuration	800
B-19	Oil shale from rock sample heated at 60°C/min	800
B-20	Oil shale, total organic content 1.2%, pulse heated	800
B-21	Beeswax	500
B-22	Silk	675
Manufactured goods		
G-1	Unprinted newspaper	650
G-2	White magazine paper	650
G-3	White bond paper	750
G-4	Bathroom cleaner product	650
G-5	Disinfectant cleaner product	650
G-6	Dishwashing liquid	700
G-7	Ink (black) ballpoint pen	650
G-8	Ink (black) ballpoint pen pyrolyzed intact with paper on which it had been	650

Pyrogram	Sample Material	Setpoint (°C)
	written. Numbered peaks are from the paper, lettered peaks from the ink.	
G-9	Ink (blue) ballpoint pen pyrolyzed intact with paper on which it had been written. Numbered peaks are from the paper, lettered peaks from the ink.	650
G-10	Printing ink formulation included linseed oil, wax, petroleum resins.	700
G-11	Petroleum resin A	700
G-12	Petroleum resin	700
G-13	Kodak photocopy, paper and toner material pyrolyzed together, peak number 1 = methyl methacrylate, 2 = styrene	650
G-14	Xerox photocopy, peak number 1 = styrene, 2 = butyl methacrylate	650
G-15	Mascara A, peaks indicated with arrows result from beeswax	750
G-16	Mascara B with acrylate, peaks indicated with arrows result from beeswax	750
G-17	Shirt thread, 50/50 cotton/polyester blend fabric.	750
G-18	*n*-Tetracontane	650
G-19	Silicone grease	900

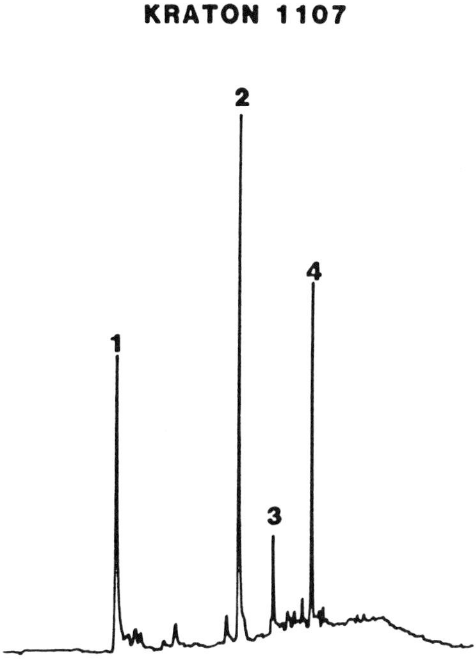

PYROGRAM S-1

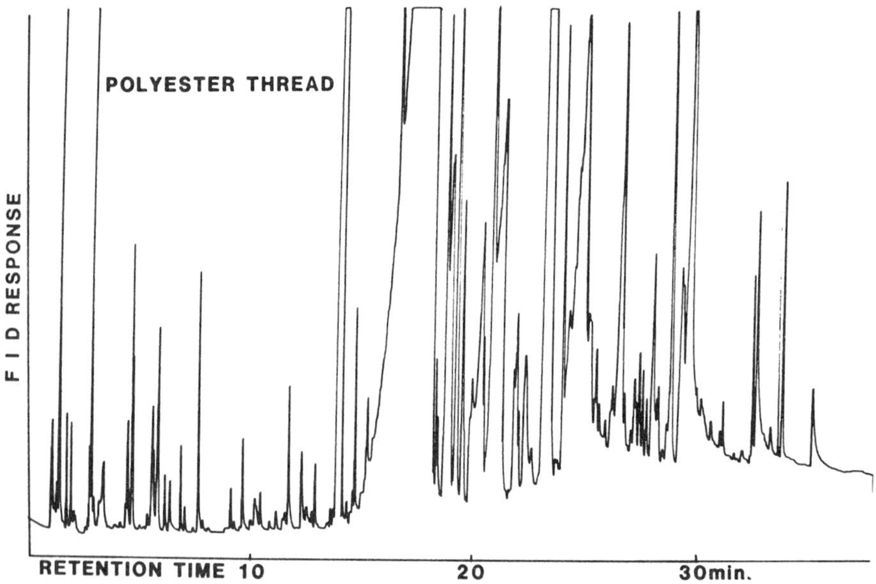

PYROGRAM S-2

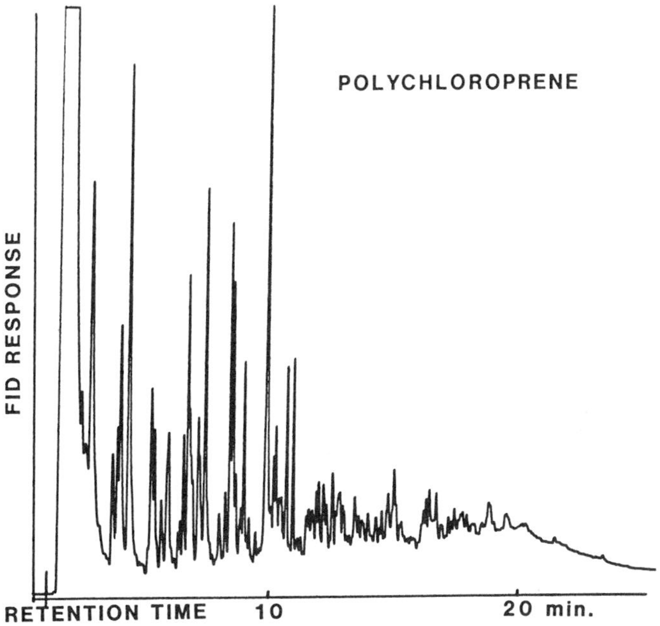
POLYCHLOROPRENE
FID RESPONSE
RETENTION TIME
10
20 min.

PYROGRAM S-3

POLYETHYL METHACRYLATE

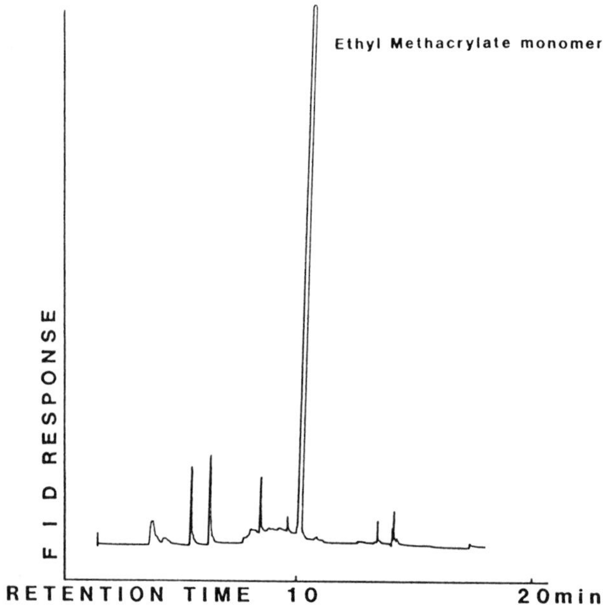

PYROGRAM S-4

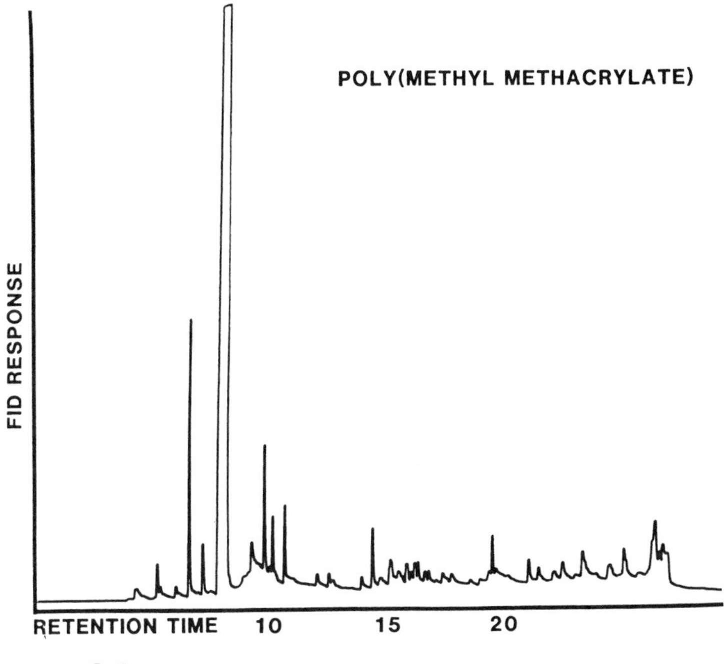
POLY(METHYL METHACRYLATE)
FID RESPONSE
RETENTION TIME
10
15
20

PYROGRAM S-5

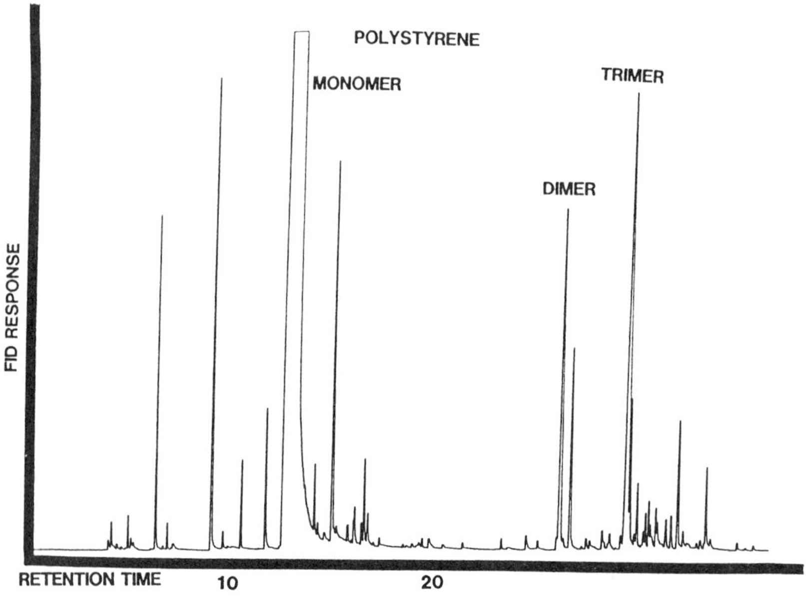
POLYSTYRENE
MONOMER
TRIMER
DIMER
FID RESPONSE
RETENTION TIME
10
20

PYROGRAM S-6

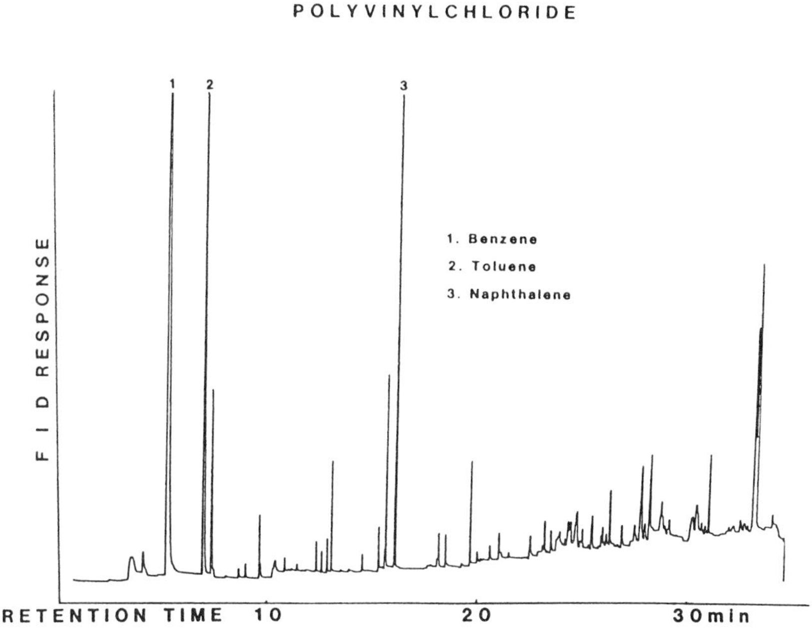
POLYVINYLCHLORIDE
1
2
3
1. Benzene
2. Toluene
3. Naphthalene
F I D RESPONSE
RETENTION TIME 10
20
30min

PYROGRAM S-7

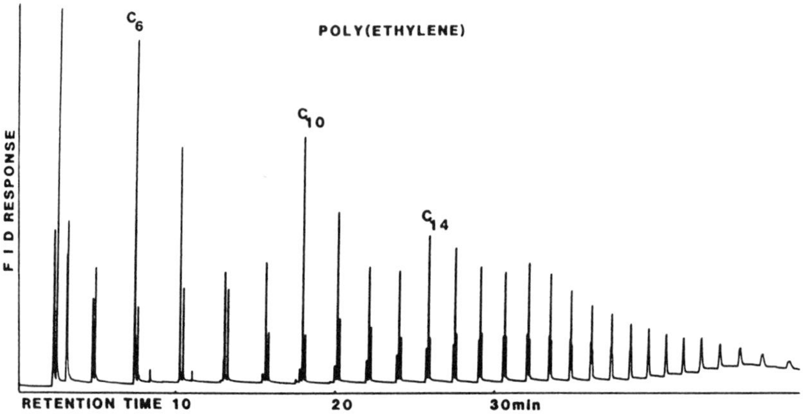

PYROGRAM O-1

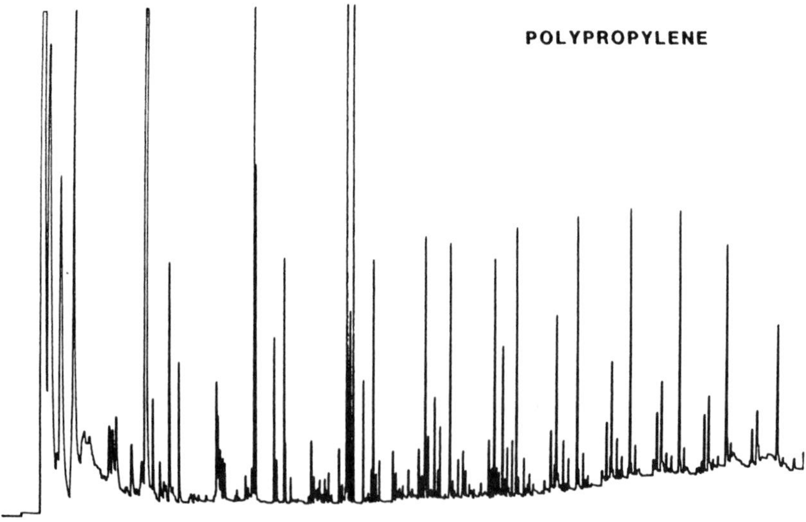

PYROGRAM O-2

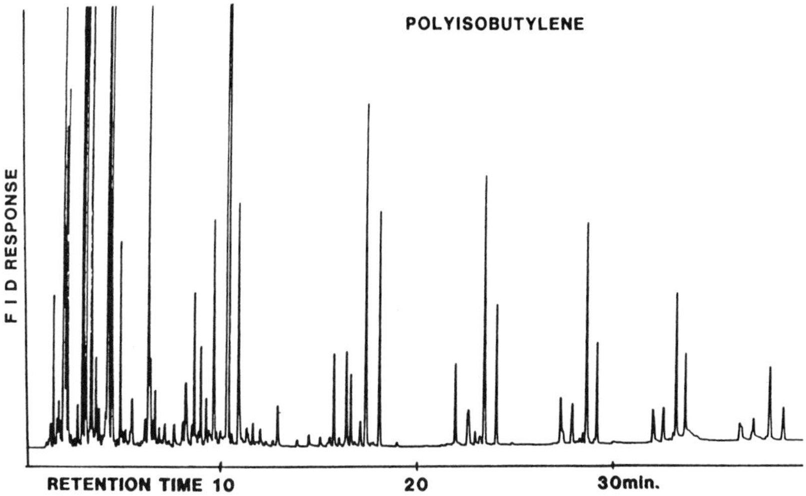

Pyrogram O-3

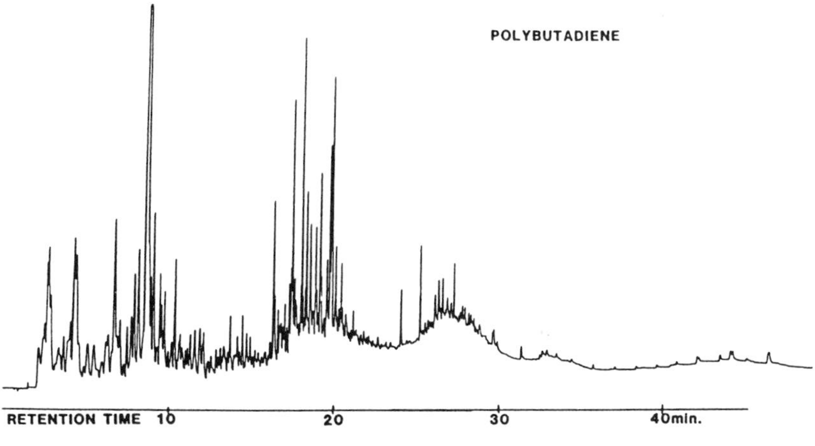

Pyrogram O-4

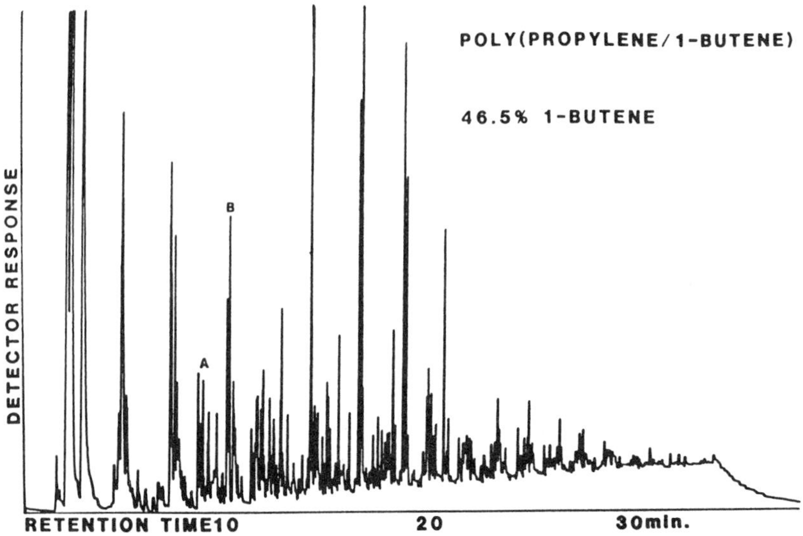

PYROGRAM O-5

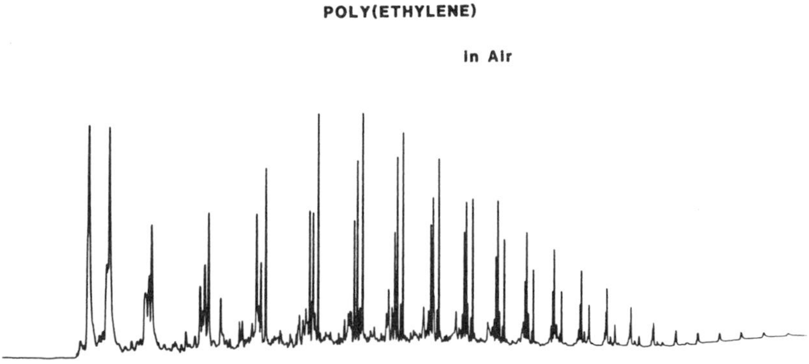

PYROGRAM O-6

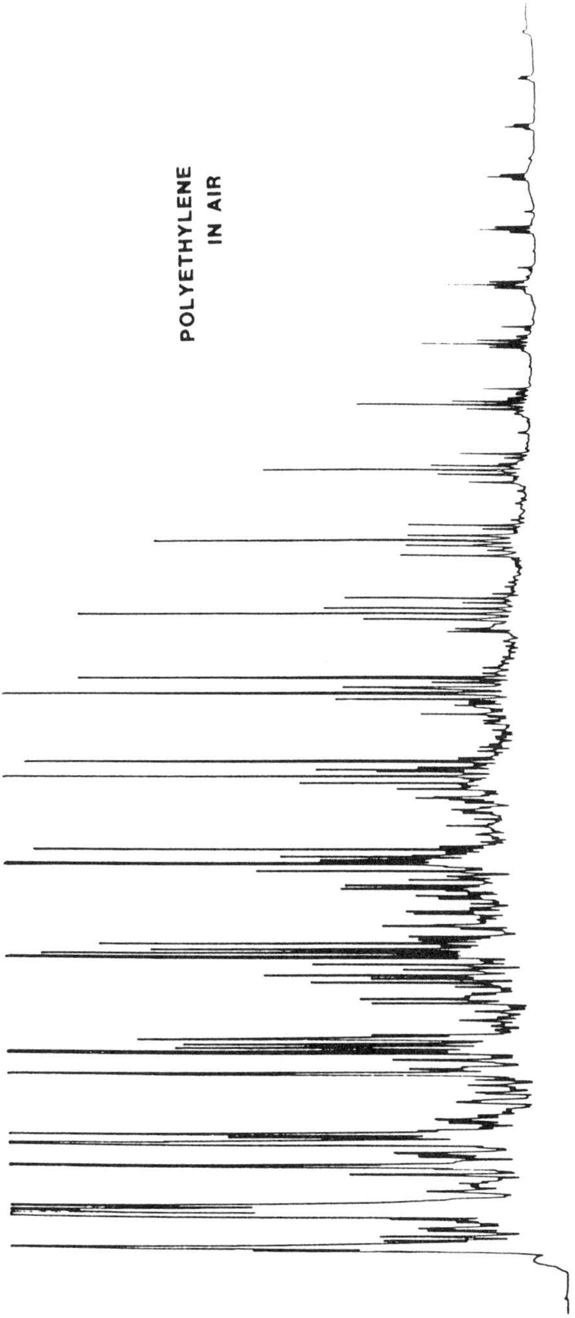

PYROGRAM O-7

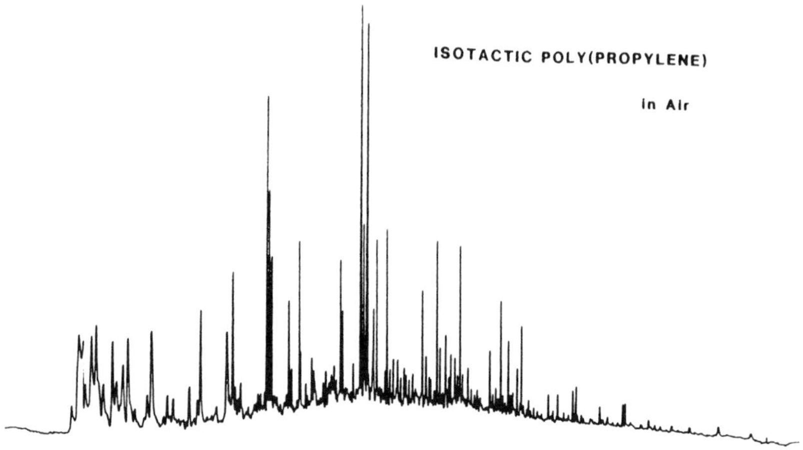

PYROGRAM O-8

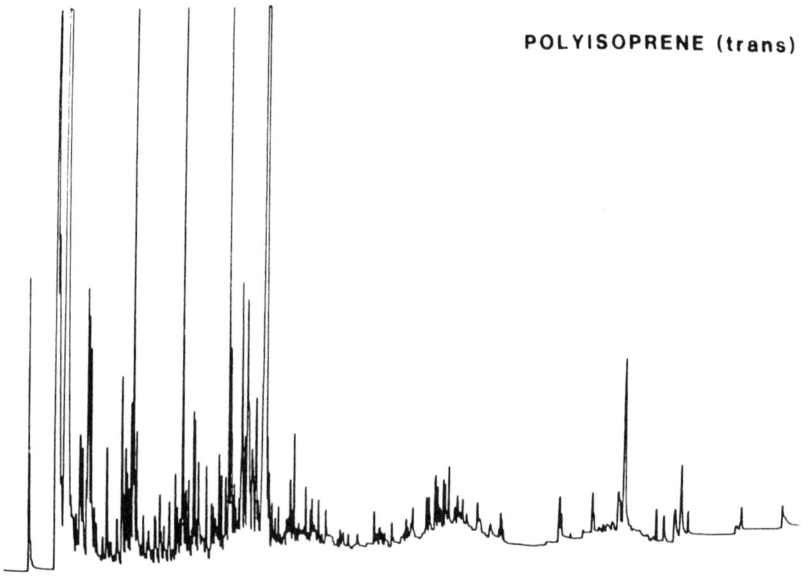

PYROGRAM O-9

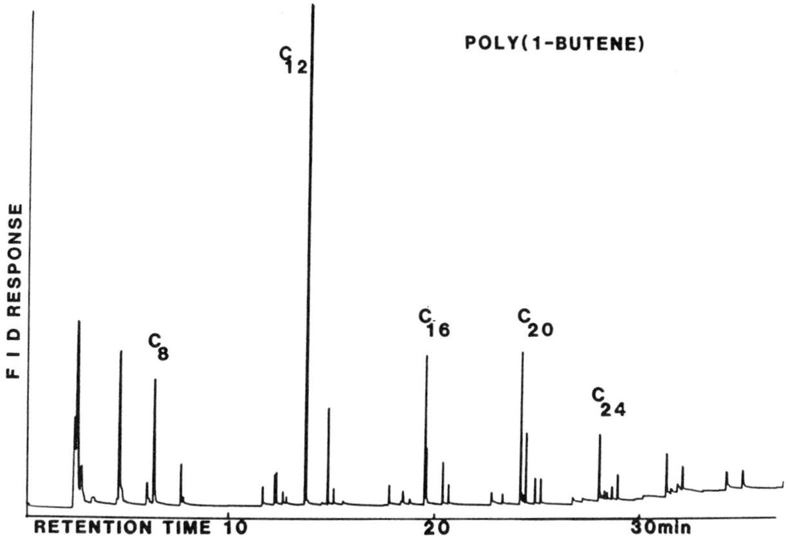

PYROGRAM O-10

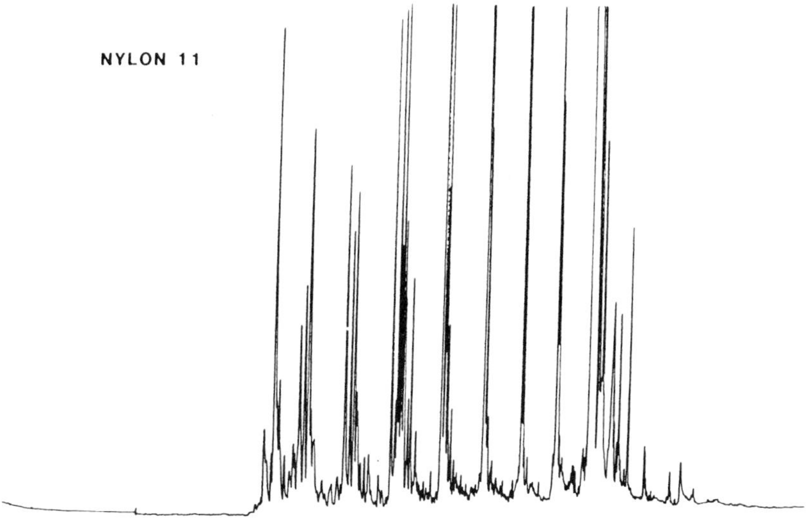

PYROGRAM N-1

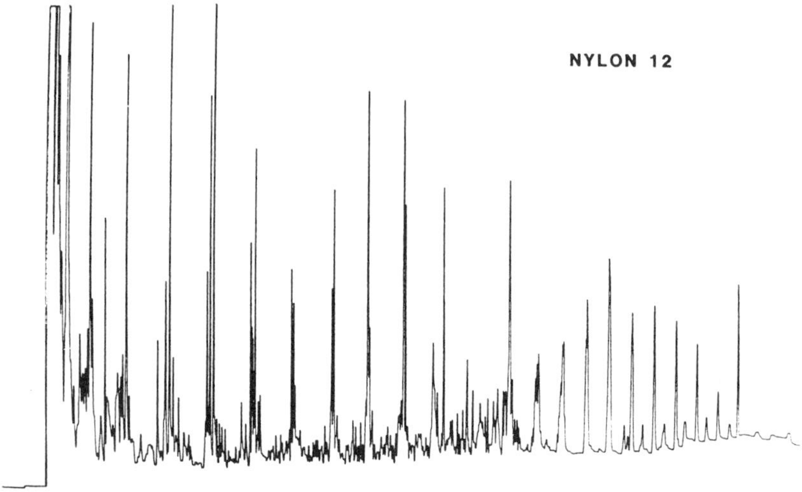

PYROGRAM N-2

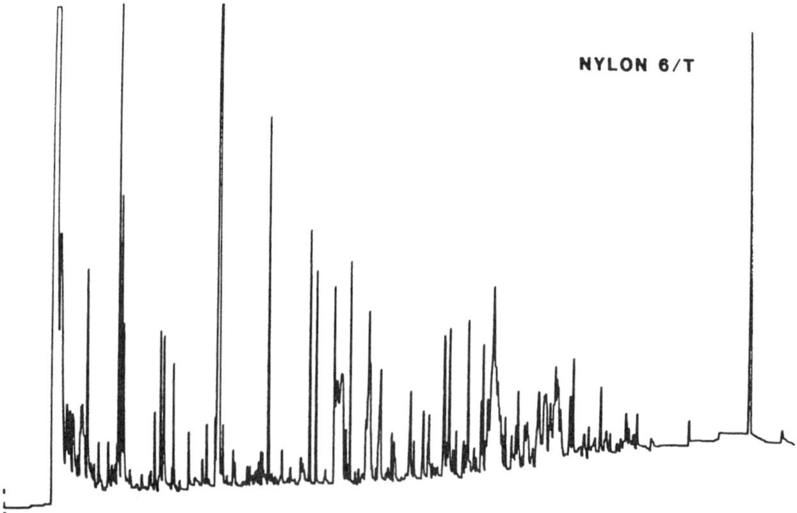

PYROGRAM N-3

SAMPLE: NYLON 6/6

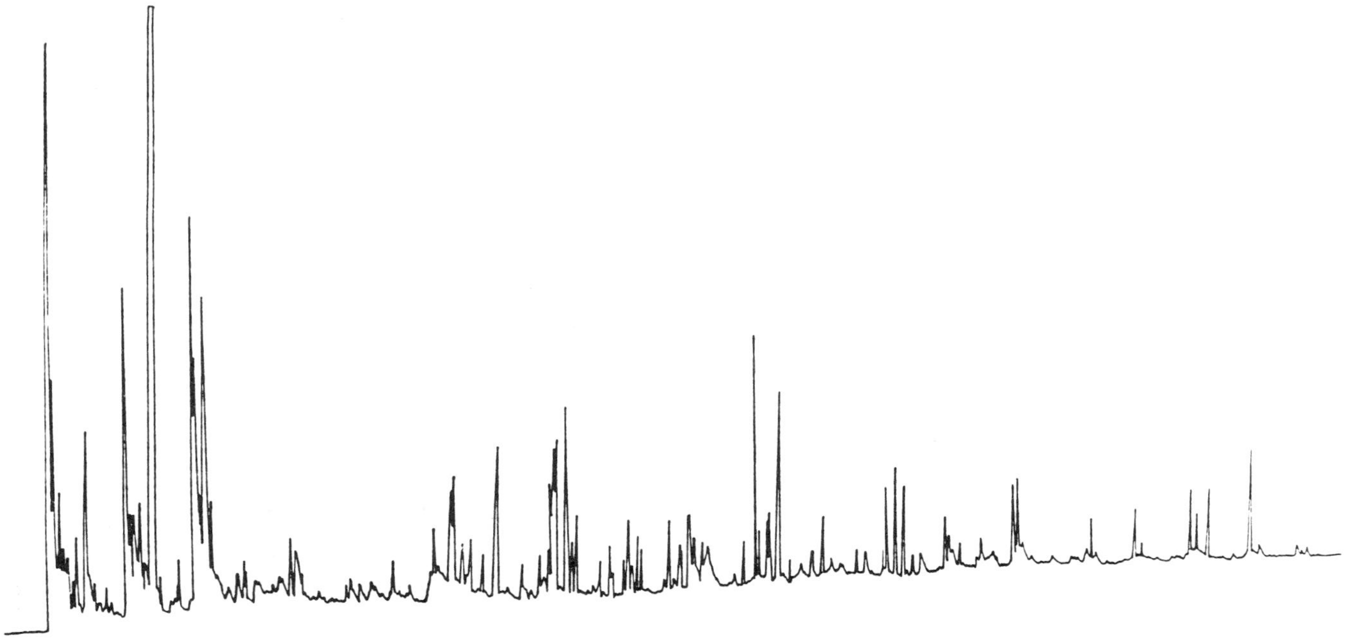

PYROGRAM N-4

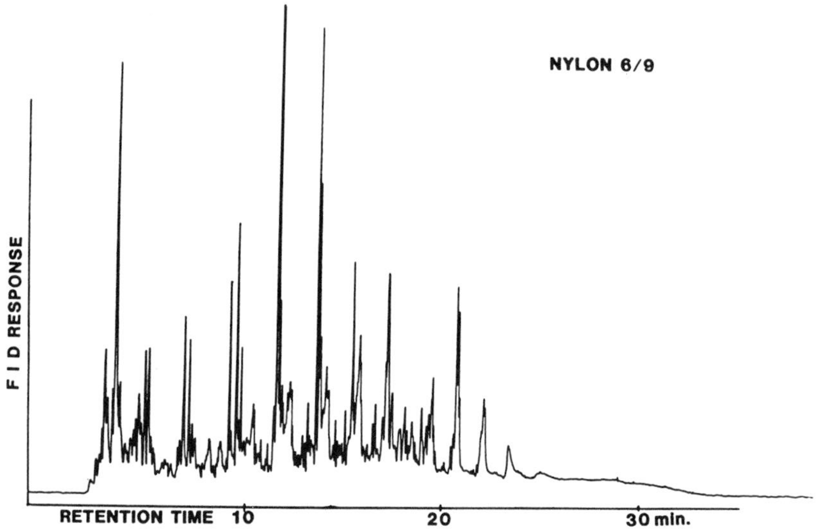

PYROGRAM N-5

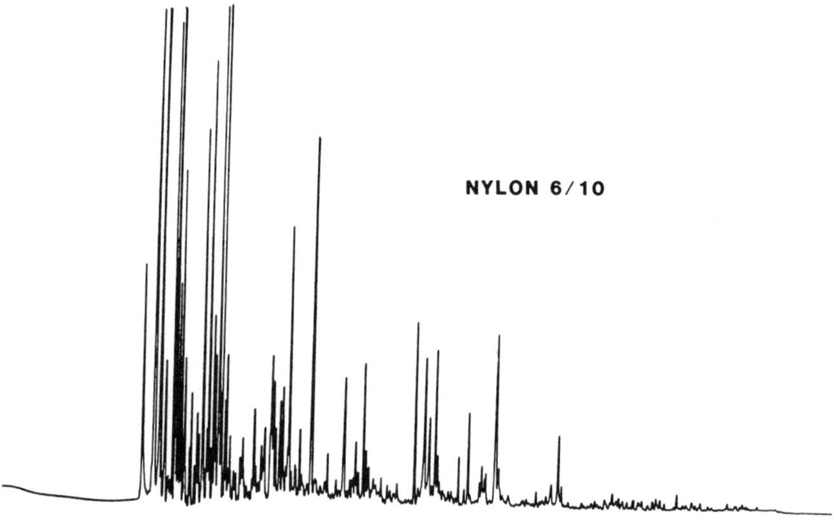

PYROGRAM N-6

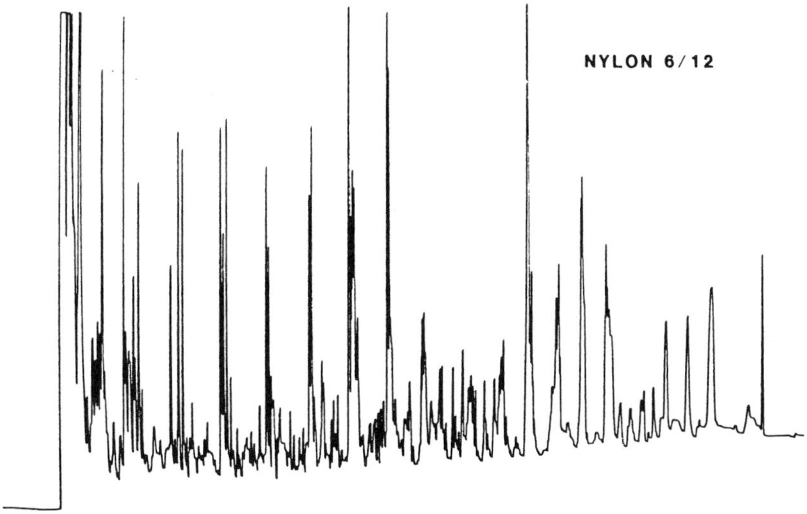

PYROGRAM N-7

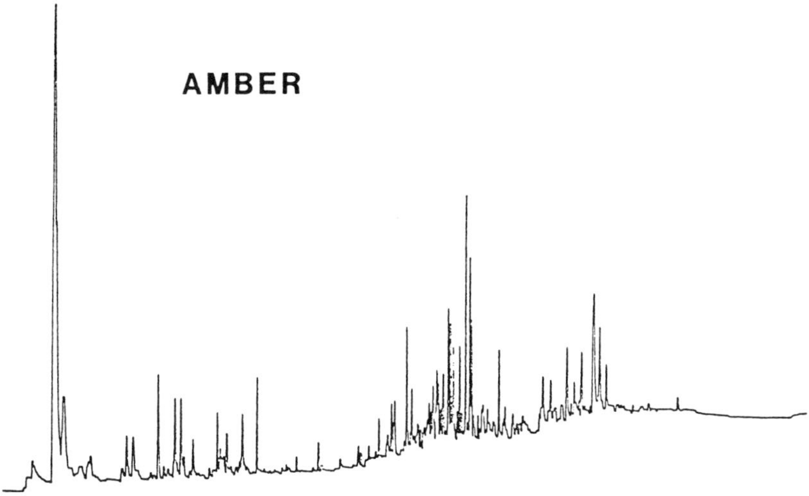

PYROGRAM B-1

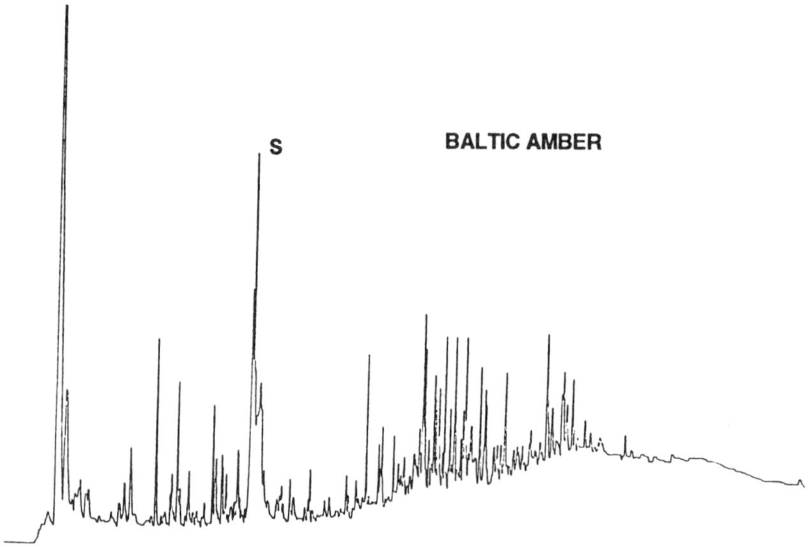

PYROGRAM B-2

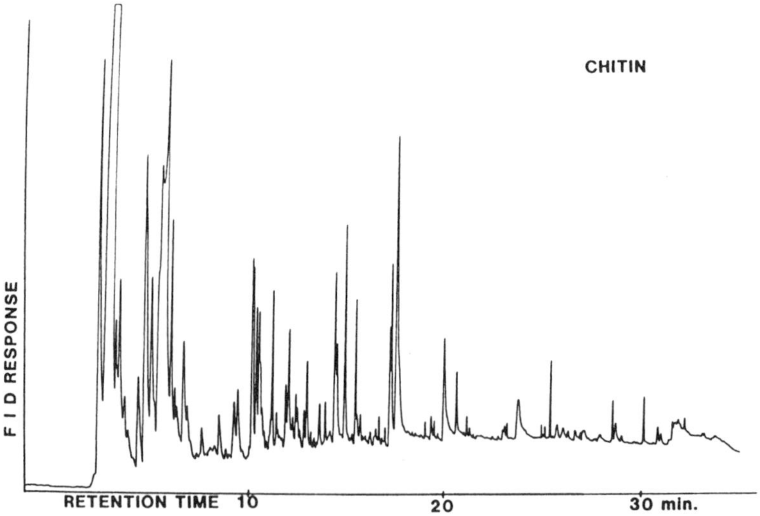

PYROGRAM B-3

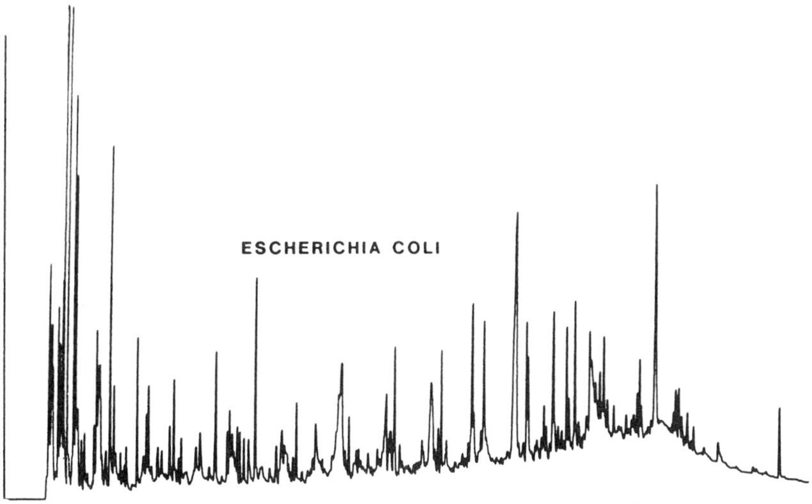

PYROGRAM B-4

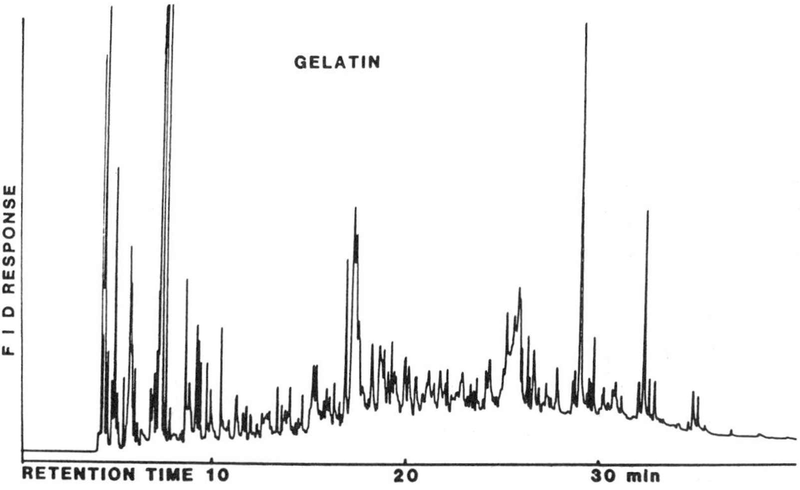

PYROGRAM B-5

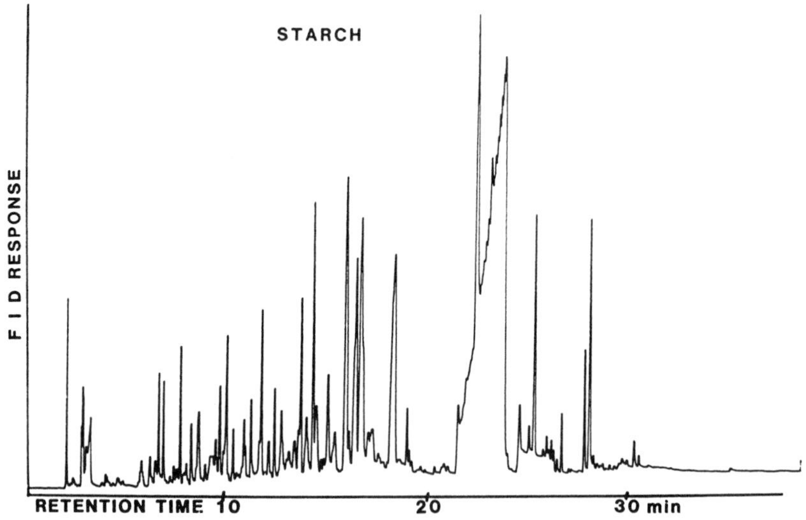

PYROGRAM B-6

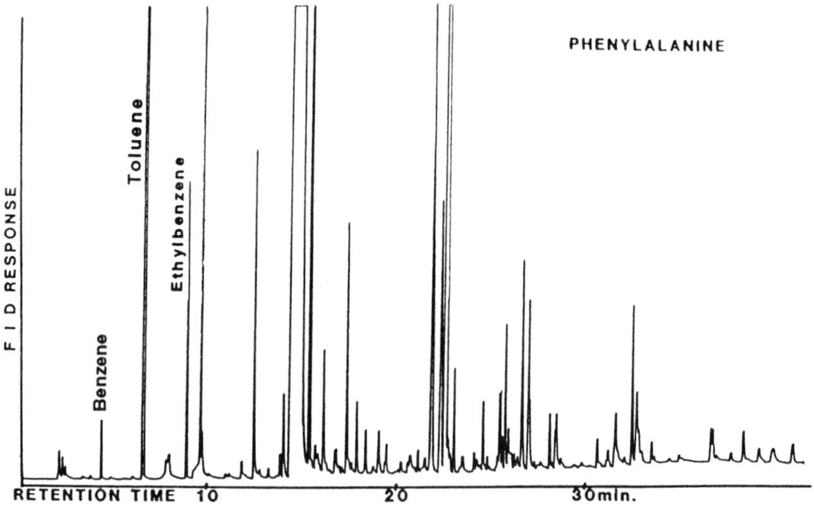

PYROGRAM B-7

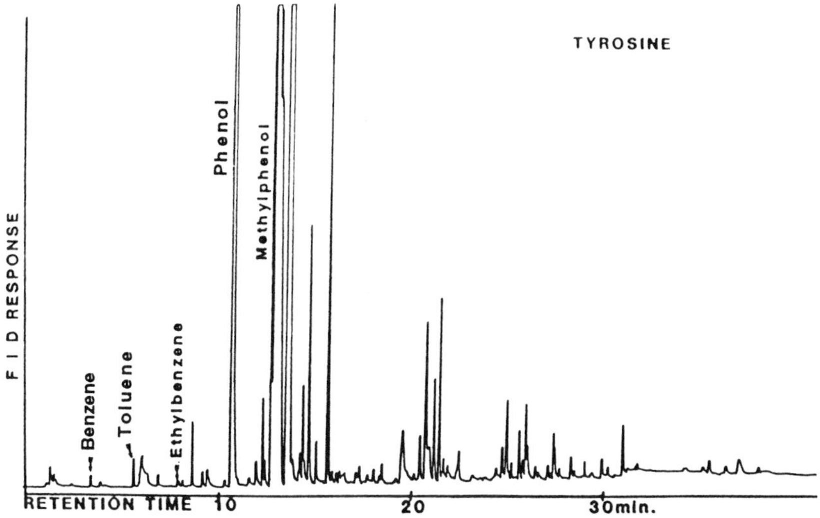
TYROSINE
Phenol
Methylphenol
Ethylbenzene
Toluene
Benzene
F I D RESPONSE
RETENTION TIME 10
20
30min.

PYROGRAM B-8

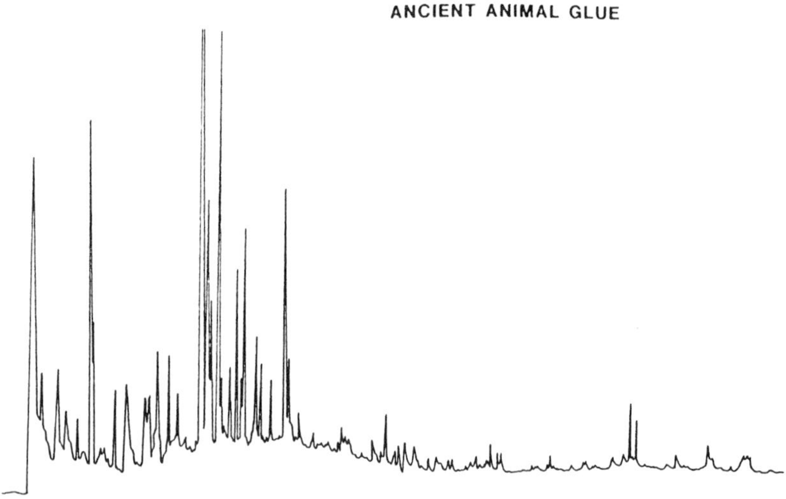
ANCIENT ANIMAL GLUE

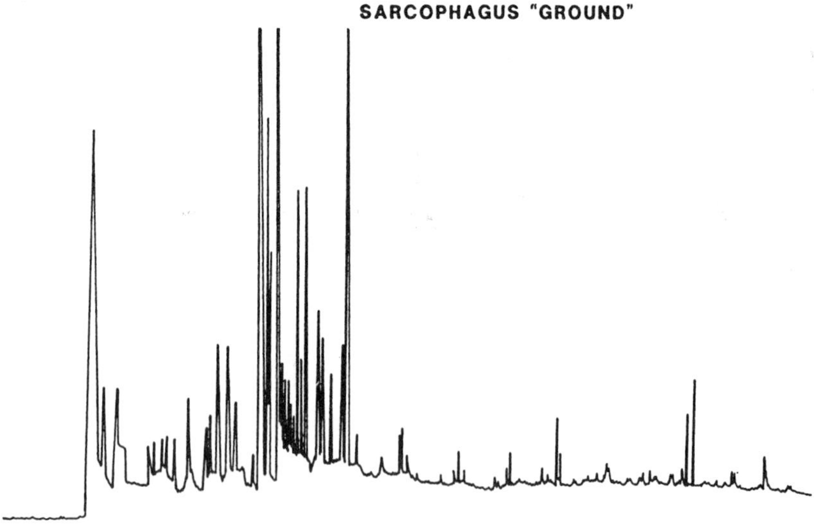
SARCOPHAGUS "GROUND"

PYROGRAM B-9

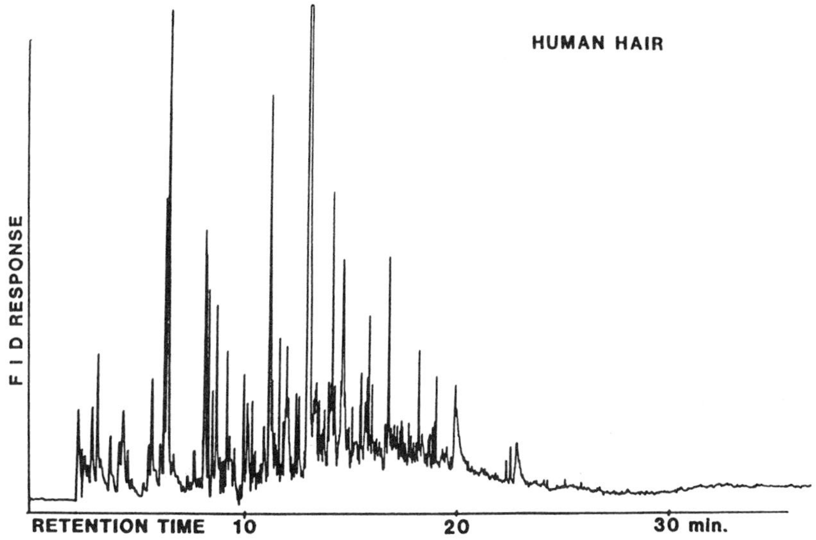

PYROGRAM B-10

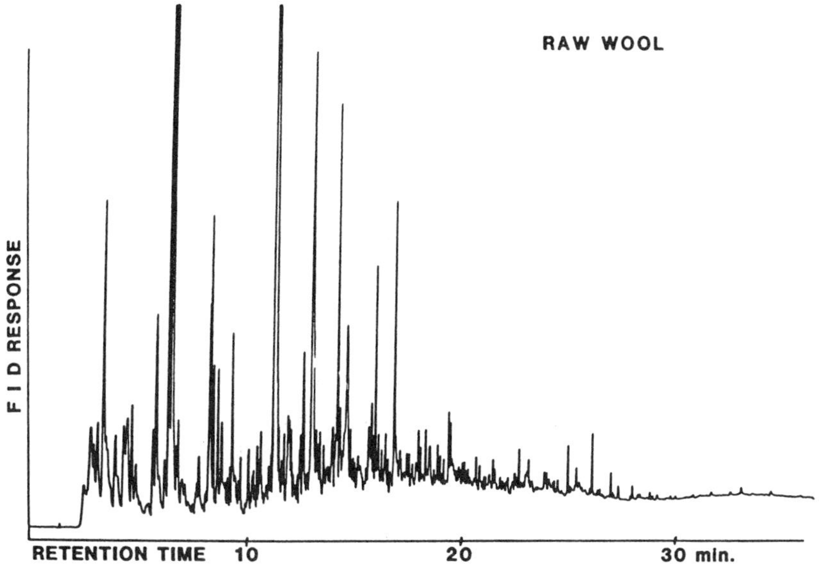

PYROGRAM B-11

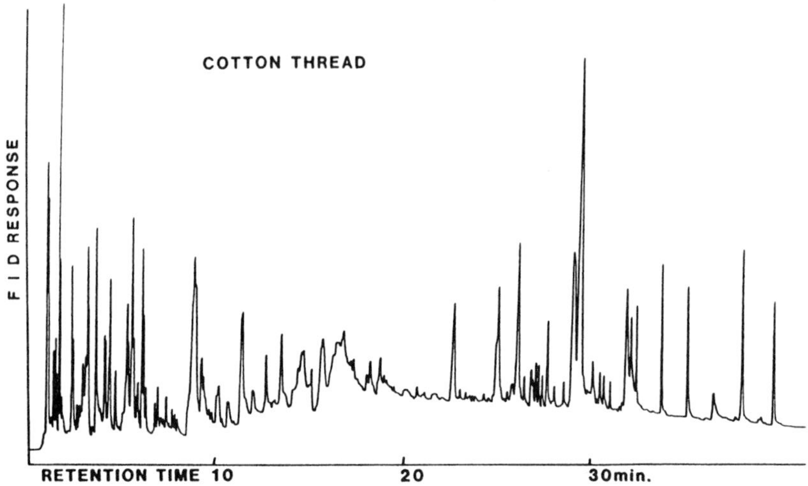

PYROGRAM B-12

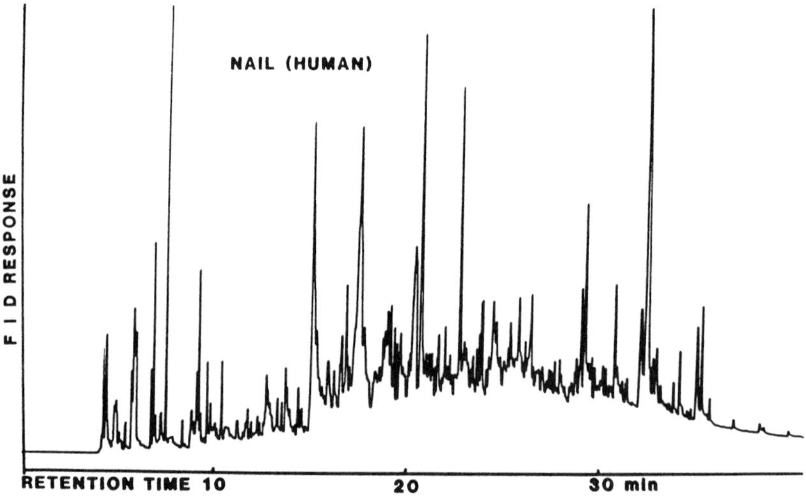

PYROGRAM B-13

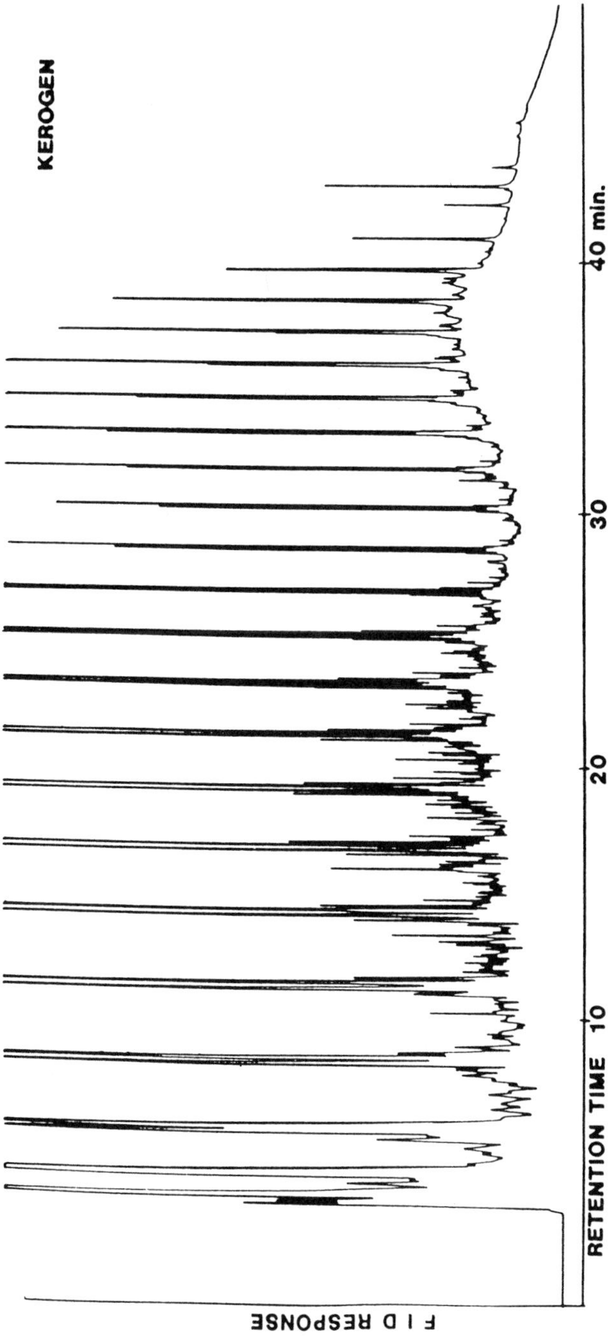

PYROGRAM B-14

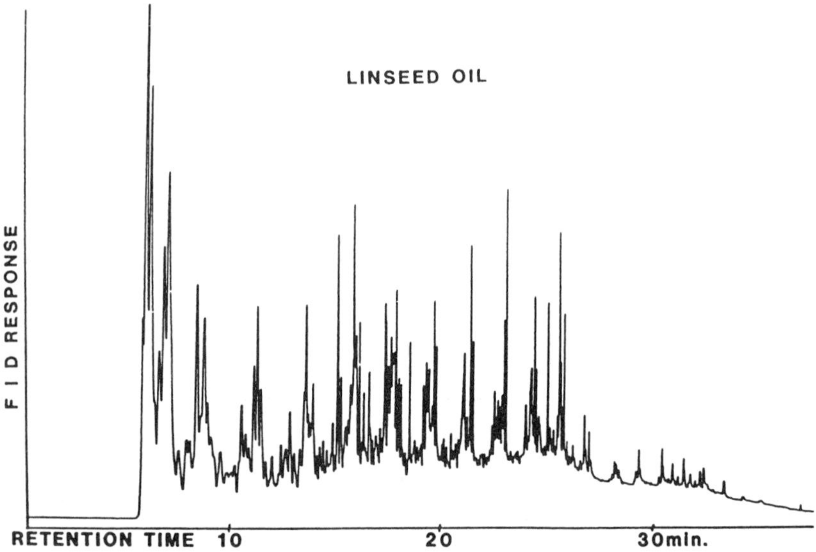

Pyrogram B-15

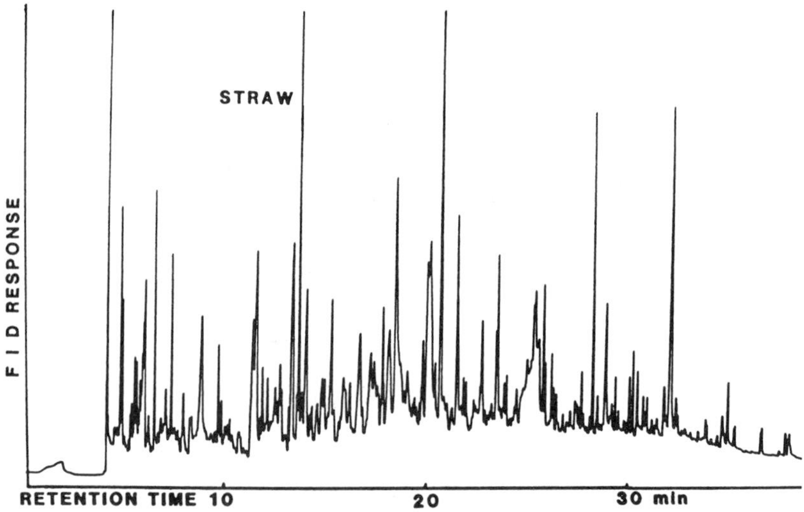

Pyrogram B-16

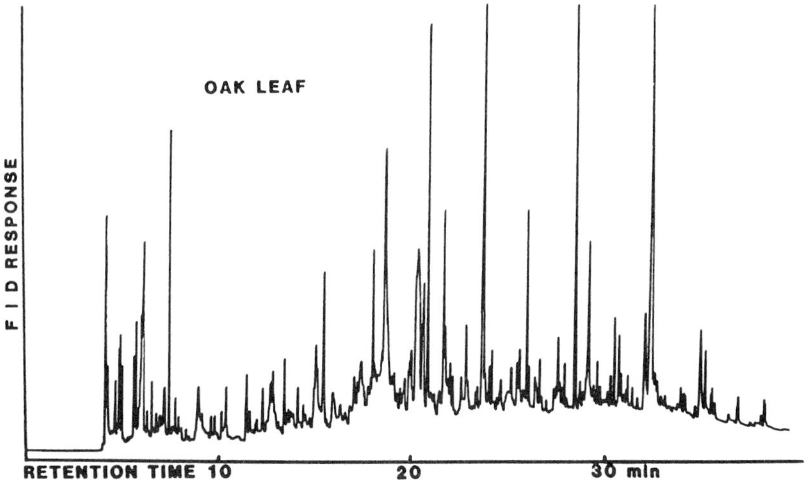

PYROGRAM B-17

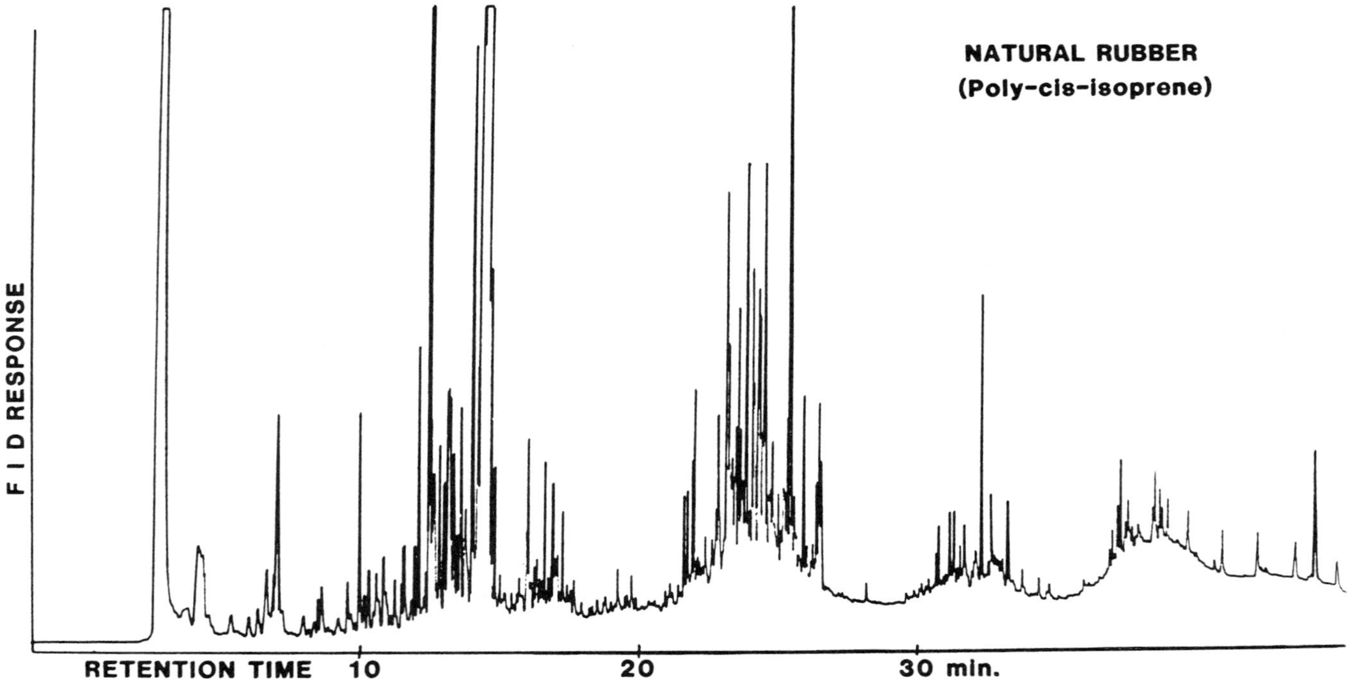

Pyrogram B-18

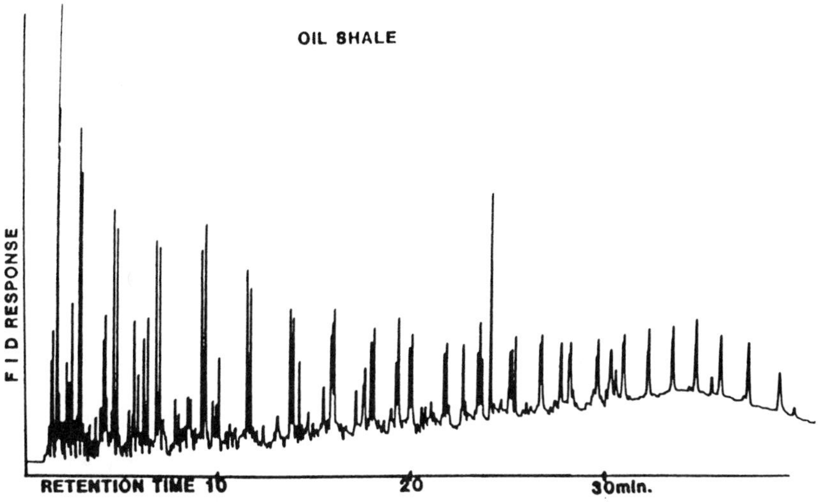

PYROGRAM B-19

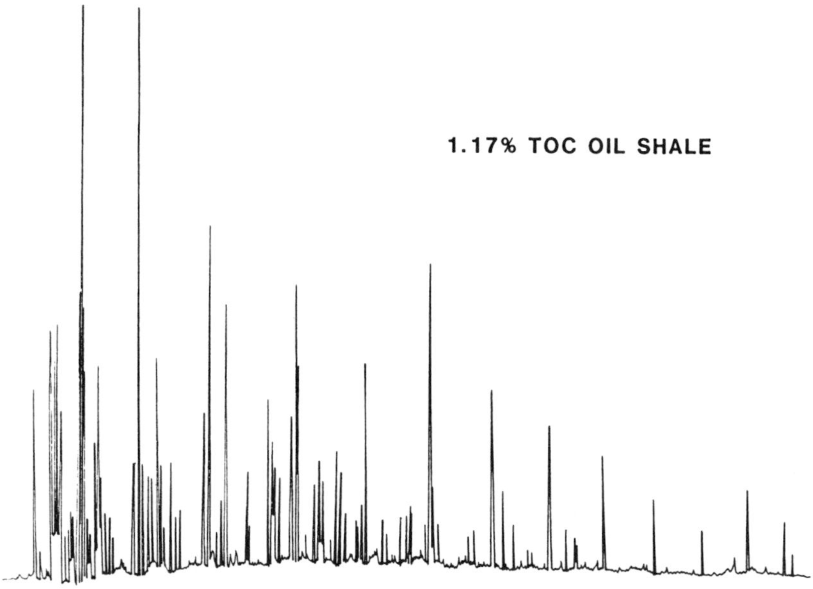

PYROGRAM B-20

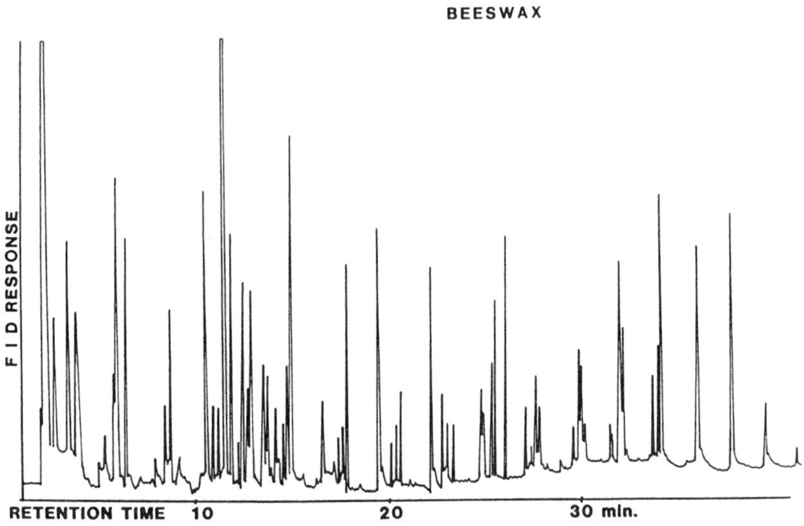

PYROGRAM B-21

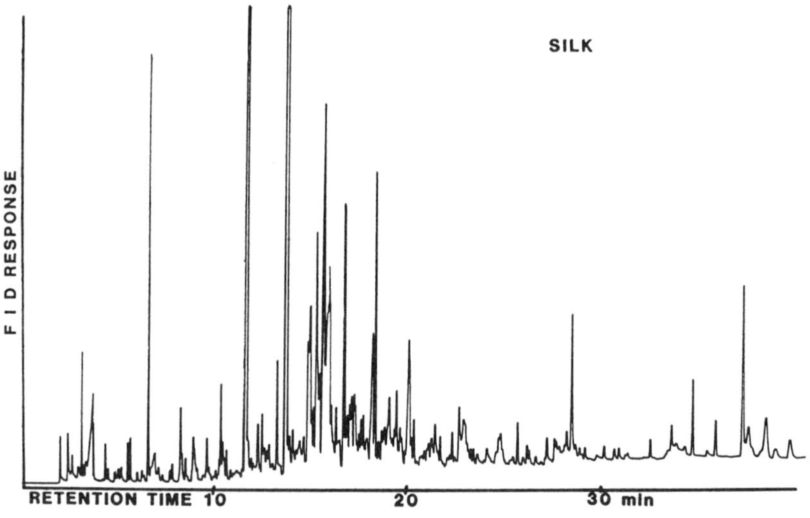

PYROGRAM B-22

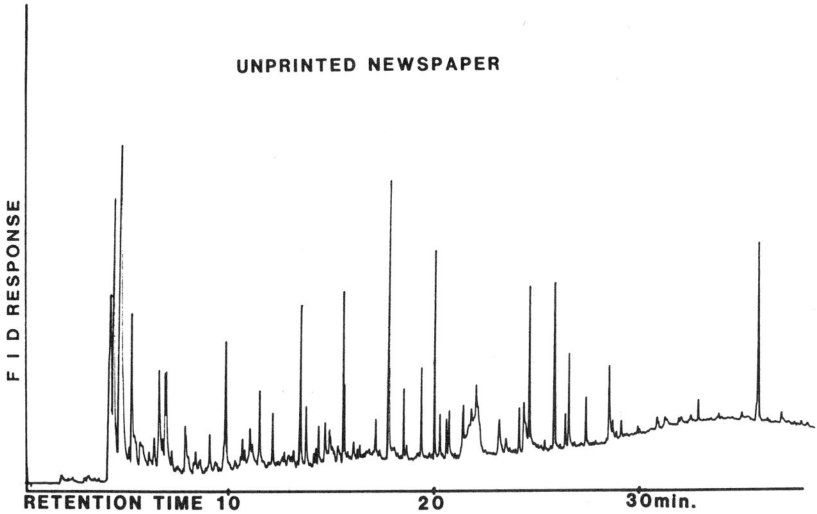

PYROGRAM G-1

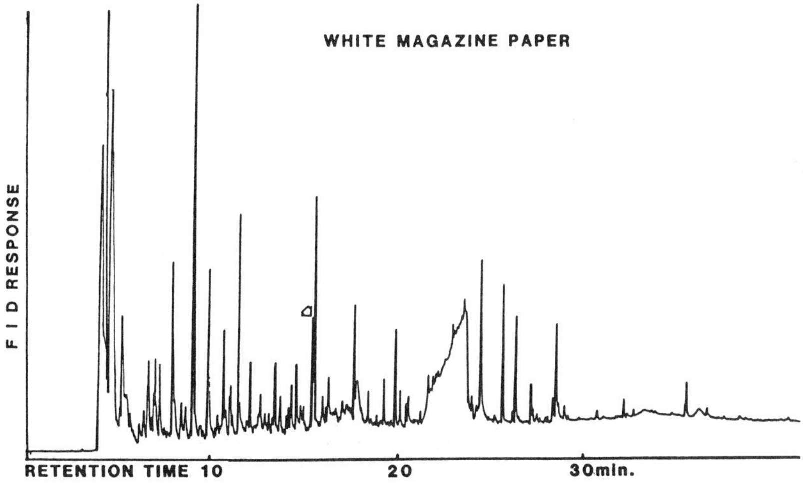

PYROGRAM G-2

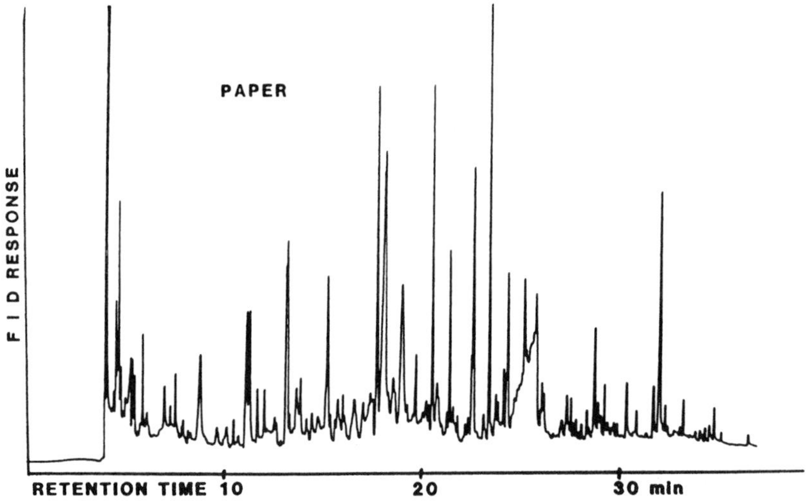

PYROGRAM G-3

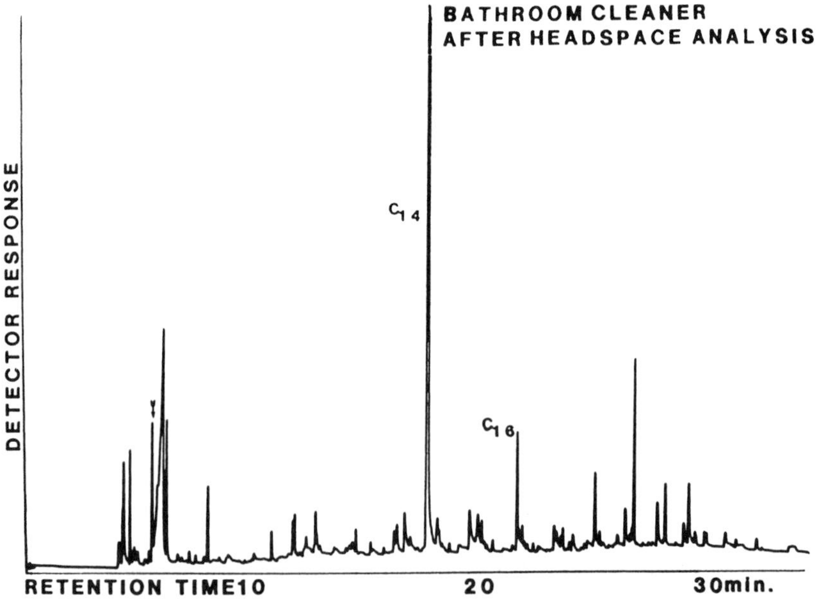

PYROGRAM G-4

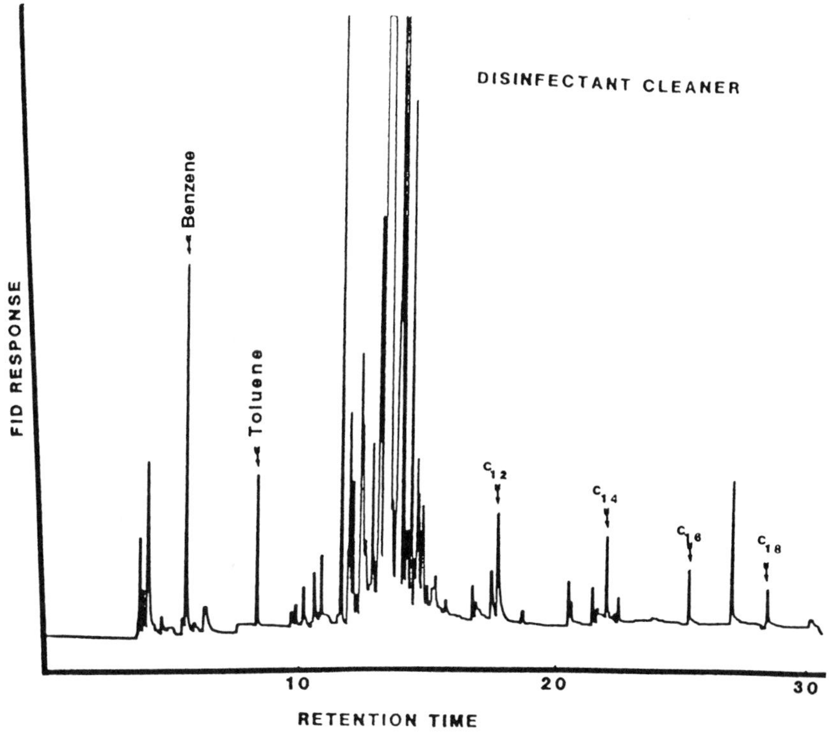
DISINFECTANT CLEANER
FID RESPONSE
Benzene
Toluene
C_{12}
C_{14}
C_{16}
C_{18}
10
20
30
RETENTION TIME

PYROGRAM G-5

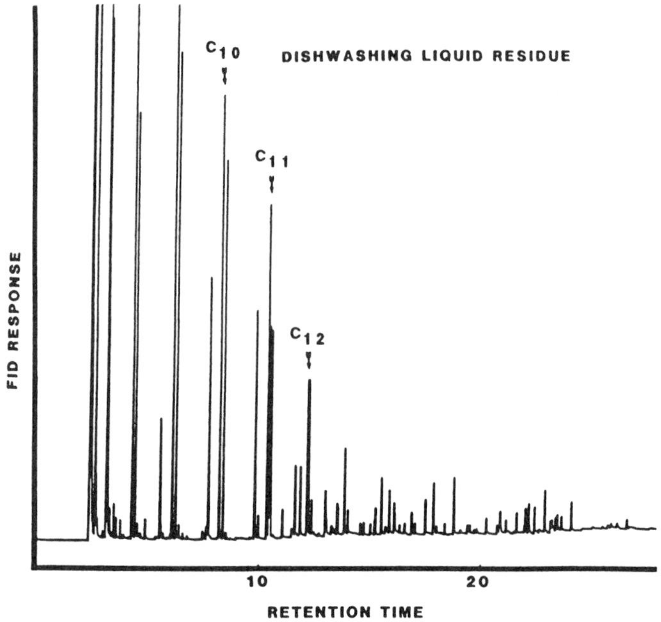

PYROGRAM G-6

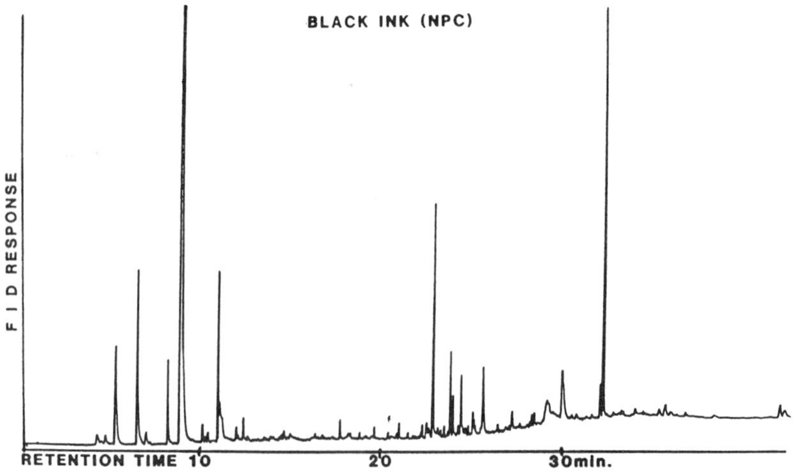

PYROGRAM G-7

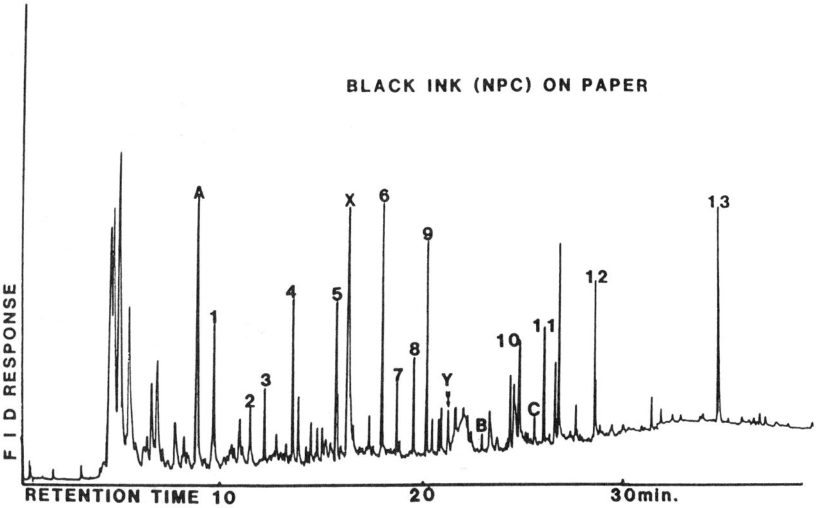

PYROGRAM G-8

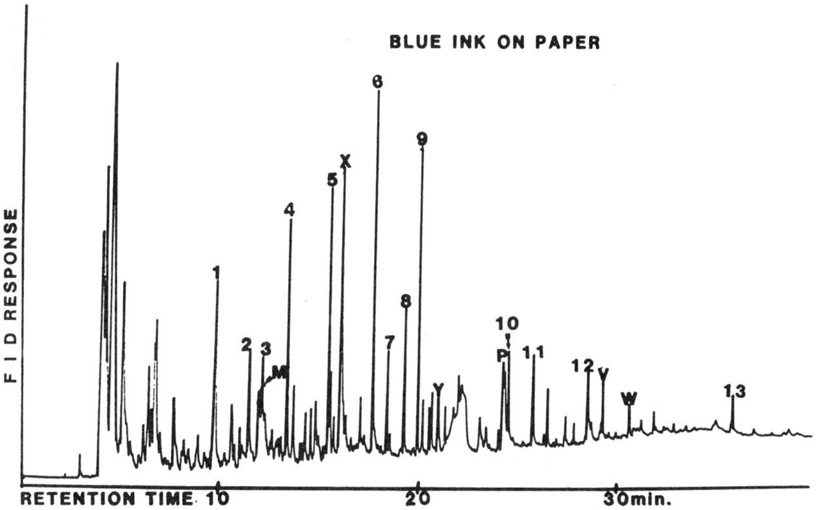

PYROGRAM G-9

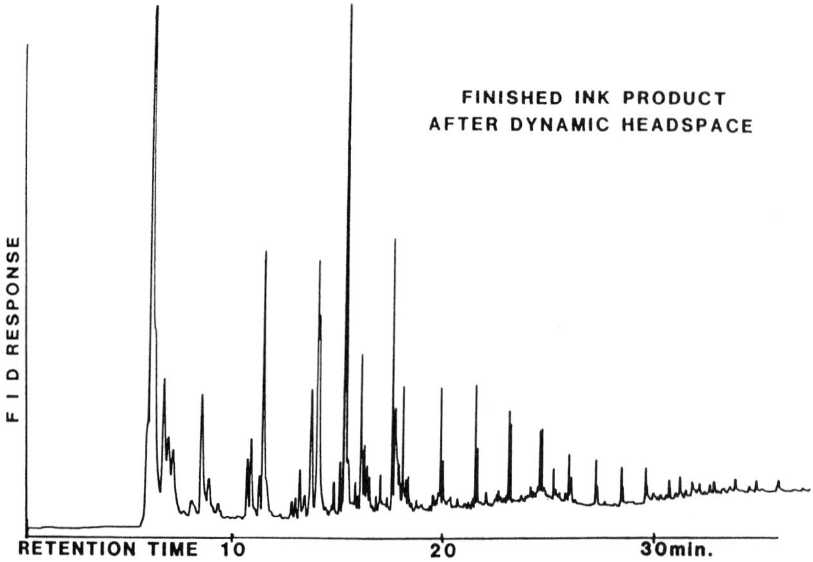

PYROGRAM G-10

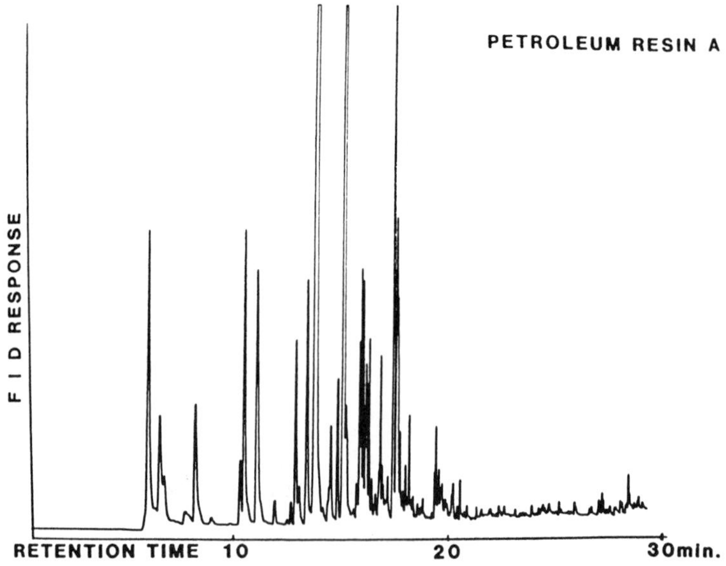

PYROGRAM G-11

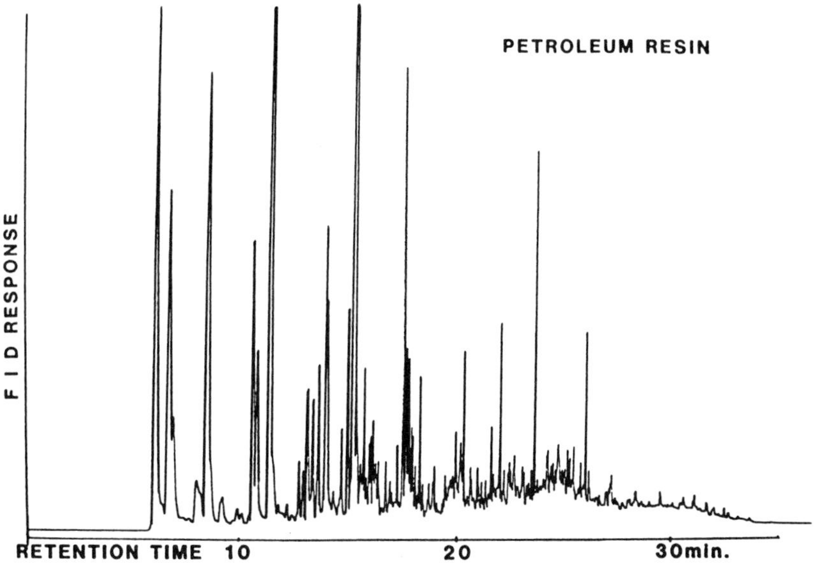

PYROGRAM G-12

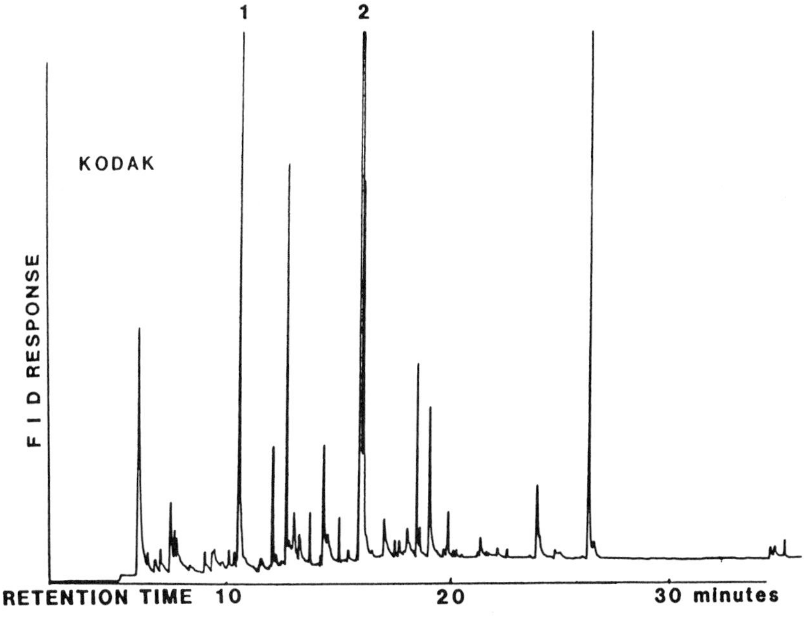

PYROGRAM G-13

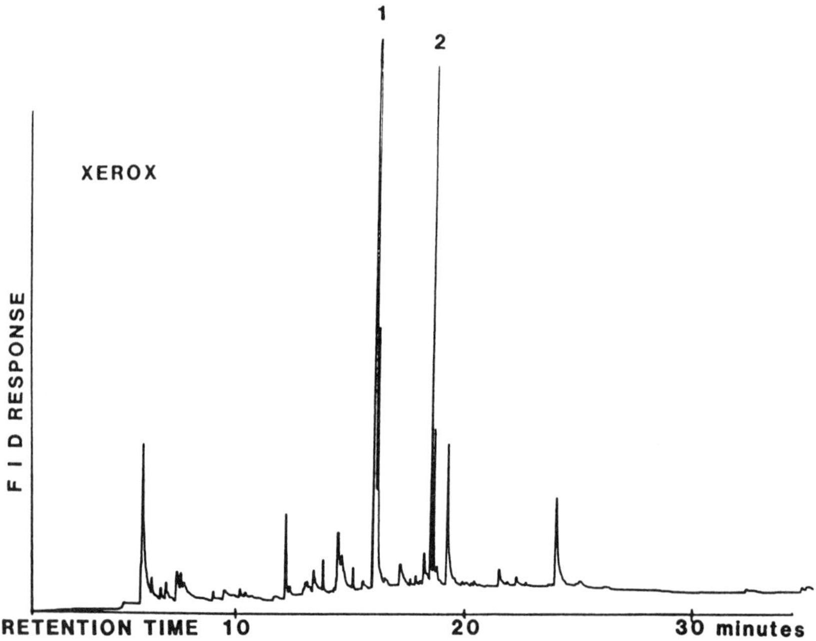

PYROGRAM G-14

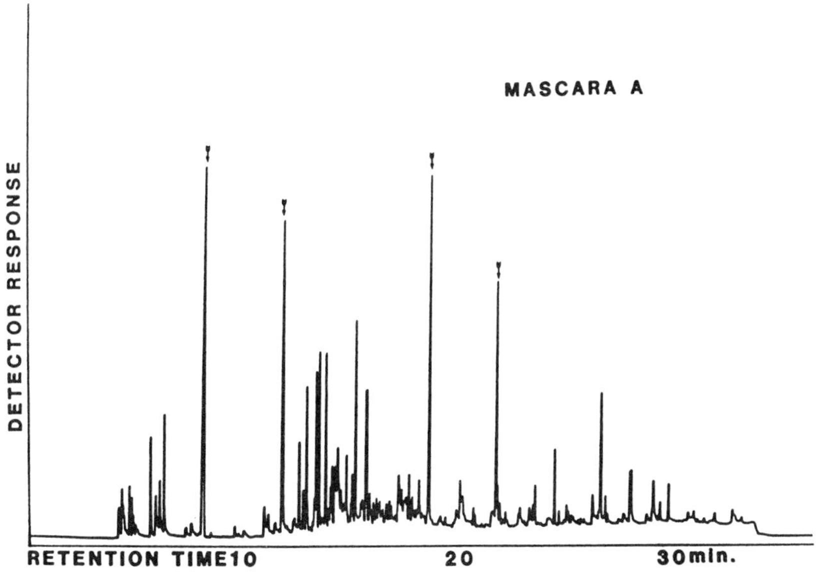

PYROGRAM G-15

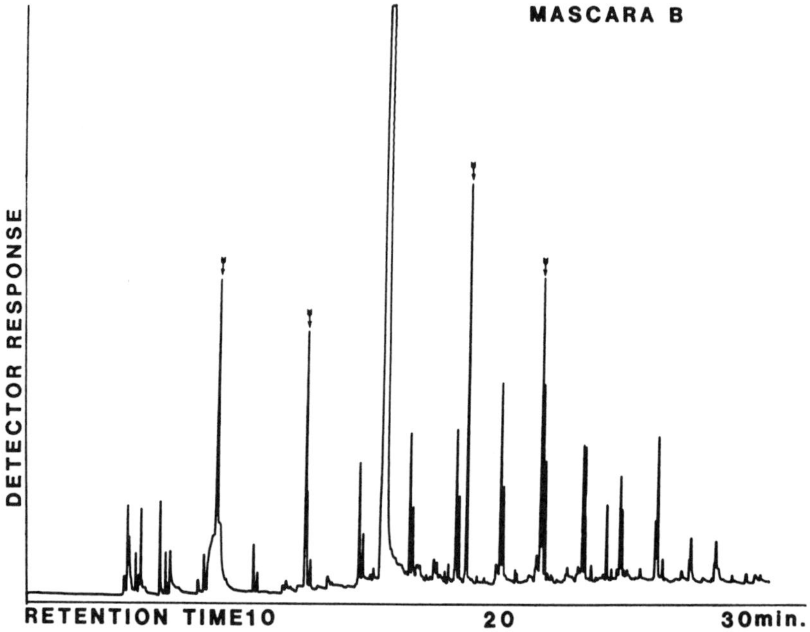

Pyrogram G-16

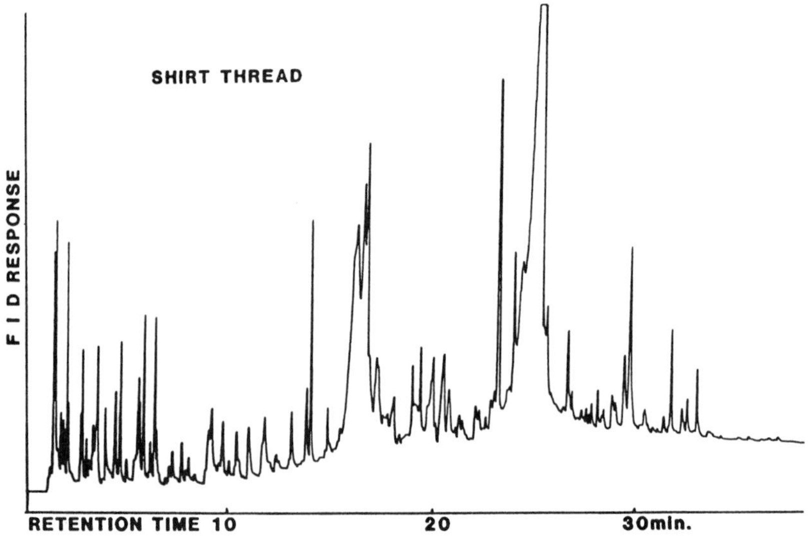

Pyrogram G-17

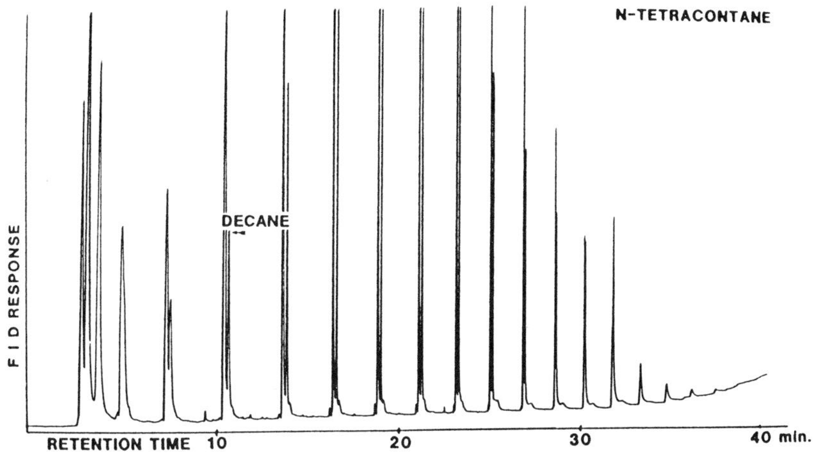

PYROGRAM G-18

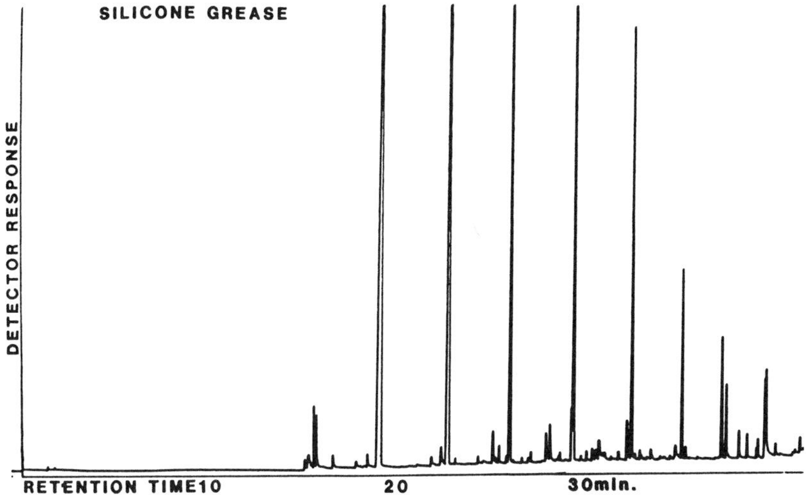

PYROGRAM G-19

Index